Advanced Materials and Nano Systems: Theory and Experiment

Part 2

Edited by

Dibya Prakash Rai

Assistant Professor
Department of Physics, Pachhunga University College
Mizoram University, Aizawl, 796001
India

Advanced Materials and Nano Systems: Theory and Experiment

(Part 2)

Editor: Dibya Prakash Rai

ISBN (Online): 978-981-5049-96-1

ISBN (Print): 978-981-5049-97-8

ISBN (Paperback): 978-981-5049-98-5

Published by Bentham Science Publishers Pte. Ltd. Singapore. All Rights Reserved.

First published in 2022.

need for a court order if at any point you breach any terms of this License Agreement. In no event will any delay or failure by Bentham Science Publishers in enforcing your compliance with this License Agreement constitute a waiver of any of its rights.

3. You acknowledge that you have read this License Agreement, and agree to be bound by its terms and conditions. To the extent that any other terms and conditions presented on any website of Bentham Science Publishers conflict with, or are inconsistent with, the terms and conditions set out in this License Agreement, you acknowledge that the terms and conditions set out in this License Agreement shall prevail.

Bentham Science Publishers Pte. Ltd.
80 Robinson Road #02-00
Singapore 068898
Singapore
Email: subscriptions@benthamscience.net

BENTHAM SCIENCE

CONTENTS

FOREWORD

First of all, I would like to congratulate **Dr Dibya Prakash Rai** for successfully publishing the first edited version of the book entitled *"Advanced Materials and Nanosystems: Theory and Experiment"*. Nothing surprised me more than the fact that, in comparison to his first release, Dr. Rai prepared the second edited book in such a short period. I know Dr. Rai since when he joined MSc. Physics at Mizoram University in 2007 and he has completed his Ph.D. degree under my supervision in 2013. He was a very hardworking student and later he emerged as a promising researcher. He has a dynamic personality with exceptional leadership qualities which reflect his intellectuality in management by coordinating the authors and the publishers, as a result, he could compile two edited books within a short period.

I'm overjoyed today for two reasons. First, my student has followed the route I've shown him. Second, for his achievements and efforts in a subject that I have always been passionate about (R & D). The title of the book makes me very happy. I wanted to read the full book, but I only had time to skim over each chapter of this edited book due to time constraints. All the chapters and the titles are diverse, very attractive, and cover most of the latest research topics. Though I would love to read every chapter thoroughly once it is published.

The content in this book is concise and thorough, and it covers current advanced materials and nanosystems research. This book takes us on a voyage to a newly found mystical realm that allows us to view atoms with the invention of transmission electron microscopy (TEM). We can now manipulate atoms to obtain useful information thanks to technological advancements. Electronics experts working with integrated circuits competed in a size-minimizing race that progressed from micro to submicron dimensions, bringing them closer and closer to the nanoscale. When scientists eventually get to the point where they can use a single electron as the basis of electronics, the entire concept of electronics will have to be rethought.

This evolution is not limited to electronics, since other disciplines of study such as mechanics, optics, chemistry, and biology have begun to develop their nanoworlds, which we now refer to as microsystems. Nanosystems, on the other hand, do not yet exist. It will take some time for them to emerge from the laboratory. As one must read this book to understand the difficulties and latest technical advancements, in the said topics, which raise the curtain from certain unclear concepts. As a result, we must be conscious of the ongoing difficulties and the stakes.

I'd like to encourage Dr. Rai for publishing a series of volumes of such books every year. I also encourage all the material scientists, engineers, and scholars to read this book and contribute to it.

Once again I would like to congratulate & wish you all the good luck !!!

Prof. R. K. Thapa
Retired Professor
Vice Chancellor
Sikkim Alpine University
Namchi, Sikkim, India

PREFACE

Nanoscience and nanotechnology have emerged out as unique and distinct disciplines in the contemporary field of science and engineering. The size-dependent phenomena of materials when their dimension is reduced below 100nm can be dealt with in Nanoscience. On the other hand, nanotechnology plays an important role in the creation and manipulation of materials at the nanometre scale, either by scaling up from single groups of atoms or by refining or reducing bulk materials. The second edition of the book *"Advanced Materials and Nanosystems"* covers the advancement of bulk to nanomaterials and their implication in the development of new technology. This book gives a solid understanding regarding the variation of the physical properties of the materials while reducing sizes from bulk to nanoscale. The book helps to give information on the various effects of nanomaterials as bio-sensors, bio-agent, nanocatalysts, nano-robot, *etc.* The book also covers the various physical, chemical and hybrid methods of nanomaterial synthesis and nanofabrication as well as advanced characterization techniques.

This book includes chapters from all fields of sciences such as; Nanosciences, Physical sciences, Chemical Sciences, Biosciences, Material Sciences, Engineering sciences *etc.*, is an integrated, multidisciplinary edited book. This book is an amalgamation of diverse chapters from different trending fields from various contributing authors. All the contributing authors systematically discuss the chemistry, physics, and biology *etc.*, aspects of nanoscience, providing a complete picture of the challenges, opportunities, and inspirations posed by each facet before giving a brief glimpse at nanoscience in action: nanotechnology. All the contributing chapters give an overview of the latest research work in their respective field, which has importance in our daily lives. This book highlighted the latest development and the significant role of different new materials for various applications. Here is a brief introduction to each chapter.

Chapter 1, Mandal *et al.*, elucidated the recent advancement of nanotechnology from a human health perspective. They have discussed the crucial points in this chapter and give a brief review of the merit and demerit of nanoparticles in human health. The development of smart nano molecules and nanodevices using molecular and supra-molecular would be a blessing for medical sciences. The nanoparticles like silver nanoparticles, gold nanoparticles, *etc.* are proved to be novel nanoparticles for their applications in biomedicine. While they have also highlighted the adverse effects, risk factors on the human body.

Chapter 2, discussed the optimised characteristics of SWIRL (Short Wave Infra-Red Light) of AlGaAs/GaAs heterogeneous type nanostructure under various GRINSLs (Graded Refractive Index Nano Scale Layers) in advanced bio-based nanotechnological systems. They reported the enhanced performances of SWIRL gain with wavelengths of photons for various GRINSLs. This behaviour can be integrated into medical devices for the treatment of wound, pain and various types of sensitive skin diseases by using the FONSCs (Fibre Optic Nano Scale Cables) through the TIRPs (Total Internal Reflection Processes) without any attenuation in dB/Km due to diminished scattering, dispersion and absorptions in the nanotechnological biosciences and medical sciences. Moreover, SWIRL of wavelength ~ 830 nm has provided the most fabulous role in the proper controlling of inflammation, edema as well as infections of various bacteria in advanced bio-based nanotechnological systems.

Chapter 3, this chapter gives the theoretical analysis of the electronic band structure of the half-Heusler alloys ScAuSn, LuAuSn and their Superlattice using density functional theory (DFT) within the full-potential linearized augmented plane waves (FP-LAPW). They have discussed the inefficiency of generalized gradient approximation (GGA) in opening the electronic bandgap. While they report that Trans and Blaha modified Becke-Johnson potential (TB-mBJ) is more appropriate for calculating the electronic bandgap. Their results revealed that LuAuSn and ScAuSn are indirect bandgap semiconductors while their superlattice is a direct bandgap semiconductor.

Chapter 4, overview the importance of nanotechnology and claim to be a multidisciplinary approach contribution from Physicists, chemists, biologists, material scientists, engineers, and computer scientists. In this chapter, they have discussed the evolving and growing interest in nanotechnology, and its implication in size scale technology to construct a computer that is smaller, faster, and more trustworthy. They prepared nanoparticles from the top-down and bottom-up approaches to have direct impact on the current computer design and architecture.

Chapter 5, herein the authors have deposited the amorphous silicon oxide (a-SiO$_x$:H) and silicon nitride (a-SiN$_x$:H) on the low substrate at 250°C -300°C by the chemical Vapour deposition technique. They have estimated the interface charge density (D_{it}) and fixed charge density (Q_f) using a high frequency (1 MHz) capacitance-voltage measurement on Metal-Insulator–Silicon structure (CV-MIS). They reported the reduction of the surface recombination velocity due to low interface charge density (D_{it}). An improved efficiency and short circuit current has been reported for a-SiO$_x$:H and a-SiN$_x$:H on the front surface of c-Si solar cells.

Chapter 6, deals with the synthesis and characterization of nanoparticles. This chapter reports the gradual development in the synthesis techniques from the bulk to nanoparticles synthesis. All types of synthesis methods have been discussed here. They found that various bottom-up and top-down approaches are appropriate for the commercial production of nanoparticles. They summarize the basic principle of solid phase, vapour phase and liquid phase synthetic techniques in detail with schematic setup. They focus on the matrix of the activated carbon for nanocomposites synthesis, with large surface area and porosity, offer vivid applications in various fields such as environmental remediation as adsorbents, suitable sorbents in analytical determination of organics, targeted drug delivery, diagnostic agents, fuel cells and sensors, to name a few.

Chapter 7 is contributed by S. Rai, and reported the effect of Nanostructure-materials on the optical properties of some rare-earth ions (Eu^{3+}, Sm^{3+} & Tb^{3+}) doped in the silica matrix. Nanoparticles of CdS incorporating in Rare Earth doped silica xerogel (RE^{3+}:SiO$_2$) matrix have been prepared by sol-gel method. The prepared materials have been characterized by physical and optical techniques, such as XRD, SEM, TEM and Photoluminescence (PL) in which he has reported the particle size of 8 nm and an average particle dimension of 5 nm. He has observed the enhanced luminescence in rare-earth (RE) ions in the presence of CdS NPs in RE^{3+}:SiO$_2$ matrix. A twenty time more intense dominating orange peaks (616 nm) from the characteristic peak of Eu^{3+} ions are observed for CdS/Eu^{3+}:SiO$_2$ matrix compared to the sample without CdS NPs.

Chapter 8, overviews a description of the **Nd$_2$Fe$_{14}$B and SmCo$_5$ based permanent magnet** nanomaterials. The **Nd$_2$Fe$_{14}$B and SmCo$_5$** nanoparticles have been studied using the first principle approach opting for the self-predictable maximum capacity linearized increased plane wave (FPLAPW) strategy as programed in the WIEN2K code. The magnetic moment of BCC Fe and HCP Nd are 2.27μ_B and 2.65μ_B, respectively.

Chapter 9, detailed a comparative study on visible-light-induced photocatalytic activity of MWCNTs decorated sulfide-based (ZnS & CdS) nano photocatalysts. ZnS and CdS of different sizes show photocatalytic activities in the visible region due to their appropriate energy bandgap (Eg). They report the multi-walled carbon nanotubes (MWCNTs) intercalated sulfide-based photocatalysts like ZnS/MWCNTs and CdS/MWCNTs composites enhance photocatalytic response in comparison to ZnS and CdS NCs.

Chapter 10, discusses the Organic solar cells and their working principle. Here, the authors have reported the new efficient type of solar cell and photovoltaic energy technology. The bulk-heterojunction (BHJ) organic solar cells (OSCs) consisting of a mixture of a conjugated donor polymer with a fullerene acceptor are considered a promising approach. They are attractive owing to their mechanical flexibility, lightweight, low cost and environmentally friendly solar cells with highly tunable electrical and chemical properties. This chapter highlights the fundamental Physics of OSCs, working mechanism, novel materials, device architectures, strategies to improve the stability of OSCs and the current status of BHJ solar cells with all critical aspects considered important to understand.

Chapter 11, final chapter which presents synthesis and characterization of Nd_2O_3 doped lithium borosilicate glasses from melt-quench technique. Electrical conductivity of produced samples was tested in frequency band of 2mHz to 20MHz at 423K to 673K, using Impedance Analyser. From this study it has been reported that conduction is based on the composition and not on the temperature. In the temperature band, 423-673 K, the variance of the dielectric loss (Tan δ), dielectric constant (ε') and ac conductivity (σ') with frequency was measured using impedance spectroscopy and discussed at length.

Chapter 12, Thomas *et al.*, shows the comprehensive quantum mechanical study of structural features, reactivity, molecular properties characteristics of capmatinib. They reported that Capmatinib is an effective medicine to fight lung cancer. They used molecular modelling using DFT and TD-DFT methods using B3LYP/CAM-B3LYP/aug-cc-pVDZ level to study the structure, reactivity and other Physico-chemical properties of this compound.

The main goal of the compilation of this edited book was to explain the underlying physics ideas, assumptions often seen in the nano literature to the learner. This book tried to demonstrate and motivate these notions by inviting all the informative chapters from enthusiastic scholars and scientists. The objective is to give the readers a foundation that will allow them to critically examine and perhaps contribute to the growing area of material sciences. It is a dream that this book will one day be turned into an introductory text for many.

Dibya Prakash Rai
Department of Physics
Pachhunga University College
Mizoram University, Aizawl
India

DEDICATION

I wholeheartedly dedicate my second edited book to my Guru (Supervisor), retired Professor R. K. Thapa, Vice Chancellor of Sikkim Alpine University, Namchi, Sikkim, who inspired and introduced me to this profession (Teaching & Research).

List of Contributors

Abeer E. Aly	Basic science department, institute of engineering and technology, Cairo, Egypt
Atul Kumar Dadhich	Department of Electrical Engineering, Vivekananda Global University, Jaipur 302012, India
Balram Tripathi	Department of Physics, S.S. Jain Subodh P.G. College, Jaipur-302004, India
Biswajit Roy	CSIR-Centre for Cellular and Molecular Biology, Hyderabad-500007, Telangana, India
Himanshu Joshi	Condensed Matter Theory Research Laboratory, Kurseong College, Darjeeling, India-734203, India St. Joseph's College, North Point, Darjeeling, India-734104, India
H.V. Ganvir	Department of Applied Physics, Yeshwantrao Chavan College of Engineering, Nagpur - 441110, India
Kumaresh Mandal	Department of Zoology, University of Calcutta, 35 Ballygunge Circular Road, Kolkata-700019, West Bengal, India
Mahesh Ram	Condensed Matter Theory Research Laboratory, Kurseong College, Darjeeling, India-734203, India Department of Physics, North-Eastern Hill University, Shillong, India-739002, India
Nihal Limbu	Condensed Matter Theory Research Laboratory, Kurseong College, Darjeeling, India-734203, India Department of Physics, North-Eastern Hill University, Shillong, India-739002, India
Pyare Lal	Department of Physics, School of Physical Sciences, Banasthali Vidyapith-304022 (Rajasthan), India
Rakesh Tamang	Department of Zoology, University of Calcutta, 35 Ballygunge Circular Road, Kolkata-700019, West Bengal, India
Renjith Thomas	Department of Chemistry, St Berchmans College, Changanaserry, Kerala, India
R. Nagarajan	Department of Electrical and Electronics Engineering, Gnanamani College of Technology, Tamilnadu, India
Romyani Goswami	Department of Physics, Surya Sen Mahavidyalaya, Surya Sen Colony, Siliguri 734004, West Bengal, India
R.S. Gedam	Department of Physics, Visvesvaraya National Institute of Technology, Nagpur - 440010, India
Rekha Garg Solanki	Department of Physics, , Dr. Harisingh Gour University, Sagar (M.P.), India
Soni Subba	Department of Zoology, University of Calcutta, 35 Ballygunge Circular Road, Kolkata-700019, West Bengal, India
S. Rai	Physics Department, Mizoram University, Mizoram, Aizawl – 796004, India
S.K. Jain	Department of Physics, School of Basic Sciences, Manipal University Jaipur, Jaipur-303007, India
Shyam Sunder Sharma	Department of Physics, Government Women Engineering College, Ajmer 305002, India

Subodh Srivastava   Department of Physics, Vivekananda Global University, Jaipur 302012, India

Shishir Tamang   Department of Zoology, University of Calcutta, 35 Ballygunge Circular Road, Kolkata-700019, West Bengal, India
Department of Zoology, Darjeeling Government College, Darjeeling-734101, West Bengal, India

Tapati Jana   Department of Physics, Sarojini Naidu College for Women, 30 Jessore Road, Kolkata 700 028, India

V.Y. Ganvir   Department of Applied Physics, Yeshwantrao Chavan College of Engineering, Nagpur - 441110, India

CHAPTER 1

Recent Advancements in Nanotechnology: A Human Health Perspectives

Kumaresh Mandal[1], Shishir Tamang[1,2], Soni Subba[1], Biswajit Roy[3] and **Rakesh Tamang[1,*]**

[1] *Department of Zoology, University of Calcutta, 35 Ballygunge Circular Road, Kolkata-700019, West Bengal, India*

[2] *Department of Zoology, Darjeeling Government College, Darjeeling-734101, West Bengal, India*

[3] *CSIR-Centre for Cellular and Molecular Biology, Hyderabad-500007, Telangana, India*

Abstract: Nanotechnology came into the limelight during the last decade of the twentieth century. It finds immense application in developing nano molecules and nanodevices using molecular, supra-molecular, and atomic level matters. Its role in biomedical engineering is proving crucial. Nanoparticles like silver nanoparticles, gold nanoparticles, *etc.* have wide implications in biomedicine. Even though there are arguments regarding the side effects, risk factors, removal from the human body, *etc.*, the regular use of nanoparticles has proven cost and time-effective solutions for several human health problems. Due to their small size, nanoparticles have an extended reach in the human body and thus have become effective tools in diagnosis and disease treatment. Most importantly the application of nanotechnology in human health includes drug and protein delivery, treating cardiovascular diseases, cancer, neurodegenerative diseases, ophthalmology, *etc.* Various nanosystems like dendrimers, nanoshells, nanocrystals, and quantum dots are effectively used to examine and cure cancer and other patients with complex health problems. Despite its wide range of applications in human health and diseases, the toxicological risk assessment of the ecosystem and human health itself is necessary for every newly developed nanomedicine. Thus, interdisciplinary understanding and evaluation of nanotechnology-based solution tools are necessary for its judicial use in human health.

Keywords: Drug delivery, Human health, Nanotechnology, Nanoparticles, Nanomedicine.

* Corresponding author Rakesh Tamang: Department of Zoology, University of Calcutta 35, Ballygunge Circular Road, Kolkata-700 019, India; Tel: +91332461544; Fax:+913324614849; E-mail: rtzoo@caluniv.ac.in

Dibya Prakash Rai (Ed.)

INTRODUCTION

Nanotechnology involves maneuvering particles of sizes less than 100 nanometers [1]. This size is several hundred folds thinner as compared to the width of human hair. Hence, nanotechnology deals with materials or devices invisible to the human eye. The strengths, conductivity, and reactivity of materials radically change when reduced to the nanoscale. These changed properties are useful in providing innovative solutions in medical science and several other industries, through application-specific engineering of nanoscale materials [2, 3]. However, the advantages of nanotechnology must be critically evaluated against potential hazards associated with its development, usage, and clearance as it may pose potential harm to the individual as well as the environmental health. This is the reason why the National Nanotechnology Initiative, USA, critically emphasizes on environmental, health, and safety impacts of nanotechnology [4]. The public acceptance of nanotechnology will depend on our liability to assess and manage its possible risks to human health and the environment.

The neologism "nano-medicine" emerged in the scientific articles at the end of the twentieth century [5]. Subsequently, the innovation and development of brand new drugs, implantable devices, molecular machinery engineered to the nanolevel have facilitated precision in drug delivery and disease diagnosis. In 2009, the National Institutes of Health, USA [http://www.clinicaltrials.gov], conducted almost 600 clinical studies with the application of nano-products following the standard protocol. Nearly 40% of these experiments are mostly in phase I or phase II. Similarly, the other nano-products are at their preclinical phase and some are *in vitro* use level. Even though there is a huge surge in the novel nano-drugs development, their detailed pharmacokinetic experiments and toxicological knowledge regarding these new drugs are fragmentary. This field lacks efficient prognostic methods and standard protocols for evaluating the toxicological properties of the designed nano molecules *in-vivo* [6]. The World Health Organization (WHO) precisely emphasized the health risk of nanomaterials and suggested that a pragmatic model of "risk governance" is necessary for its various sittings (WHO report 2012, Bonn, Germany WHO report 2010, Parma, Italy). In this chapter, we attempt to discuss the various implication of nanotechnology in human health and its possible threat to the environment.

NANOMATERIALS AND NANOPARTICLES IN MEDICINE

Nanotechnology is promising because it advocates the improvement of existing products and the creation of newer ones with brand new features with massive implications in clinical practices. Biochemical interactions within an organism take place mostly at DNA, RNA, and protein levels. Interventions at these biological units for their efficient functioning can be better comprehended using nanotechnology. Its main applications in medical fields are primarily in disease diagnosis and imaging, disease monitoring, and innovative drug-delivery systems for drugs with possible risks. It is an area with the ability to detect molecules associated with diseases like cancer, neurodegenerative diseases, diabetes mellitus, and detect harmful microorganisms. For instance, in cancer treatment, effective novel nanoparticles are expected to respond to externally applied stimuli making them proper therapeutics or drug delivery systems [2, 3]. Sufficient knowledge on the associated toxicological risk needs extensive research for the nanotechnology-based product available in markets. This is why the risk assessment strategy is a prerequisite for the biomedical and technological application of nanotechnology. As nanoparticles are small in size, they can easily pass through the blood-brain barrier and can migrate through cell membranes. This characteristic of the nanoparticle is exploited to develop nanoscaled vehicles transporting high potential pharmaceutics precisely to the targeted region. Liposomes are used in delivering the desired genes and drugs. Polymer nanoparticles are used in DNA examination [7]. The "Nanorobots" and other "nano-devices" are the future devices with great benefits for health. The artificially designed spherical red blood cell called "respirocyte" with a 1nm diameter delivers more oxygen in comparison with the natural red blood cells [8], as well as to well to manage carbonic acidity. Blood transfusion, controlling anaemia and lung diseases to a certain extent, artificial breathing, preventing asphyxia, *etc.*, will become more efficient and effective with the application of respirocytes [9].

Some nano-level molecules can be applied as tags and labels. They make biological tests more sensitive and flexible. Two types of nanomedicine that have been already experimented within the mice model and will be tested for human trials are 1) gold nanoshells implicated in cancer diagnosis and cure, and 2) liposomes used as a vaccine adjuvant and as mediators in drug transport [10]. Similarly, another application of nanomedicine is drug detoxification. Small and less invasive devices can be developed and accurately implanted inside the body with the help of nanotechnology. The biochemical reaction time of those nanoparticles is much shorter and more sensitive [11].

Nanotechnology in Drug Delivery

The drug delivery system facilitates transporting drugs to the targeted region in the body, its delivery, and absorption in the site of action. Nanotechnology associated drug delivery depends on three major points: i) proficient drug encapsulation, ii) effective drug delivery to the specific sites in the body, and iii) efficient release of delivered drug at the desired region. Nanoparticles find their application in target-specific drug delivery where the side effects are considerably reduced due to their high accuracy. This ultimately reduces the cost of the drugs and pain for the patients. Thus, various dendrimers and nanoporous materials are used in drug delivery systems. The application of micelles from block co-polymers is effective in encapsulating the drug. It helps in the transportation of smaller drug molecules to the target location. Iron nanoparticles or gold shells are used in the treatment of cancer due to their efficiency, success rate, and minimal side effects. A targeted medicine may reduce drug consumption, side effects, and in turn expenses for disease treatment. Nanomedicines are nanoscaled particles or molecules capable of improving the bioavailability of desired drugs. Molecular targeting is carried out by nano-engineered devices called nanorobots for maximizing bioavailability [12]. *In vivo* imaging is another broad-spectrum field where nanotools and devices are being illuminated for high-resolution imaging. In MRI and ultrasound, nanoparticles are introduced as contrasting agents. Biocompatible and self-assembled nanodevices can be used for the detection of cancerous cells and disease examination mechanically.

Using lipid and polymer-based nanoparticles, the therapeutic and pharmacological characteristics of drugs can be enhanced with specific drug delivery systems [13, 14]. The potency of the drug delivery system is its ability to modify the pharmacokinetics and bio-distribution of the drug. As the nanoparticles evade the defence mechanisms of the body [15], they are used as the preferred drug delivery vehicle. Novel drug delivery systems are being designed for providing better treatment opportunities. These systems can transport drugs through cellular membranes and cytoplasm precisely enhancing the efficiency of the drugs. The triggered response is one of the major approaches for drugs to be applied more effectively where the drugs which are administered in the body can be activated only with the proper signal or stimuli specified for them. Inadequately soluble drugs will be substituted by a nano-engineered system that results in better solubility owing to the occurrence of both hydrophilic and hydrophobic surroundings [16] (Fig. **1**). By regulated drug release, tissue damage can be prevented.

Thus, the uses of nanoparticles in diagnostic sensors and bioimaging have resulted in the improvement of a well-organized drug delivery system. The biodistribution of those nanoparticles is still flawed because of the interactions between host and engineered nanoparticles, and the complexities in targeting only the desired tissues in the body accurately. Efforts are being made for optimization and understanding of the advantages and disadvantages of nanoparticle based systems. For studying the excretory system in mice, dendrimers are used as an encapsulation for drug transportation. These were seen to go into the kidneys precisely. However, the negatively charged gold nanoparticles stayed in vital organs. The reason behind this difference is that the positively charged surface of the nanoparticle reduces the nanoparticles' opsonization rate in the liver which in turn leads to hampering the pathways of the excretory system. Nanoparticles can be stored in the peripheral tissues because of their smallness in size (5 nm), and therefore can accumulate in the body in due course of time. Further research is needed on the toxicity of nanoparticles so that their application in medical science can be enhanced to the maximum level [17].

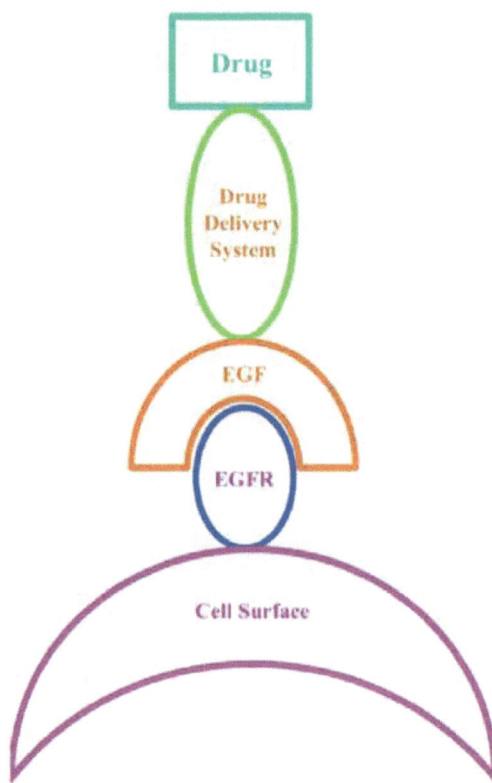

Fig. (1). Schematic outline of drug delivery systems designed using nanotechnology. The drug delivery system may be a nanoparticle or a nanodevice attached to the drug. It facilitates the binding of the drug with the epidermal growth factor (EGF) which in turn binds to its receptor (EGFR) on the cell surface thereby assisting the drug's entry into the target cell.

Another type of nanoparticles used in medicine is minicell nanoparticles used in early clinical trials as a drug transporting medium to treat complex and inoperable cancers. The membranes of various mutant bacteria are used to develop minicells. They are loaded with a cetuximab coat and paclitaxel. When minicells enter the tumour cells, the anti-cancer drug loaded in it destroys the tumour cells. Mostly the large-sized minicells provide desired results. The minicell drug delivery system can be used at a lower dose of the drug and thus will have lesser side effects [18]. One more nano-system used in drug delivery is nanosponges [19]. Due to their minute size and porosity, nanosponges can attach less to insoluble drugs within their matrix and recover their bioavailability which was unavailable previously. Hence, they become helpful in preventing drug and protein degradation and facilitating the guided discharge of drugs.

Nanotechnology in Proteins and Peptide Delivery

Protein and peptides are a group of macromolecules with comparatively longer and shorter chains of amino acid, respectively. These macromolecules are used in the treatment of a variety of diseases and malfunctions as they are part of various biological phenomena inside the body. Nanomaterials like nanoparticles and dendrimers noted as nano-biopharmaceuticals are important for the site-specific delivery of needed molecules. Myelin antigens may trigger immune activity in relapsing multiple sclerosis. Their delivery inside the body becomes much more effective with the help of nanomaterials. In this, the biodegradable myelin sheath coated polystyrene microparticles is thought to reorganize the immune system of mouse and thus check the reappearance of disorders. This is because the shielding myelin sheath forms a coat over the nerve fibers of the central nervous system and is useful in treating various autoimmune diseases [20, 21].

Nanopore Sequencing Nanotechnology

Deoxyribonucleic acid (DNA) is the heritable unit that governs the development, functioning, growth, and propagation of an organism. It consists of four nucleotides namely, Adenine, Guanine, Cytosine, and Thymine arranged to form a polynucleotide [22, 23]. The variation in their arrangement in a polynucleotide stretch may change the genetic information that may ultimately change the respective phenotype. DNA sequencing is one of the most powerful techniques to know the exact order of the nucleotides within a stretch of a DNA molecule [24]. DNA sequencing can be categorized into 3 groups [25] viz., First, Second, and Third generation sequencing. First-generation sequencing is mainly amplification-based Sanger sequencing. Second-generation sequencing includes high throughput sequencing techniques such as arrays of microbeads, massively parallelized chips, DNA clusters, Illumina [26 - 30]. Third-generation sequencing includes nanopore sequencing [31 - 33], single-molecule sequencing by synthesis [34], single-molecule motion sequencing [35 - 37], sequencing by tip-enhanced Raman scattering, *etc* [38 - 42]. Oxford Nanopore Technologies (ONT) launched the first nanopore sequencer, recognized as MinION in 2012 [43]. Then, ONT launched the project on nanopore sequencing – the MinION Access Program (MAP) [44 - 46]. In the MinION device, an enzyme unwinds DNA first, feeding one strand inside a protein nanopore [47 - 49]. The distinct shape of each base produces a typical alteration in the electrical current which provides a readout of the underlying sequence [50 - 53]. With the impact of nanopore sequencing, the genome sequence of the resistance-causing element in the multidrug-resistant (MDR) strain of *E.coli* was illustrated [54 - 56].

Nanotechnology in Cardiovascular Disease

Global health estimates by WHO on 9th December 2020 stated that cardiovascular disease (Ischemic heart disease and stroke) plays the main role in human mortality globally. Minimal invasive treatments for cardiovascular disease are the desirable goal for health workers across the globe. The advancement in nanotechnology to the lesser invasive methods has brought a ray of hope for the cardiovascular patient. The cardiovascular gene therapy system can be understood through the detection of a protein that leads to the formation of blood vessels, packaging, and development of strands of DNA which comprises the gene responsible for the production of the right protein, and delivery of the DNA in heart muscle [9]. Nanotechnology has immense significance in the interventional therapeutics of atherosclerosis and coronary artery disease (CAD). Applications of various nanoparticles improved the biocompatibility of intracoronary stents and regulation of the chief limit factors for *Percutaneous Transluminal Coronary Angioplasty* (PTCA). It was noted that overexpression of nanotized PPARα (Peroxisomal Proliferator-Activated Receptor alpha) can ameliorate pathological hypertrophy and improve cardiac activities. Overexpression of myocardium-targeted nanotized PPARα is carried out by using a conjugated carboxymethyl-chitosan nanopolymer (CMC) modified with stearic acid and this reduces apoptosis *via* downregulation of the p53 acetylation [57].

Nanotechnology in Cancer

Nanoparticles have promising applications in oncology, mostly in imaging. For this, Quantum dots are used. These are nanoparticles that contain quantum confinement characters, like size-tunable light emission. These can be applied in magnetic resonance imaging for the production of very minute and advanced images of the tumour [58]. The fluorescent quantum dots produce a high-resolution image at a cheaper price as compared to the traditional method. Several toxic elements in quantum dots are the only drawback of this method of imaging.

Nanoparticles are comprised of a distinct character of high surface area to volume ratio. This property of nanoparticles facilitates various functional groups to attach to a nanoparticle and thus efficiently fix with specific tumour cells. Multifunctional nanoparticles can be manufactured that would be helpful in detection, imaging, and then treatment of a tumour in the future [59]. In Kanzius RF therapy used in killing cancer cells, the nanoparticles are attached to the cancerous cells that are destroyed through radio waves. Nanowires are used in preparing sensor test chips used in detecting cancer biomarkers. They are also capable of providing an accurate cancer diagnosis at an early stage from blood samples [60, 61].

The various types of nanosystems (Fig. **2**) used in cancer therapy [62] are **1) Carbon nanotubes:** these are of 0.5 nm to 3 nm in diameter, and 20 nm to 1000 nm length and have a usage in the detection of DNA mutation and protein biomarkers associated with diseases, **2) Dendrimers:** their size is smaller than 10 nm and are useful in controlled delivery of the drug, and as contrast particles in imaging, **3) Nanocrystals:** their size ranges is 2 nm to 9.5 nm. They are helpful in enhanced expression of very less soluble drugs, labelling of HeR2, a marker for breast cancer in the cancer cell surface, **4) Nanoparticles:** these are of 10 nm to 1000 nm in size and find their application in ultrasound and MRI as image contrast particles, and for site-specific drug delivery **5) Nanoshells:** they are used in imaging associated with tumour, deep tissue thermal ablation, **6) Nanowires:** these are helpful for detection of protein biomarker, detection of DNA mutation and gene expression **7) Quantum dots:** these are of 2-9.5 nm in size and assists in optically detecting proteins and genes in model animals and cellular experiments, and imaging of tumour and lymph node.

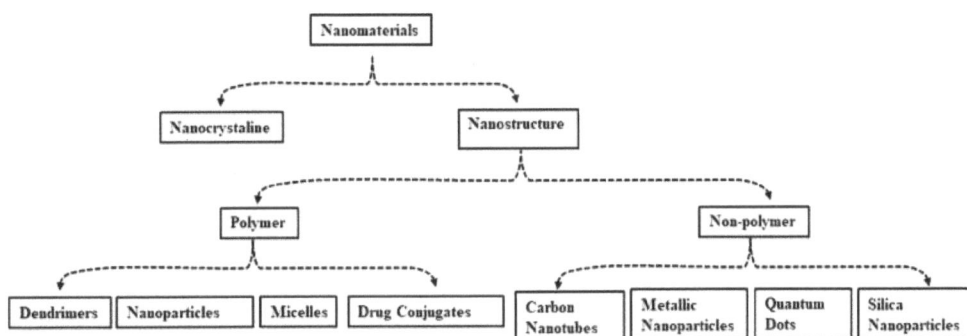

Fig. (2). Classification of Nanomaterials.

The major areas where nanomedicine is being designed in cancer biology are *early-stage detection of tumours and cancer treatment.* Early-stage detection of tumours can be facilitated by the development of "smart" tissue collection platforms for synchronized investigation of markers related to cancer. This is followed by the treatment process *via* the creation of nanodevices that are capable of releasing chemotherapeutic agents precisely to the target region. For overcoming cancer, preventing it is the best option. If it happens, early tumour diagnosis and its timely eradication will significantly augment better survival. Nanowires are used in the detection of molecules associated with cancer and thereby contribute to diagnosis at an early stage and detection of tumours [63]. For tumour detection, nanoparticle contrast agents have already been designed. Both labelled and non-labelled nanoparticles have been experimented with as agents for imaging tumours to aid in better diagnosis. Tumour treatment can become effective with silica-coated micelles, liposomes, dendrimers, and ceramic nanoparticles. These particles may be used in vehicles for site-specific drug delivery that carries therapeutic genes or chemotherapeutic agents towards malignant cells [64].

Nanotechnology in the Treatment of Neurodegenerative Disorders

For treating neurodegenerative disorders, application of the nanotechnology is proving advantageous [65]. Several nano-vehicles like dendrimers, nanoemulsions, nano gels, liposomes, solid lipid nanoparticles, polymeric nanoparticles, and nanosuspensions have been greatly studied for the delivery of the central nervous system (CNS) therapeutics. Efficient transportation of these nanomedicines across a range of *in vitro* and *in vivo* BBB (blood-brain barrier) experiments through endocytosis or transcytosis, has shown efficiency at an early stage in preclinical trials for the supervision of various CNS situations like Alzheimer's disease, brain tumours, and HIV encephalopathy. Improvement of their permeability through BBB and reduction in their neurotoxicity are major areas in nanomedicine research in the future with a great promise for combating neurological diseases.

With the application of nanotechnology, a significant improvement in the current approaches in Parkinson's disease (PD) therapy has been achieved. After Alzheimer's disease, Parkinson's disease (PD) which affects persons above 65 years of age at a rate of 0.01, is the second most familiar neurodegenerative disorder. PD is a disease of the central nervous system (CNS). Currently, the available therapies try to improve the functional capacity of the PD patient though they are not able to alter the succession of the neurodegenerative processes. The goal of functional nanotechnology is neuroprotection and regeneration of the CNS to check the neurodegeneration which is a challenging task. A multidisciplinary approach combining nanotechnology, neurophysiology, neuropathology, and cell biology will solve the intricacies of its progression and treatment of neuro-degenerative disease.

Applications of Nanotechnology in Ophthalmology

With the application of nanobiotechnology, extensive progress in several sophisticated fields of ophthalmology has become possible now. Application of nanomedicines, various nanodevices, and regenerative nanomedicine have fetched a new horizon in managing oxidative stress, intraocular pressure measurement, healing choroidal new vessels, and prevention of scarring in patients operated for glaucoma. For the treatment of rigorous evaporative dry eye, a new nanoscale-dispersed eye ointment (NDEO) has been effectively developed [66].

Nanotechnology in Immunological Perspective

The nanodevice buckyballs find their application to modify the allergy or immune reaction. Because of their better attachment with free radicals than vitamin E or any available anti-oxidant, they can check mast cells from secreting histamine in the body especially in blood and tissue [67]. Through the interference with different proteins implicated in the resistance to antibiotics and pharmacological processes of drugs, zinc oxide nanoparticles are capable of decreasing antibiotic resistance and augmenting the antibacterial action of Ciprofloxacin over other microorganisms [68]. With the application of nanotechnology in tissue engineering, the reproduction or repair of damaged tissues can be possible. In organ transplants, cell proliferation due to artificial stimuli in or simulated implant therapy, nanotechnology may help provide suitable nanomaterial-based scaffolds and growth factors. This ultimately may lead to the patient's life extension.

Nanopharmaceuticals

Nanopharmaceuticals are useful in detecting diseases at early stages. It is an emerging field where nanoscaled drug particle or nano-level therapeutic delivery systems are used. the delivery of a particular active agent with a suitable dose to a specific disease region is vital and complicated in the pharmaceutical industry. Nanopharmaceuticals have huge potential in increasing precision and accuracy in site-specific targeting of active agents. Also, it plays a significant role in the reduction of noxious side effects. Further to delivering high-quality products to patients after maintaining profitability, the pharmaceutical industry faces enormous pressure. Thus, to improve drug target discovery and drug formulation, companies are taking advantage of nanotechnology. Nanopharmaceuticals are vital in making the drug discovery procedure cost-effective.

POSSIBLE RISKS FOR HUMAN HEALTH AND ETHICAL QUESTIONS

The literature on toxicological risks of nanotechnology in the field of medicine is inadequate. Size reduction of structures to nanolevel alters various distinct characteristics [69]. For the toxic effects of particles, the dominant indicator is the smallness in size. Hence, nano-formulation requires a proper evaluation in terms of its activity, reactivity, and toxicity. Chemical property dictating the fundamental toxic natures of the chemical is important in the determination of the particles' toxicity. The detailed mechanism of nanoparticle elimination from the human body is still not well-known. The studies so far indicate the way of elimination of nanoparticles *via* liver sinusoids, space of Disse, hepatocytes, bile ducts, and intestines. However, their transport processes are not well studied [70].

NANOTECHNOLOGY IN THERAPEUTICS: DRAWBACK AND PROSPECTIVE

Theoretically, many regular substances can be used as medicines. Whether they precisely reach the unhealthy organs or tissues in the body [71] is still investigated. These substances are hardly soluble in water. They are susceptible to breakage or get inactivated before reaching the target region. Their capability to pass through several biological barriers (blood-brain barrier, placenta, cell membranes, *etc.*) is lower, and is generally released non-specifically to almost all types of organs and tissues. There are several requirements that a delivery system has to fulfill, such as the residence time of the delivery systems in blood should be longer for the deposition in the specific site, they should have the capability to contain adequate active material, the systems, and the degraded products should comprise a complimentary toxicity outline, their shelf-life should be long enough to permit proper distribution and storage, the efficiency of delivery systems must be in proportion to the cost of production and treatment [72 - 75].

CONCLUSION

Nanomaterials have increased surface area and are comprised of nanoscale effects. These properties make them a potential tool for drug development, gene delivery systems, imaging in biomedical fields, and biosensor particles. Nanomaterials pose distinctive biological and physicochemical characteristic features than regular materials. The characteristics of nanomaterials can significantly control their connections with biomolecules and cells. These interactions occur due to their unusual shape, size, surface property, chemical structure, charge, and solubility. With the usage of nanoparticles, exceptional images of tumour sites can be constructed. Due to their high efficiency in delivery and transportation, single-walled carbon nanotubes can be used in transporting molecules inside the cellular bodies. Several highly sophisticated biological technologies are intertwined with nanotechnology for better results. Nanotechnology is capable to engineer a phenomenon at the smallest scale and thus has a strong role in the advancement of broad-spectrum areas like information technology, cognitive science, biotechnology, and other integrative biological fields. With the advancement of futuristic research in nanotechnology, its influence on human health is prominent. Personalization in various highly developed fields in biomedical technology, regenerative medicine, stem cell, and nutraceuticals will be materialized and glorified by nanotechnology innovations and progressions.

CONSENT FOR PUBLICATION

Not applicable.

CONFLICT OF INTEREST

The authors declare no conflict of interest, financial or otherwise.

ACKNOWLEDGEMENTS

This study was supported by SERB, the Government of India through SERB-Core grant (CRG/2018/001727) and SERB-Empowerment and Equity Grant (EEQ/2019/000750). KM was supported by UGC JRF (649/CSIR-UGC NET JUNE 2019). ST was supported by a grant (Memo No. 215(Sanc.)/ST/P/S&T/5G-9/2018) from the Department of Science and Technology, Government of West Bengal. Financial assistance to SS from SERB, the Government of India is acknowledged.

REFERENCES

[1]   Bayley, H. Holes with an edge. *Nature,* **2010**, *467*(7312), 164-165.
        [http://dx.doi.org/10.1038/467164a]

[2]   Logothetidis, S. Nanotechnology in medicine: The medicine of tomorrow and nanomedicine. *Hippokratia,* **2006**.

[3]   Logothetidis, S.; Gioti, M.; Lousinian, S.; Fotiadou, S. Haemocompatibility studies on carbon-based thin films by ellipsometry. *Thin Solid Films,* **2005**, *482*(1-2), 126-132.
        [http://dx.doi.org/10.1016/j.tsf.2004.11.131]

[4]   Navalakhe, R.M.; Nandedkar, T.D. Application of nanotechnology in biomedicine. *Indian J. Exp. Biol.,* **2007**, *45*(2), 160-165.

[PMID: 17375555]

[5]   Freitas, R.A., Jr What is nanomedicine? *Nanomedicine*, **2005**, *1*(1), 2-9.
         [http://dx.doi.org/10.1016/j.nano.2004.11.003] [PMID: 17292052]

[6]   World Health Organisation. Nanotechnology and human health: Scientific evidence and risk governance. In: *Report of the WHO expert meeting*; Bonn, Germany, **2013**.

[7]   Highlights of the first annual meeting of the American Academy of Nanomedicine. *Nanomedicine Nanotechnology, Biol. Med*, **2005**, *1*

[8]   Zhu, R.; Avsievich, T.; Popov, A.; Bykov, A.; Meglinski, I. In vivo nano-biosensing element of red blood cell-mediated delivery. *Biosens. Bioelectron.*, **2021**, *175*, 112845.
         [http://dx.doi.org/10.1016/j.bios.2020.112845] [PMID: 33262059]

[9]   Korpanty, G.; Chen, S.; Shohet, R.V.; Ding, J.; Yang, B.; Frenkel, P.A.; Grayburn, P.A. Targeting of VEGF-mediated angiogenesis to rat myocardium using ultrasonic destruction of microbubbles. *Gene Ther.*, **2005**, *12*(17), 1305-1312.
         [http://dx.doi.org/10.1038/sj.gt.3302532] [PMID: 15829992]

[10]   Boisseau, P.; Loubaton, B. Nanomedicine, nanotechnology in medicine. *C. R. Phys.*, **2011**, *12*(7), 620-636.
         [http://dx.doi.org/10.1016/j.crhy.2011.06.001]

[11]   LaVan, D.A.; McGuire, T.; Langer, R. Small-scale systems for in vivo drug delivery. *Nat. Biotechnol.*, **2003**, *21*(10), 1184-1191.
         [http://dx.doi.org/10.1038/nbt876] [PMID: 14520404]

[12]   Cavalcanti, A.; Shirinzadeh, B.; Freitas, R.A., Jr; Hogg, T. Nanorobot architecture for medical target identification. *Nanotechnology*, **2008**, *19*(1), 015103.
         [http://dx.doi.org/10.1088/0957-4484/19/01/015103]

[13]   Allen, T.M.; Cullis, P.R. Drug delivery systems: entering the mainstream. *Science*, **2004**, *303*(5665), 1818-1822.
         [http://dx.doi.org/10.1126/science.1095833] [PMID: 15031496]

[14]   Allen, T.M.; Cullis, P.R. Supplementary Material Drug delivery systems: entering the mainstream. *Science*, **2004**, *303*(5665), 1818-1822.
         [http://dx.doi.org/10.1126/science.1095833]

[15]   Bertrand, N.; Leroux, J.C. The journey of a drug-carrier in the body: An anatomo-physiological perspective. *J. Control. Release*, **2012**, *161*(2), 152-163.
         [http://dx.doi.org/10.1016/j.jconrel.2011.09.098] [PMID: 22001607]

[16]   Nagy, Z.K.; Balogh, A.; Vajna, B.; Farkas, A.; Patyi, G.; Kramarics, Á.; Marosi, G. Comparison of electrospun and extruded Soluplus®-based solid dosage forms of improved dissolution. *J. Pharm. Sci.*, **2012**, *101*(1), 322-332.
         [http://dx.doi.org/10.1002/jps.22731] [PMID: 21918982]

[17]   Minchin, R. Sizing up targets with nanoparticles. *Nat. Nanotechnol.*, **2008**, *3*(1), 12-13.
         [http://dx.doi.org/10.1038/nnano.2007.433]

[18]   Brambilla, D.; Le Droumaguet, B.; Nicolas, J.; Hashemi, S.H.; Wu, L.P.; Moghimi, S.M.; Couvreur, P.; Andrieux, K. Nanotechnologies for Alzheimer's disease: diagnosis, therapy, and safety issues. *Nanomedicine*, **2011**, *7*(5), 521-540.
         [http://dx.doi.org/10.1016/j.nano.2011.03.008] [PMID: 21477665]

[19]   Ahmed, R.Z.; Patil, G.; Zaheer, Z. Nanosponges – a completely new nano-horizon: pharmaceutical applications and recent advances. *Drug Dev. Ind. Pharm.*, **2013**, *39*(9), 1263-1272.
         [http://dx.doi.org/10.3109/03639045.2012.694610] [PMID: 22681585]

[20]   Nikalje, A.P. Nanotechnology and its Applications in Medicine. *Med. Chem. (Los Angeles)*, **2015**, *5*(2)
         [http://dx.doi.org/10.4172/2161-0444.1000247]

[21]   Ou, A.; H, L.; Oo, O.; Bi, I.; Pi, A.; M, P. Utility of Nanomedicine for Cancer Treatment. *J. Nanomed. Nanotechnol.,* **2018**, *9*(1)
[http://dx.doi.org/10.4172/2157-7439.1000481]

[22]   Simpson, J.T.; Workman, R.E.; Zuzarte, P.C.; David, M.; Dursi, L.J.; Timp, W. Detecting DNA cytosine methylation using nanopore sequencing. *Nat. Methods,* **2017**, *14*(4), 407-410.
[http://dx.doi.org/10.1038/nmeth.4184] [PMID: 28218898]

[23]   Kono, N.; Arakawa, K. Nanopore sequencing: Review of potential applications in functional genomics. *Dev. Growth Differ.,* **2019**, *61*(5), 316-326.
[http://dx.doi.org/10.1111/dgd.12608] [PMID: 31037722]

[24]   Jilsha, Nanoscience and Nanotechnology : An Introduction Why Nano Rather than Something Else. *Sci. Technol,* **2015**, *1*(2)

[25]   Schadt, E.E.; Turner, S.; Kasarskis, A. A window into third-generation sequencing. *Hum. Mol. Genet.,* **2010**, *19*(R2), R227-R240.
[http://dx.doi.org/10.1093/hmg/ddq416] [PMID: 20858600]

[26]   Heng, J.B.; Aksimentiev, A.; Ho, C.; Dimitrov, V.; Sorsch, T.W.; Miner, J.F.; Mansfield, W.M.; Schulten, K.; Timp, G. Beyond the gene chip. *Bell Labs Tech. J.,* **2005**, *10*(3), 5-22.
[http://dx.doi.org/10.1002/bltj.20102] [PMID: 18815623]

[27]   Erlich, Y.; Mitra, P.P.; delaBastide, M.; McCombie, W.R.; Hannon, G.J. Alta-Cyclic: a self-optimizing base caller for next-generation sequencing. *Nat. Methods,* **2008**, *5*(8), 679-682.
[http://dx.doi.org/10.1038/nmeth.1230] [PMID: 18604217]

[28]   Saunders, C.J.; Miller, N.A.; Soden, S.E.; Dinwiddie, D.L.; Noll, A.; Alnadi, N.A.; Andraws, N.; Patterson, M.L.; Krivohlavek, L.A.; Fellis, J.; Humphray, S.; Saffrey, P.; Kingsbury, Z.; Weir, J.C.; Betley, J.; Grocock, R.J.; Margulies, E.H.; Farrow, E.G.; Artman, M.; Safina, N.P.; Petrikin, J.E.; Hall, K.P.; Kingsmore, S.F. Rapid whole-genome sequencing for genetic disease diagnosis in neonatal intensive care units. *Sci. Transl. Med.,* **2012**, *4*(154), 154ra135.
[http://dx.doi.org/10.1126/scitranslmed.3004041] [PMID: 23035047]

[29]   Minervini, C.F.; Cumbo, C.; Orsini, P.; Anelli, L.; Zagaria, A.; Specchia, G.; Albano, F. Nanopore Sequencing in Blood Diseases: A Wide Range of Opportunities. *Front. Genet.,* **2020**, *11*, 76.
[http://dx.doi.org/10.3389/fgene.2020.00076] [PMID: 32140171]

[30]   Schneider, G.F.; Dekker, C. DNA sequencing with nanopores. *Nat. Biotechnol.,* **2012**, *30*(4), 326-328.
[http://dx.doi.org/10.1038/nbt.2181] [PMID: 22491281]

[31]   Branton, D. W. Deamer; A., Marziali; J. A, Schloss The potential and challenges of nanopore sequencing, in Nanoscience and Technology: A Collection of Reviews. *Nature Journals,* **2009**.

[32]   Branton, D.; Deamer, D.W.; Marziali, A.; Bayley, H.; Benner, S.A.; Butler, T.; Di Ventra, M.; Garaj, S.; Hibbs, A.; Huang, X.; Jovanovich, S.B.; Krstic, P.S.; Lindsay, S.; Ling, X.S.; Mastrangelo, C.H.; Meller, A.; Oliver, J.S.; Pershin, Y.V.; Ramsey, J.M.; Riehn, R.; Soni, G.V.; Tabard-Cossa, V.; Wanunu, M.; Wiggin, M.; Schloss, J.A. The potential and challenges of nanopore sequencing. *Nat. Biotechnol.,* **2008**, *26*(10), 1146-1153.
[http://dx.doi.org/10.1038/nbt.1495]

[33]   Schreiber, J.; Wescoe, Z.L.; Abu-Shumays, R.; Vivian, J.T.; Baatar, B.; Karplus, K.; Akeson, M. Error rates for nanopore discrimination among cytosine, methylcytosine, and hydroxymethylcytosine along individual DNA strands. *Proc. Natl. Acad. Sci. USA,* **2013**, *110*(47), 18910-18915.
[http://dx.doi.org/10.1073/pnas.1310615110]

[34]   Harris, D.; Buzby, P. R.; Babcock, H.; Beer, F.; Bowers, J.; Xie, Z. Single-molecule DNA sequencing of a viral genome. *Science (80-),* **2008**.
[http://dx.doi.org/10.1126/science.1150427]

[35]   Greenleaf, J.; Block, S. M. Single-molecule, motion-based DNA sequencing using rna polymerase.*Science (80-),* **2006**.

[http://dx.doi.org/10.1126/science.1130105]

[36]   Ding, F.; Manosas, M.; Spiering, M.M.; Benkovic, S.J.; Bensimon, D.; Allemand, J.F.; Croquette, V. Single-molecule mechanical identification and sequencing. *Nat. Methods,* **2012**, *9*(4), 367-372.
[http://dx.doi.org/10.1038/nmeth.1925] [PMID: 22406857]

[37]   Bell, D.C.; Thomas, W.K.; Murtagh, K.M.; Dionne, C.A.; Graham, A.C.; Anderson, J.E.; Glover, W.R. DNA base identification by electron microscopy. *Microsc. Microanal.,* **2012**, *18*(5), 1049-1053.
[http://dx.doi.org/10.1017/S1431927612012615] [PMID: 23046798]

[38]   Cover Picture: Tip-Enhanced Raman Spectroscopy of Single RNA Strands: Towards a Novel Direct-Sequencing Method (Angew. Chem. Int. Ed. 9/2008). , **2008**.
[http://dx.doi.org/10.1002/anie.200890029]

[39]   Bailo, E.; Deckert, V. Tip-enhanced Raman spectroscopy of single RNA strands: Towards a novel direct-sequencing method. *Angew. Chem. Int. Ed.,* **2008**, *47*(9), 1658-1661.
[http://dx.doi.org/10.1002/anie.200704054]

[40]   Treffer, X. Lin, E. Bailo, T. Deckert-Gaudig, and V. Deckert, Distinction of nucleobases - A tip-enhanced Raman approach. *Beilstein J. Nanotechnol.,* **2011**.
[http://dx.doi.org/10.3762/bjnano.2.66] [PMID: 22003468]

[41]   Hong, J.; Lee, Y.; Chansin, G.A.T.; Edel, J.B.; deMello, A.J. Design of a solid-state nanopore-based platform for single-molecule spectroscopy. *Nanotechnology,* **2008**, *19*(16), 165205.
[http://dx.doi.org/10.1088/0957-4484/19/16/165205] [PMID: 21825639]

[42]   Wang, Q.Y.; Wang, Z. The evolution of nanopore sequencing. *Front. Genet.,* **2014**.
[http://dx.doi.org/10.3389/fgene.2014.00449] [PMID: 25610451]

[43]   Fanget, A.; Traversi, F.; Khlybov, S.; Granjon, P.; Magrez, A.; Forró, L.; Radenovic, A. Nanopore integrated nanogaps for DNA detection. *Nano Lett.,* **2014**, *14*(1), 244-249.
[http://dx.doi.org/10.1021/nl403849g] [PMID: 24308689]

[44]   Goodwin, S.; McPherson, J.D.; McCombie, W.R. Coming of age: ten years of next-generation sequencing technologies. *Nat. Rev. Genet.,* **2016**, *17*(6), 333-351.
[http://dx.doi.org/10.1038/nrg.2016.49] [PMID: 27184599]

[45]   Magi, A.; Semeraro, R.; Mingrino, A.; Giusti, B.; D'Aurizio, R. Nanopore sequencing data analysis: state of the art, applications and challenges. *Brief. Bioinform.,* **2017**.
[http://dx.doi.org/10.1093/bib/bbx062] [PMID: 28637243]

[46]   Rohrandt, N. Kraft; B., Brandl; B. M., Schuldt; F. J., Muller Nanopore SimulatION - A raw data simulator for Nanopore Sequencing. **2019**.
[http://dx.doi.org/10.1109/BIBM.2018.8621253]

[47]   Chen, X.; Rungger, I.; Pemmaraju, C.D.; Schwingenschlögl, U.; Sanvito, S. First-principles study of high-conductance DNA sequencing with carbon nanotube electrodes. *Phys. Rev. B Condens. Matter Mater. Phys.,* **2012**, *85*(11), 115436.
[http://dx.doi.org/10.1103/PhysRevB.85.115436]

[48]   Li, W. Si; J., Sha; Y, Chen Molecular dynamics study of DNA translocation through graphene nanopores with controllable speed. **2015**.
[http://dx.doi.org/10.1115/IMECE2015-50858]

[49]   Schneider, G.F.; Kowalczyk, S.W.; Calado, V.E.; Pandraud, G.; Zandbergen, H.W.; Vandersypen, L.M.K.; Dekker, C. DNA translocation through graphene nanopores. *Nano Lett.,* **2010**, *10*(8), 3163-3167.
[http://dx.doi.org/10.1021/nl102069z] [PMID: 20608744]

[50]   Translocation of Single-Stranded DNA Through Single-Walled Carbon Nanotubes. *Science (80),* **2010**.

[51]   Liu, L.; Yang, C.; Zhao, K.; Li, J.; Wu, H.C. Ultrashort single-walled carbon nanotubes in a lipid

bilayer as a new nanopore sensor. *Nat. Commun.,* **2013**, *4*(1), 2989.
[http://dx.doi.org/10.1038/ncomms3989] [PMID: 24352224]

[52]   Liu, L.; Xie, J.; Li, T.; Wu, H.C. Fabrication of nanopores with ultrashort single-walled carbon nanotubes inserted in a lipid bilayer. *Nat. Protoc.,* **2015**, *10*(11), 1670-1678.
[http://dx.doi.org/10.1038/nprot.2015.112] [PMID: 26426500]

[53]   Ling, X.S. DNA sequencing using nanopores and kinetic proofreading. *Quant. Biol.,* **2020**, *8*(3), 187-194.
[http://dx.doi.org/10.1007/s40484-020-0201-x]

[54]   Fouda, M.M.G.; Abdel-Halim, E.S.; Al-Deyab, S.S. RETRACTED: Antibacterial modification of cotton using nanotechnology. *Carbohydr. Polym.,* **2013**, *92*(2), 943-954.
[http://dx.doi.org/10.1016/j.carbpol.2012.09.074] [PMID: 23399115]

[55]   Orsini, P.; Minervini, C.F.; Cumbo, C.; Anelli, L.; Zagaria, A.; Minervini, A.; Coccaro, N.; Tota, G.; Casieri, P.; Impera, L.; Parciante, E.; Brunetti, C.; Giordano, A.; Specchia, G.; Albano, F. Design and MinION testing of a nanopore targeted gene sequencing panel for chronic lymphocytic leukemia. *Sci. Rep.,* **2018**, *8*(1), 11798.
[http://dx.doi.org/10.1038/s41598-018-30330-y] [PMID: 30087429]

[56]   Kuo, F.C.; Mar, B.G.; Lindsley, R.C.; Lindeman, N.I. The relative utilities of genome-wide, gene panel, and individual gene sequencing in clinical practice. *Blood,* **2017**, *130*(4), 433-439.
[http://dx.doi.org/10.1182/blood-2017-03-734533] [PMID: 28600338]

[57]   Rana, S.; Datta, R.; Chaudhuri, R.D.; Chatterjee, E.; Chawla-Sarkar, M.; Sarkar, S. Nanotized PPARα overexpression targeted to hypertrophied myocardium improves cardiac function by attenuating the p53-gsk3β-mediated mitochondrial death pathway. *Antioxid. Redox Signal.,* **2019**, *30*(5), 713-732.
[http://dx.doi.org/10.1089/ars.2017.7371] [PMID: 29631413]

[58]   Gao, D.; Guo, X.; Zhang, X.; Chen, S.; Wang, Y.; Chen, T.; Huang, G.; Gao, Y.; Tian, Z.; Yang, Z. Multifunctional phototheranostic nanomedicine for cancer imaging and treatment. *Mater. Today Bio,* **2020**, *5*, 100035.
[http://dx.doi.org/10.1016/j.mtbio.2019.100035] [PMID: 32211603]

[59]   Nie, S.; Xing, Y.; Kim, G.J.; Simons, J.W. Nanotechnology applications in cancer. *Annu. Rev. Biomed. Eng.,* **2007**, *9*(1), 257-288.
[http://dx.doi.org/10.1146/annurev.bioeng.9.060906.152025] [PMID: 17439359]

[60]   Zheng, G.; Patolsky, F.; Cui, Y.; Wang, W.U.; Lieber, C.M. Multiplexed electrical detection of cancer markers with nanowire sensor arrays. *Nat. Biotechnol.,* **2005**, *23*(10), 1294-1301.
[http://dx.doi.org/10.1038/nbt1138] [PMID: 16170313]

[61]   Roszek, W. H. De Jong; R. E, Geertsma Nanotechnology in Medical Applications : State-of-the-Art in Materials and Devices RIVM report 265001001 / 2005 Nanotechnology in medical applications : state-of-the-art in materials and devices. **2005**.

[62]   Nahar, M.; Dutta, T.; Murugesan, S.; Asthana, A.; Mishra, D.; Rajkumar, V.; Tare, M.; Saraf, S.; Jain, N.K. Functional polymeric nanoparticles: an efficient and promising tool for active delivery of bioactives. *Crit. Rev. Ther. Drug Carrier Syst.,* **2006**, *23*(4), 259-318.
[http://dx.doi.org/10.1615/CritRevTherDrugCarrierSyst.v23.i4.10] [PMID: 17341200]

[63]   Kawasaki, E.S.; Player, A. Nanotechnology, nanomedicine, and the development of new, effective therapies for cancer. *Nanomedicine,* **2005**, *1*(2), 101-109.
[http://dx.doi.org/10.1016/j.nano.2005.03.002] [PMID: 17292064]

[64]   Zharov, V.P.; Kim, J.W.; Curiel, D.T.; Everts, M. Self-assembling nanoclusters in living systems: application for integrated photothermal nanodiagnostics and nanotherapy. *Nanomedicine,* **2005**, *1*(4), 326-345.
[http://dx.doi.org/10.1016/j.nano.2005.10.006] [PMID: 17292107]

[65]   Wong, H.L.; Wu, X.Y.; Bendayan, R. Nanotechnological advances for the delivery of CNS

therapeutics. *Adv. Drug Deliv. Rev.,* **2012**, *64*(7), 686-700.
[http://dx.doi.org/10.1016/j.addr.2011.10.007] [PMID: 22100125]

[66] Li, Y. *Zhang, J. Yang, K. Bi, Z. Ni, D. Li, Y. Chen, Molecular dynamics study of DNA translocation through graphene nanopores*; Phys. Rev. E - Stat. Nonlinear, Soft Matter Phys, **2013**.
[http://dx.doi.org/10.1103/PhysRevE.87.062707]

[67] Bayda, M. Adeel, T. Tuccinardi, M. Cordani, and F. Rizzolio, "The history of nanoscience and nanotechnology: From chemical-physical applications to nanomedicine. *Molecules,* **2020**.
[http://dx.doi.org/10.3390/molecules25010112] [PMID: 31892180]

[68] Merchant, C. DNA Translocation Through Graphene Nanopores. *Biophys. J.,* **2011**, *100*(3), 521a.
[http://dx.doi.org/10.1016/j.bpj.2010.12.3046]

[69] Ebbesen, M.; Jensen, T.G. Nanomedicine: techniques, potentials, and ethical implications. *J. Biomed. Biotechnol.,* **2006**, *2006*(5), 1-11.
[http://dx.doi.org/10.1155/JBB/2006/51516] [PMID: 17489016]

[70] Poon, W.; Zhang, Y.N.; Ouyang, B.; Kingston, B.R.; Wu, J.L.Y.; Wilhelm, S.; Chan, W.C.W. Elimination Pathways of Nanoparticles. *ACS Nano,* **2019**, *13*(5), 5785-5798.
[http://dx.doi.org/10.1021/acsnano.9b01383] [PMID: 30990673]

[71] Shrivastava, S.; Dash, D. Applying Nanotechnology to Human Health: Revolution in Biomedical Sciences. *J. Nanotechnol.,* **2009**, *2009*, 1-14.
[http://dx.doi.org/10.1155/2009/184702]

[72] Misra, R.; Acharya, S.; Sahoo, S.K. Cancer nanotechnology: application of nanotechnology in cancer therapy. *Drug Discov. Today,* **2010**, *15*(19-20), 842-850.
[http://dx.doi.org/10.1016/j.drudis.2010.08.006] [PMID: 20727417]

[73] Sahoo, S.K.; Labhasetwar, V. Nanotech approaches to drug delivery and imaging. *Drug Discov. Today,* **2003**, *8*(24), 1112-1120.
[http://dx.doi.org/10.1016/S1359-6446(03)02903-9] [PMID: 14678737]

[74] Sahoo, S.; Dilnawaz, F.; Krishnakumar, S. Nanotechnology in ocular drug delivery. *Drug Discov. Today,* **2008**, *13*(3-4), 144-151.
[http://dx.doi.org/10.1016/j.drudis.2007.10.021] [PMID: 18275912]

[75] Agarwal, ; Huang, D.; Thakur, S.S.; Rupenthal, I.D. Nanotechnology for ocular drug delivery *Design of Nanostructures for Versatile Therapeutic Applications,* **2018**.

CHAPTER 2

An Exploratory Study on Characteristics of SWIRL of AlGaAs/GaAs in Advanced Bio based Nanotechnological Systems

Pyare Lal[1,*]

[1] *Department of Physics, School of Physical Sciences, Banasthali Vidyapith-304022 (Rajasthan), India*

Abstract: The most prominent aim of this innovative research book chapter has been to study an optimised exploration of characteristics of SWIRL (Short Wave Infra-Red Light) of AlGaAs/GaAs heterogeneous type nanostructure under various GRINSLs (Graded Refractive Index Nano Scale Layers) in advanced bio-based nanotechnological systems. Under this optimised exploration, the simulating performances of SWIRL gain enhancement with wavelengths of light photons for various GRINSLs have been systematically calculated. Other important parameters like SWIRL modal confinement gain with current densities per unit cm^2, SWIRL differential gain and parameter of anti-guiding with carriers per unit cm^3 have been computed. In the innovative investigation through the results, the highest value of SWIRL gain has been achieved at the wavelength ~ 830 nm. The SWIR light of wavelength ~ 830 nm has mostly been utilised in the optimization of a proper combination of higher penetrating abilities and cellular type interactions. Hence, this wavelength's SWIRL source has also been used in the treatment of wound, pain and various types of sensitive skin diseases by using the FONSCs (Fibre Optic Nano Scale Cables) through the TIRPs (Total Internal Reflection Processes) without any attenuation in dB/Km due to diminished scattering, dispersion and absorptions in the nanotechnological biosciences and medical sciences. Moreover, this SWIRL of wavelength ~ 830 nm has provided the most fabulous role in the proper controlling of inflammation, oedema as well as infections of various bacteria in advanced bio-based nanotechnological systems.

Keywords: AlGaAs, Change in indices of refraction, Differential SWIRL gain, GaAs, GRINSLs, Modal SWIRL confinement gain, Net peak SWIR gain, SWIRL gain, SWIRL output power, SWIRL parameter of antiguiding, SWIRL loss.

---

* **Corresponding author Pyare Lal:** Department of Physics, School of Physical Sciences, Banasthali Vidyapith-304022 (Rajasthan), India; Tel:+919829587431; E-mails: drpyarephysics@gmail.com and pyarelal@banasthali.in

Dibya Prakash Rai (Ed.)

INTRODUCTION

The importance of heterogeneous nanostructure in the advanced bio-based nano-technological sciences has been found to be increasing in recent times. At the present time under the nanoscale-technological and engineering sciences, the performances of heterogeneous nanostructures are very critical because of their unique optical light properties. The various types of experimental and theoretical research-based innovative work all over the world have been done by the researchers. In several fields like medical science, industries, radar system, aerospace, photovoltaic and detectors areas, lasing type devices, *etc*. The nanoscale heterogeneous structures provide a great role due to their several optical performances. The several types of optical properties of various nanoscale type heterogeneous nanostructures [1 - 7] have been investigated by researchers. In general, heterogeneous structures are formed by the process of the combination of multiple hetero type junctions. The hetero-junctions are those junctions that are formed by the interface between the dissimilar band gap nano-materials. Among various nanoscale heterostructures, AlGaAs/GaAs material based nano-heterogeneous structure has been very popular due to the emission of radiations of ~ 830 nm wavelength. These types of wavelengths have been of high concern due to their potential performances in fibre optic appliances based telecommunications due to diminishing attenuation. These nanotechnological materials have been set up with some additional rewards such as gaining stability at a higher temperature, improved linewidth enhancement factor and photonic wavelength. The heterogeneous nanostructure based on the materials AlGaAs and AlGaAsIn/InP [8 - 10] has also been reported as a platform on which the nanotechnological devices can be fabricated. Recently, the properties of SWIRL, NIRL, and UVL for AlGaAs/GaAs, InAlGaAs/InP and AlGaN/AlN have been investigated and presented by authors at several national and international conferences. For example, the electrical results such as the I-V and C-V curves of the Schottky-diodes, which were fabricated on AlGaAs/GaAs in nanotechnology, have been studied under the various nanoscale barrier layers. In this innovative chapter, an investigative study on the growth of modal confinement light gain of proposed nanostructure in nanotechnological life sciences has been done by computational nano techniques under the number of NGILs (Nano Graded Index Layers). This chapter provides a substantial contribution to nano-biological sciences because of their unique utilities. The spectral performances of growth of modal type light gain with energies of light photons in eV of proposed nanostructure have been calculated and computed by spectral performances. In these spectral performances the peaks of spectra have been achieved at energy of photons ~1.5 eV correspondence wavelengths of light photons ~ 830 nm have been illustrated by appropriate graphical curves under various types of NGILs. Next, the modal type behaviours of transparency energies of light photons with different types of

NGILs for the proposed nanostructure have been calculated by graphical results. Next, this light of wavelength ~ 830 nm has been utilised in the optimization of a proper combination of higher penetrating abilities and cellular interactions.

SIMULATED HETEROGENEOUS TYPE NANOSTRUCTURE AND THEORETICAL DETAILS

The proposed and simulated structure is an innovative heterogeneous nanoscale structure because its dimensions are in nanoscale order. The proposed nanoscale structure has total 21 nanoscale layers in which SQWNSL (Single Quantum Well Nanoscale Layer) is sandwiched among the 20 GRINSLs in the form of 18 BNSLs (Barriers Nanoscale Layers) and 2 CNSLs (Cladding Nanoscale Layers). Hence, the entire system is simulated and grown on the substrate. In the proposed nanostructures, nanoscale layers are such that the value of QWOCPs (Quantum Well Optical Confinement Parameters) is diminished from the QW-NSL to CNSLs. In other words, the QW-NSL has a maximum value of QWOPs while the CNSLs have a minimum value of QWOPs. Parameter details of the simulated heterogeneous nanostructure have been shown in below Table **1**.

Table 1. Parameter details of proposed and simulated heterogeneous type nanostructures.

Nanoscale Layers Specified	Parameters of Nanoscale Layers ($Al_xGa_{1-x}As$) Value of x	Thickness of Nanoscale Layers (nm)	Values in eV of Energy of C-BOs (Conduction-Band Offsets)	Values in eV of Energy of V-BOs (Valences- Band Offsets)
1-GRINSL	0.20	0.5511	0.16267	-0.10500
2-GRINSL	0.24	0.5511	0.19359	-0.12630
3-GRINSL	0.28	0.5511	0.22467	-0.14789
4-GRINSL	0.32	0.5511	0.25639	-0.16987
5-GRINSL	0.36	0.5511	0.28976	-0.18765
6-GRINSL	0.40	0.5511	0.31999	-0.21211
7-GRINSL	0.44	0.5511	0.35131	-0.24611
8-GRINSL	0.48	0.5511	0.39543	-0.27987
9-GRINSL	0.52	0.5511	0.44986	-0.31432
10-GRINSL (CNSL)	0.56	10.000	0.49576	-0.35678
(QWNSL)	0.10	06.000	----------	---------

Since the last few decades it has been clear that, in advanced bio-based nanotechnological and engineering sciences for optical communication-based treatment, the SWIR light gain [11 - 14] has been provided with a substantial role due to its unique light properties. Commonly, the enhancement in SWIR light is given by the incremental variation of SWIR light per unit of original SWIR light intensity and per unit length of the active region of the nanostructure. In the nano-optoelectronics SWIR light yielding has been achieved when transitions towards upward direction enhance than transitions towards the downward direction, while in the equal condition of upward and downward transitions the achieved yielding is occurred negligible this type of condition is termed as a condition of transparent. When the power of absorption is higher than stimulated type emission then the loss is obtained in the SWIR light. An expression related to the yielding of SWIR light gain as a function of photon's energy is given by the below mathematical equation. This SWIR light gain equation is given by the ref [15].

$$G(E) = \frac{q^2 h}{2n_{eff}\, h\nu m_0^2 \varepsilon_0 c} \times \left[1 - \exp\left(\frac{E - \Delta f}{k_b T}\right)\right]$$

$$\times \sum_{nc,nv} \frac{|M_b|^2 f_c f_v}{4\pi^2 L_W} \times \frac{(h/2\pi\tau)dk_x dk_y}{\pi(\{E_{nc} + E_{nv} + E_{sg}\} - E)^2 + (h/2\pi\tau)^2}$$

(1)

In the above equation (1), the value of SWIR light gain is exponentially enhanced as a decrease in temperature values. In other words, the gain value tends to zero as the temperature value increases and tends to infinity. At absolute zero temperature, the gain value becomes always positive and maximum. Moreover, the SWIR light gain has an inverse relationship with the width value of QWWNLs (Quantum Well Width Nanoscale Layers) due to diminished energy separation between various energy levels in the active region. It has also been observed by the gain relationship that gain value has proportional behaviours concerning energy values of separation between quasi-Fermi energies of levels. In this relationship, it has been predicted that the gain enhancement value has also been affected by bulk momentum matrix elements. Sometimes gain value depends on the modal confinement parameter with the reduced amount in laser processing.

The modal confinement parameter has a critical role in the computing of yielding of modal confinement light gain per cm in advanced bio-based nanosystems. The fundamental equation of the modal confinement parameter is given as the following relation (2).

$$\Gamma = \frac{\int_{-W/2}^{W/2} \left[|\varepsilon(z)|^2\right]dz}{\int_{-\infty}^{\infty} \left[|\varepsilon(z)|^2\right]dz}$$

(2)

When photon light power related to the active region is divided by the photon light related to the entire region then the modal confinement parameter is achieved. It has been verified that the modal confinement gain is basically a product of modal confinement SWIRL parameter and material based light gain enhancement. It is approximately equal to one but always less than one. It can be determined by the ratio of modal confinement gain coefficient and light gain coefficient. It depends on the geometry of the wave function and active region as well as the number of NGILs and the number of QWWNLs.

The accurate rate at which light gain varies with respect to carriers per unit volume is termed as differential light gain. It can also be determined by the change in SWIRL gain per unit charge carrier per unit volume. The equation of differential gain as a function of the energy of photons is expressed as the below numerically integral relation. In the above differential SWIRL gain equation (3), it has been observed that the value of differential SWIRL gain depends on the thicknesses of QWWNLs as well as bulk momentum matrix element and Lorentzian lineshape function.

$$G'(E) = \frac{dG(E)}{dN} = \frac{8\pi^2 m_r E}{c\varepsilon h^3 L_w} \times \int_{E'}^{\infty} |M_b|^2$$
$$\times \left(\frac{df_c(E)}{dN} - \frac{df_v(E)}{dN} \right) \times L(E')dE' \tag{3}$$

In general, the value of the index of refraction has been enhanced from CNLs (Cladding Nano Layers) to The QWWNLs in the structure. Next, the rate at which the index of refraction has been changed with respect to carriers is called the differential type index of refraction and it is shown by the below mathematical equation (4).

$$n'(hv) = \frac{dn(E)}{dN} = \frac{4\pi^2 m_r E \lambda \tau}{c\varepsilon h^4 L_w}$$
$$\times \int_{E'}^{\infty} |M_b|^2 \times \left(\frac{df_c(E)}{dN} - \frac{df_v(E)}{dN} \right) \times (E' - E) \times L(E')dE' \tag{4}$$

The fractional ratio of the differential index of refraction and differential SWIRL gain is termed as a parameter of antiguiding. It is a substantial parameter that provides supporting behaviours with SWIRL gain in lasing action. Its value can be occurred negative and zero as well as positive values due to changes in values of the index of refraction. It is also called AP (Alpha Parameter) and LWEP (Line Width Enhancement Parameter). The parameter of antiguiding is expressed in terms of differential index of refraction and differential SWIRL gain by the below mathematical equation (5).

$$G' = \frac{dG}{dN} = \frac{4\pi}{\lambda} \times \left(\frac{1}{\alpha}\right) \times \left(-\frac{dn}{dN}\right) \tag{5}$$

The frequency value that can be affected by differential SWIRL gain and OOP (Optical Output Power) is called RRFs (Relaxation Resonance Frequencies). In general, RRFs can be measured in GHz. The RRFs as well as OOP have a very crucial role in the determination of threshold conditions for the lasing process. In the enhanced study of differential type gain, RRFs provide vital contributions. The equation of relaxation oscillation type frequency is exhibited by the following equation (6). In this equation, the brief details of appropriate terms can be exhibited in refs [16 - 18].

$$f_r = \frac{1}{2\pi} \times \left(\frac{(cP)}{(n\tau_p)} \times \{G'(E,N)\}\right)^{1/2} \tag{6}$$

The exact positive value of injection current in mA at which the value of RRFs tends to zero is known as threshold current and this situation is called a case of a threshold condition. At and above the threshold condition, the rate stimulated emission has been greater than that of the rate of absorption, hence SWIRL gain is achieved. But below the threshold condition *i.e.* at transparency situations the achieved value of gain is negligible because the rate of stimulated emission is approximately equal to the rate of absorption. In this condition, the positive and exact values of current densities per unit area of the cross-section are termed as transparent current densities. Sometimes the value of achieved SWIRL gain is negative because, in this phenomenon, the rate at which stimulated emission occurred is less than the rate at which absorption occurred, hence this loss is called net modal loss per cm. The value of net modal loss per cm is basically the sum of mirror loss per cm and internal loss per cm. The expression of threshold current in mA is given as below appropriate exponential relation (7).

$$I_{th} = \left(\frac{nJ_0 L_w L}{\eta}\right) \times \exp\left[\left(\frac{1}{n\Gamma G_0(J)}\right) \times \left(\alpha_{int} + \frac{1}{2L}\ln\frac{1}{R_1 R_2}\right) - 1\right] \tag{7}$$

The attenuation loss dB/km has been found maximum at a wavelength range of photons ~ 1401 nm but it has been found minimum at wavelengths ~ 831 nm, 1331 nm and 1551 nm, when SWIR light propagates through the NFOCs (Nano Fibre Optic Cables) by process of TIRs (Total Internal Reflections). SWIRL attenuation depends on various factors like the length of wire of fiber optic as well as output optical power and input optical power. In general, it is measured in dB/km. The proper and exact attenuation equation in terms of OPs (Optical Powers) is given by the following mathematical and logarithmic formula (8).

$$A = \left(\frac{10}{L}\right) \times \log_{10}\left(\frac{P_{in}}{P_{out}}\right)^{-1} \tag{8}$$

OPTIMIZED COMPUTATIONAL RESULTS AND DISCUSSIONS

Basically, SWIR light increments are the net amount of the stimulated emission that a typical SWIRL photon generates as it travels at a given appropriate distance of propagation. In the hetero nanostructures, the SWIR light amplification has been caused by photon induced transition of electrons from the c-band to the v-band in the QW active region. If the rate of downward transitions exceeds the rate of upward transitions, there will be a net generation of photons and enhancement or profit in SWIRL gain *i.e.* optical type gain [11, 12] can be achieved. The SWIR light gain is exponentially enhanced as a decrease in temperature values. In other words, the gain value tends to be zero as the temperature value increases and tends to be infinity. At absolute zero temperature, the gain value becomes always positive and maximum.

Moreover, the SWIR light gain has an inverse relationship with the width value of QWWNLs (Quantum Well Width Nano Layers) due to diminished energy separation between various energy levels in the active region. It has also been observed by the gain relationship that gain value has proportional behaviours with respect to energy values of separation between quasi-Fermi energies of levels. In this relationship, it has been predicted that the gain enhancement value has also been affected by bulk momentum matrix elements. Sometimes gain value depends on the modal confinement parameter with the reduced amount in laser processing. The modal confinement parameter has a critical role in the computing of yielding of modal confinement light gain per cm in advanced bio-based nanosystems. The modal properties of SWIRL, NIRL, and UVL for AlGaAs/GaAs, InAlGaAs/InP and AlGaN/AlN have been investigated by Pyare Lal [19 - 32] in recent years.

The SWIRL modal gain enhancement per cm versus photonic wavelengths for various NGILs and peak modal SWIRL gain enhancement in the intensity of light per cm versus the number of GRINSLs or NGI-Layers of ternary material AlGaAs/GaAs type heterogeneous nanostructure under the various number of GRINSLs or NGILs have been illustrated by the left y-axis and bottom x-axis; and right y-axis and top x-axis, respectively in Fig. (1). The exact and proper value of the wavelength of photons at which modal SWIRL gain is equal to approximately zero is called the transparency wavelength of photons. In this phenomenon, the stimulated and absorption rates both are approximately equal hence obtained achievement in gain is zero.

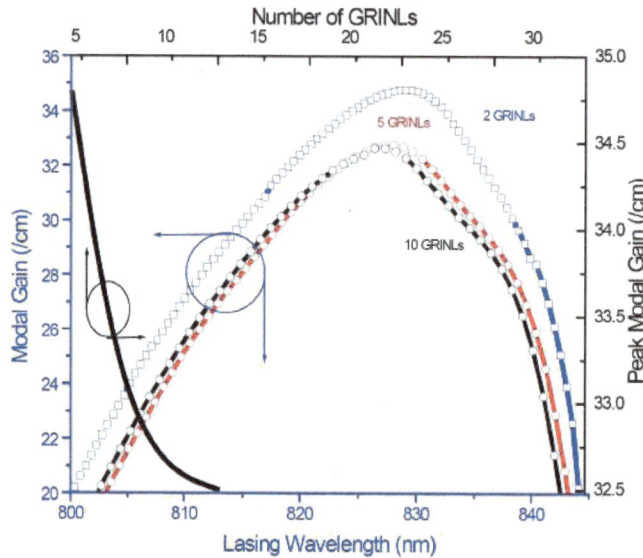

Fig. (1). Modal SWIR light gain intensity with wavelength of photons and peak modal SWIR light gain intensity with various GRINSLs for AlGaAs/GaAs heterogeneous nanostructure.

It has been predicted experimentally and theoretically, in the condition of transparency, that the energy separation between quasi-Fermi energies is equal to bandgap energy. Above the transparent condition *i.e.* at the threshold condition and above the threshold condition the net SWIRL gain has been achieved due to an increase in the rate of stimulated emission as well as energy separation between quasi-Fermi energies. The value of peak modal SWIRL gain enhancement tends to be higher as a reduction in the several numbers of NGILs due to an increase in the value of the parameter of mode confinement. The highest value of enhancement in the intensity of modal type SWIR light gain per cm is achieved at the wavelength of 830 nm.

It has also been observed that by three spectral curves in Fig. (1) the peak SWIR light modal gain at wavelength ~ 830 mm for 2 GRINSLs is maximum and approximately equal to 35 /cm, for 5 GRINSLs the SWIRL modal gain is reduced and that is approximately equal to 33 /cm and for 10 GRINSLs the SWIRL modal gain is also reduced and that is approximately equal to 32 /cm because the QW modal confinement parameter has been high for 2 GRINSLs but the QW confinement parameter for 10 GRINSLs is lowest hence peak value of SWIRL modal gain is compressed as enhancement in the number of NGILs or GRINSLs. This type SWIR light of wavelength ~ 830 mm has critical importance in the utilisation of SWIRL and NIRL applications to achieve the combination of higher penetration power and appropriate cellular interaction performances without any type absorption losses of light signals in the cable of fibre optic. The achieved modal SWIRL gain through the spectra correspondence to the maximum modal light gain of wavelength (~830 nm) for the lasing phenomenon has an essential contribution in current days for the applications of EM radiations as well as this wavelength range has been also utilised in fibre optic telecommunications by the method of TIRs with diminished loss and attenuation in dB/km of light signals. Moreover, the emitted light of 830 nm wavelength range has been used in the treatment of skin type diseases and this range also provides the contribution to the determination of correlation between the higher value of penetration abilities and interaction of cellular in the medical sciences and bio based nanotechnological systems in daily life applications.

In general, the value of the index of refraction has been enhanced from CNLs (Cladding Nano Layers) to The QWWNLs in the structure. The rate at which the index of refraction has been changed with respect to carriers is called the differential type index of refraction. The fractional ratio of the differential index of refraction and differential SWIRL gain is termed as a parameter of antiguiding. It is a substantial parameter that provides supporting behaviours with SWIRL gain in lasing action. Its value can be negative and zero as well as positive values due to changes in values of the index of refraction. It is also called AP (Alpha Parameter) and LWEP (Line Width Enhancement Parameter). The parameter of antiguiding is expressed in terms of differential type index of refraction and differential type SWIRL gain

The proper performances in the value of the index of refraction with current densities per unit area and changeable behaviours in the parameter of anti-guiding with charge carriers concentration per cubic cm of AlGaAs/GaAs type heterogeneous nanostructure under the various numbers of GRINSLs have been illustrated graphically in Fig. (**2**). It has been observed by curves that at the lowest value of current densities the value of change in the index of refraction is obtained maximum and at the largest value of current densities, the value of change in the index of refraction is minimum and saturated.

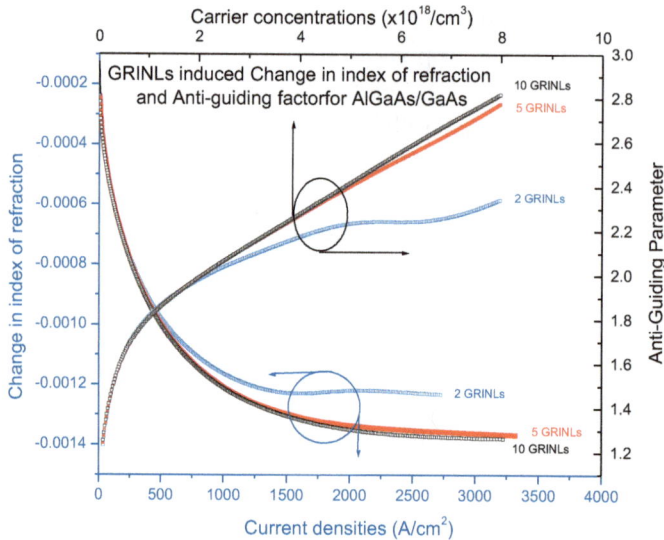

Fig. (2). The proper performances in the value of index of refraction with current densities per unit area and changeable behaviours in parameter of anti-guiding with charge carrier's concentration per cubic cm of AlGaAs/GaAs heterogeneous nanostructure.

It means the change in the index of refraction has an inverse relationship with current densities per unit cross-section area for various GRINSLs. The saturated values of change in the index of refraction have been diminished as an increment in the number of GRINSLs. In other words, it can be said that for 2 GRINSLs the value of change in the index of refraction has been found maximum.

Next, the changing behaviours of a parameter of antiguiding with carrier concentration have been expressed by the right y-axis and top x-axis that have been shown in Fig. (2). It has been also observed by graphs in Fig. (2) that the antiguiding parameter has proportional behaviours with respect to carriers per cubic cm as well as a number of GRINSLs.

Due to various types of the burning of holes like spatial and spectra, the value of SWIR light gain has been reduced, this type of reduction in gain is generally known as SWIRL gain compression in cubic cm. In general, in physical sciences, the reduction in SWIR light gain per unit saturated SWIR light gain and per unit carrier concentration per cubic cm is called SWIR light gain compression. In other words, it is equal to the ratio of differential SWIR light gain and fundamental SWIR light gain. While the substantial type parameter of antiguiding is approximately equal to the ratio of derivative of the index of refraction with respect to carriers and derivative of modal SWIR light gain with respect to carriers. The compression behaviours in peak SWIRL gain per cubic cm with various numbers of GRINSLs and range of parameters of anti-guiding with maximum change in the index of refraction for AlGaAs/GaAs heterogeneous nanostructure are shown in Fig. (3). It has been observed by Fig. (3)., both SWIRL gain compression and rate of anti-guiding parameter have inversely behaviours with GRINSLs and peak change in the index of refraction respectively.

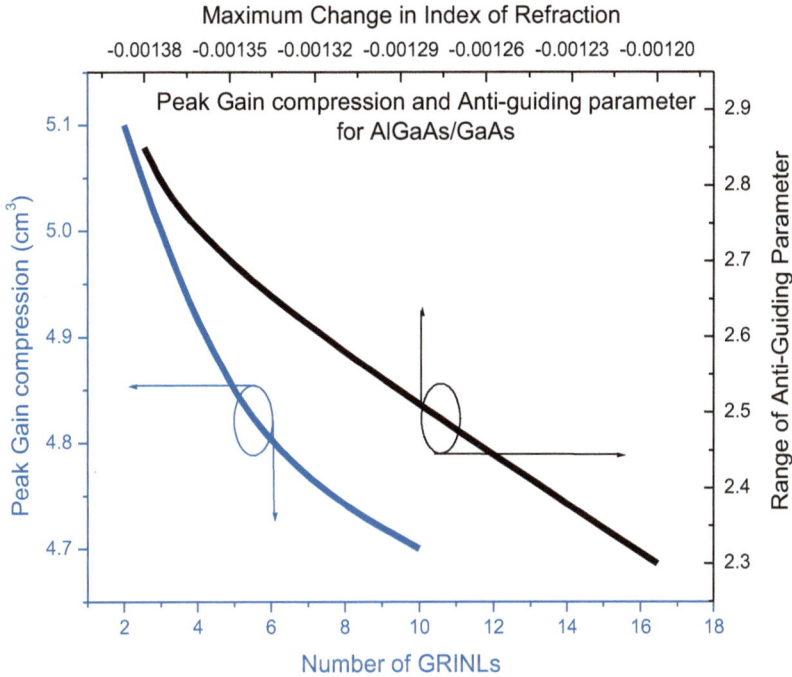

Fig. (3). The appropriate performances in the value of peak SWIR light gain compression with various numbers of GRINSLs and rate of the parameter of antiguiding with the value of peak change in the index of refraction of AlGaAs/GaAs heterogeneous nanostructure.

In Fig. (4) the proper changing behaviours and performances of the value of peak SWIR light output power with various numbers of GRINSLs and saturated SWIR light modal gain of with value of peak leakage current per unit cross-section area of AlGaAs/GaAs type heterogeneous nanostructure have been illustrated by computational and appropriate graphical curves. It has been observed by a blue graph that the peak light output power has proportional behaviours with the number of GRINSLs while it has been observed by a black curve that the saturated SWIR light modal gain has inversely behaviours with a peak value of leakage current per unit cross-section area.

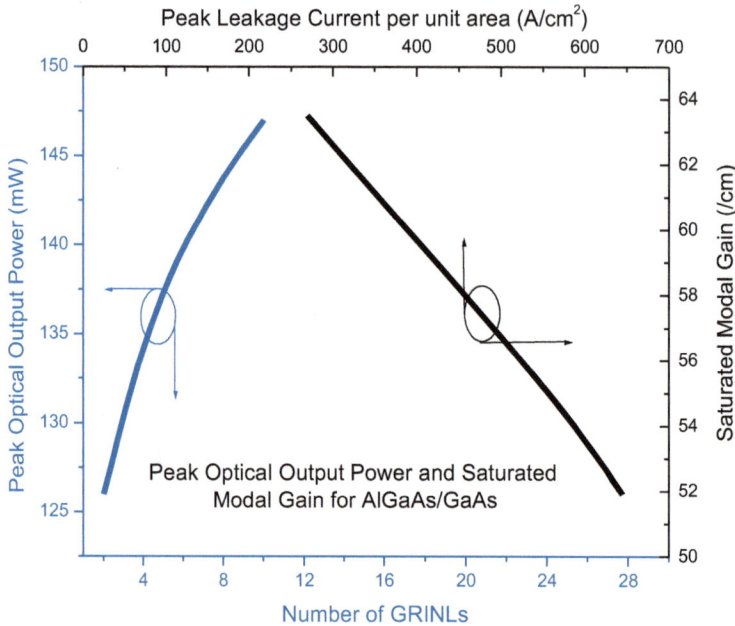

Fig. (4). The proper changing behaviours and performances of the value of peak SWIR light output power with various GRINSLs and saturated SWIR light modal gain of with value of peak leakage current per unit cross section area of AlGaAs/GaAs heterogeneous nanostructure.

The graphical behaviours of ROFs (Relaxation Oscillation Frequencies) in Hz and OOPs (Optical Output Powers) in mW with injection current per unit cross-section area in mA for several types of GRINSLs for proposed AlGaAs/GaAs type heterogeneous nanostructures have been exhibited and computed by Fig. (5). It has been observed by the exact positive value of injection current in mA at which the value of RRFs and OOPs both tends to zero is known as threshold type current (~ 25 mA) and this type situation is called a case of a threshold condition. At and above the threshold condition, the rate stimulated emission has been greater than that of the rate of absorption, hence SWIRL gain is achieved.

But below the threshold condition *i.e.* at transparency situations the achieved value of gain is negligible because the rate of stimulated emission is approximately equal to the rate of absorption. In this condition, the positive and exact values of current densities per unit area of the cross-section are termed as transparent current densities. Sometimes the value of achieved SWIRL gain is negative because, in this phenomenon, the rate at which stimulated emission occurred is less than the rate at which absorption occurred, hence this type of loss is called net modal loss per cm. The value of net modal loss per cm is basically the sum of mirror type loss per cm and internal type loss per cm.

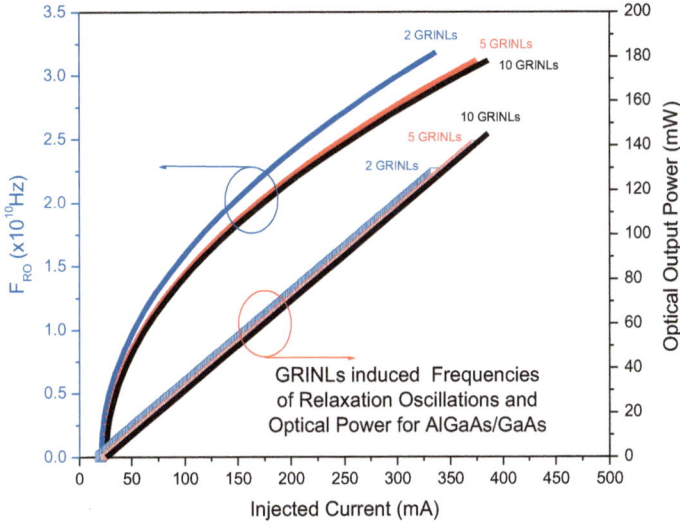

Fig. (5). The appropriate graphical performance of ROFs in Hz and OOPs in mW with injection current per unit cross-section area in mA for several types of GRINSLs for proposed AlGaAs/GaAs type heterogeneous nanostructure.

Above the threshold current (~25 mA), values of RRFs and OOPs both are enhanced as enhancement of injection current for several types of GRINSLs. The frequency value that can be affected by differential SWIRL gain and OOP (Optical Output Power) is called RRFs (Relaxation Resonance Frequencies). In general, RRFs can be measured in GHz. The RRFs as well as OOP have a very crucial role in the determination of threshold conditions for the lasing process. In the enhanced study of differential type gain, RRFs provide vital contributions. Moreover, the maximum and saturated values of RRFs have been highest for 2 GRINSLs and lowest for 10 GRINSLs while peak values of OOPs have been lowest for 2 GRINSLs and highest for 10 GRINSLs.

Hence in the last, it has been said by the author that in the innovative investigation through the results the highest value of SWIRL gain has been achieved at the wavelength ~ 830 nm. The SWIR light of wavelength ~ 830 nm has mostly been utilised in the optimization of a proper combination of higher penetrating abilities and cellular type interactions. Hence, this wavelength's SWIRL source has also been used in the treatment of wound, pain and various types of sensitive skin diseases by using the FONSCs (Fiber Optic Nanoscale Cables) through the TIRPs (Total Internal Reflection Processes) without any attenuation in dB/Km due to diminished scattering, dispersion and absorptions in the nanotechnological biosciences and medical sciences. Moreover, this SWIRL of wavelength ~ 830 nm has the most fabulous role in the proper controlling of inflammation, oedema as well as infections of various bacteria in advanced bio based nanotechnological engineering systems and nanosciences.

CONCLUSION

Under the several numbers of GRINSLs (Graded Refractive Index Nanoscale Layers), in advanced bio based nanotechnological sciences and nanosystem an investigative and exploratory study on characteristics of SWIR light of AlGaAs/GaAs type heterogeneous nanostructure has been done by computational type nano techniques. This type of innovative book chapter provides a substantial contribution in nano type advanced biological sciences because of their unique utilities. The spectral performances of growth of modal SWIR light gain with energies of light photons in eV of AlGaAs/GaAs have been calculated and computed by spectral performances. In these spectral performances, the peaks of spectra have been achieved at the energy of SWIR light photons ~1.5 eV correspondence wavelengths of light photons ~830 nm have been illustrated by appropriate graphical curves under various types of GRINSLs. The SWIRL modal type behaviours of transparency energies of light photons with different types of GRINSLs for proposed nanostructure have been calculated by graphical and simulation results. This type of SWIR light of wavelengths ~ 830 nm has mostly been utilised in the optimization of a proper combination of higher penetrating abilities and cellular type interactions. This wavelength's SWIR light source has also been used in the treatment of various sensitive types of skin diseases in life sciences and advanced bio-based nanotechnological systems and engineering sciences.

CONSENT FOR PUBLICATION

Not applicable.

CONFLICT OF INTEREST

The author declares no conflict of interest, financial or otherwise.

ACKNOWLEDGEMENTS

Author is very grateful to Banasthali Vidyapith for providing computer related appropriate facilities in the School of Physical Sciences.

REFERENCES

[1]   Alvi, P.A.; Pyare Lal, S. Dalela, M. J. Siddiqui, "An Extensive Study on Simple and GRIN SCH based $In_{0.71}Ga_{0.21}Al_{0.08}As$/InP Lasing heterostructure. *Phys. Scr.,* **2012**, *85*, 035402. [http://dx.doi.org/10.1088/0031-8949/85/03/035402]

[2]   Alvi, P.A.; Lal, P.; Yadav, R.; Dixit, S.; Dalela, S. Modal gain characteristics of GRIN-InGaAlAs/InP

lasing nano-heterostructures. *Superlattices Microstruct.,* **2013**, *61*, 1-12.
[http://dx.doi.org/10.1016/j.spmi.2013.05.019]

[3]   Alvi, P.A. Strain-induced non-linear optical properties of straddling-type indium gallium aluminum arsenic/indium phosphide nanoscale-heterostructures. *Mater. Sci. Semicond. Process.,* **2015**, *31*, 106-115.
[http://dx.doi.org/10.1016/j.mssp.2014.11.016]

[4]   Ramam, A.; Chua, S.J. Features of InGaAlAs/InP heterostructures. *J. Vac. Sci. Technol. B,* **1998**, *16*(2), 565.
[http://dx.doi.org/10.1116/1.589864]

[5]   Rybalko, D.A.; Polukhin, I.S.; Solov'ev, Y.V.; Mikhailovskiy, G.A.; Odnoblyudov, M.A.; Gubenko, A.E.; Livshits, D.A.; Firsov, A.N.; Kirsyaev, A.N.; Efremov, A.A.; Bougrov, V.E. Model of mode-locked quantum-well semiconductor laser based on InGaAs/InGaAlAs/InP heterostructure. *J. Phys. Conf. Ser.,* **2016**, *741*, 012079.
[http://dx.doi.org/10.1088/1742-6596/741/1/012079]

[6]   Lal, Pyare; Bhardwaj, Garima; Kattayat, Sandhya Tunable Anti-Guiding Factor and Optical Gain of InGaAlAs/InP Nano-Heterostructure under Internal Strain. *Journal of Nano- and Electronic Physics,* *12*(2), 02002(3pp).**2020,**

[7]   Lal, P.; Gupta, S. G-J study for GRIN InGaAlAs/InP lasing nano-heterostructures *AIP Conference Proceedings, 1536*(1), 53-54.**2013,**

[8]   Sandra, R. Selmic, Tso-Min Chou, JiehPing Sih, Jay B. Kirk, Art Mantie, Jerome K. Butler, David Bour, and Gary A. Evans, "Design and Characterization of 1.3-μm AlGaInAs–InP Multiple-Quantu--Well Lasers. *IEEE J. Sel. Top. Quantum Electron.,* **2001**, *7*(2).

[9]   Yoshitomi, S.; Yamanaka, K.; Goto, Y.; Yokomura, Y.; Nishiyama, N.; Arai, S. Continuous-wave operation of a 1.3 μm wavelength npn AlGaInAs/InP transistor laser up to 90 °C. *Jpn. J. Appl. Phys.,* **2020**, *59*(4), 042003.
[http://dx.doi.org/10.35848/1347-4065/ab7ef2]

[10]   Joachim Piprek, J. Kenton White, and Anthony J. SpringThorpe "What Limits the Maximum Output Power of Long-Wavelength AlGaInAs/InP Laser Diodes?". *IEEE J. Quantum Electron.,* **2002**, *38*(9).

[11]   Chow, W.W.; Zhang, Z.; Norman, J.C.; Liu, S.; Bowers, J.E. On quantum-dot lasing at gain peak with linewidth enhancement factor $\alpha_H = 0$. *APL Photonics,* **2020**, *5*(2), 026101.
[http://dx.doi.org/10.1063/1.5133075]

[12]   Lal, P.; Alvi, P.A. Strain induced gain optimization in type-I InGaAlAs/InP nanoscale-heterostructure. *AIP Conf. Proc.,* **2020**, *2220*, 020060.
[http://dx.doi.org/10.1063/5.0001124]

[13]   Karachinsky, L.Ya.; Novikov, I.I.; Babichev, A.V.; Gladyshev, A.G.; Kolodeznyi, E.S. Optical Gain in Laser Heterostructures with an Active Area Based on an InGaAs/InGaAlAs Superlattice *ISSN 0030-400X, Optics and Spectroscopy,* **2019**, *127*(6), 1053-1056.

[14]   Geuchies, J.J.; Brynjarsson, B.; Grimaldi, G.; Gudjonsdottir, S.; van der Stam, W.; Evers, W.H.; Houtepen, A.J. Quantitative Electrochemical Control over Optical Gain in Quantum-Dot Solids. *ACS Nano,* **2021**, *15*(1), 377-386.
[http://dx.doi.org/10.1021/acsnano.0c07365] [PMID: 33171052]

[15]   Chuang, S.L. *Physics of optoelectronic devices*; Wiley: New York, **1995**.

[16]   Henry, C. Theory of the linewidth of semiconductor lasers. *IEEE J. Quantum Electron.,* **1982**, *18*(2), 259-264.
[http://dx.doi.org/10.1109/JQE.1982.1071522]

[17]   Vahala, K.; Yariv, A. Semiclassical theory of noise in semiconductor lasers - Part II. *IEEE J. Quantum Electron.,* **1983**, *19*(6), 1102-1109.
[http://dx.doi.org/10.1109/JQE.1983.1071984]

[18] Lal, P.; Yadav, R.; Sharma, M.; Rahman, F.; Dalela, S.; Alvi, P.A. Qualitative analysis of gain spectra of InGaAlAs/InP lasing nano-heterostructure. *Int. J. Mod. Phys. B,* **2014**, *28*(29), 1450206. [http://dx.doi.org/10.1142/S0217979214502063]

[19] Lal, P. Gain enhancement study of nanomaterial AlGaAs/GaAs under GRINLs. In: *CONIAPS XXVI*; Department of Physics, Manipal University Jaipur: Rajasthan, **2021**.

[20] Lal, P. An investigative study on growth of light of AlGaAs/GaAs in nanotechnological life sciences In: *NCTLS-2021, 29 January, 2021. 3rd National Seminar on Current Trends in Life Sciences*; Amity Institute of Biotechnology, Amity University Madhya Pradesh: Gwalior, **2021**.

[21] Lal, P. Gain enhancement study of nanomaterial AlGaAs/GaAs under GRINLs In: *CONIAPS XXVI*; 26th International Conference of International Academy of Physical Sciences (CONIAPS XXVI) on "Advances in Applied Physics & Earth Sciences" at Department of Physics, Manipal University Jaipur: Rajasthan, **2020**; pp. 18-20.

[22] Lal, P. An Innovative Study on UV Gain of AlGaN/AlN for Water Purification under Nanotechnology In: *International Conference on the "Soil and Water Resource Management [ICSWRM-2021]*; College of Technology and Engineering, Maharana Pratap University of Agriculture and Technology: Udaipur, Rajasthan, INDIA, **2021**.

[23] Lal, P. An Exploring Study on Enhancement of Light Emitted by AlGaAs/GaAs in Nanotechnology In: *NCESTM-2021*; I.G.I.T: Sarang, Odisha, **2021**.

[24] Lal, P. An optimised exploration on SWIR Gain of InAlGaAs/InP in Mathematical Sciences In: *CPMV-2021*; BITS Pilani: Hyderabad, **2021**; pp. 20-22.

[25] Lal, P. An Innovative SWIRL Gain Studies of InAlGaAs/InP for Optical Communications under QWWLs *Innovation and Technological Developments in Electronics, Computers, and Communication (ITDECC- 2021),* **2021**.

[26] Lal, P. An Innovative Computation on Optical Properties of InAlGaAs/InP in Nanotechnology *NCICT-21,* **2021**.

[27] Lal, Pyare An Optimised Study on Gain of AlGaN/AlN under Simulation Techniques In: *ICSDMO-2021*; ORSI: Meerut (UP), **2021**; pp. 16-17.

[28] Lal, Pyare Nanomaterials InAlGaAs/InP Based Gain Studies Under Normal Strains In: *Applied Materials and Technology – 2020, An Approach to Trans-Disciplinary Research*; Department of Physics, KLE Society's S Nijalingappa College: Bengaluru, **2020**; pp. 9-10.

[29] Lal, P. An Investigative Optimization on NIRL Gain of InAlGaAs/InP in Engineering Sciences In: *MASE-2021*; Govt. College Satnali: Mahendergarh, **2021**.

[30] Lal, P. GRINLs Influenced Several Light Properties of AlGaAs/GaAs Type Futuristic Material In: *ICFM-20*; DUGU: Gorakhpur, UP (INDIA), **2020**; pp. 18-20.

[31] Lal, P. Light Profits Studies of AlGaAs/GaAs in Nano Technological Engineering Sciences In: *9th National Conference on Mathematics Education Organised by NCERT*; Regional Institute of Education: Bhopal, **2020**.

[32] Lal, P. An innovative investigation of science teaching by nanotechnological process In: *Indian Science Techno Festival (ISTF-2020)*; , **2021**; pp. 26-28.

CHAPTER 3

Electronic Structure of the Half-Heusler ScAuSn, LuAuSn and their Superlattice: A Comparative GGA, mBJ and GGA+SOC Study

Himanshu Joshi[1,2,*], **Mahesh Ram**[1,3] and **Nihal Limbu**[1,3]

[1] *Condensed Matter Theory Research Laboratory, Kurseong College, Darjeeling, India-734203*

[2] *St. Joseph's College, North Point, Darjeeling, India-734104*

[3] *Department of Physics, North-Eastern Hill University, Shillong, India-739002*

Abstract: A detailed analysis of electronic band structure of the ScAuSn, LuAuSn and their Superlattice have been performed using the full potential linearized augmented plane waves (FP-LAPW). The exchange-correlation between the electrons was treated with three schemes, generalized gradient approximation (GGA), Trans and Blaha modified Becke-Johnson potential (TB-mBJ) and Spin-Orbit Coupling (SOC) incorporated with GGA. The GGA method reveals an indirect spin-gapless semiconducting nature for LuAuSn, an indirect band gap semiconducting nature for ScAuSn and direct semiconducting nature for their superlattice whereas under mBJ scheme, the band gap values are found to be enhanced. The inclusion of Spin-Orbit Coupling effects in GGA predicts the materials to be semi-metallic. The density of states is mainly dominated by the Sc and Lu atom near the vicinity of Fermi energy level and in the conduction region in ScAuSn and LuAuSn alloys, respectively whereas in superlattice the density of states is mainly dominated by Sc atom with significant contribution from Sn atoms.

Keywords: Half-metal, Spin-gapless semiconductor, Semi-metal, Superlattice.

INTRODUCTION

Half-Heusler materials have been a subject of research interest due to their increasing utility in modern technological applications. The advantages of the compound over other class of materials lies in their mechanical and thermal stability, variety in electronic properties [1], elemental abundance with environmentally friendly constituents [2] and minimal production cost.

---
* **Corresponding author Himanshu Joshi:** Condensed Matter Theory Research Lab, Kurseong college, Kurseong, Darjeeling, West-Bengal, India; Tel: 8257803359; E-mail: himanshujoshi09@gmail.com

Named after the discoverer Friedrich Heusler, these compounds have more than 1500 elemental combinations and are categorized into various sub-groups. Generally, four structural categories of Heusler alloys are commonly based upon their elemental arrangement and stoichiometry, *viz.*, the Full-Heusler compounds (X_2YZ), Half-Heusler compounds (XYZ), Inverse-Heusler compounds (X_2YZ) and the Quaternary-Heusler compounds (XX'YZ) [3]. This ternary intermetallic consists usually of two *d*-block elements XY (transition elements) and one *p*-block element Z (main group element). Regular or inverse Full-Heusler follows the stoichiometry ratio 2:1:1 and has 4 *fcc* sublattices. They crystallize in cubic *fcc* structure with space group symmetry. Our compound of choice ScAuSn and LuAuSn belongs to the half- Heusler category following 1:1:1 stoichiometry ratio and crystallizing in cubic symmetry within space group orientation [4]. Indeed, the half-Heusler crystal orientation is derived from the full-Heusler one by adding a vacancy in the tetrahedral atomic site of one of the transition metal atoms, such as to reduce the stoichiometric ratio from 2:1:1 to 1:1:1. Their electronic properties, determined by the energy band structure, strongly depend on the valence electron number or concentration in the primitive cell, which in turn affects the physical properties of these compounds. The Slater-Pauling rule which governs the saturation magnetization based on the valence electron number (VEN) of an alloy, has sheer relevancy in Heusler material class, allowing to fore predict its material property. Compounds with VEN of 21 and 22 are predicted to have half-metallic ferromagnetic properties [5]. Heusler alloys with VEN = 20 are found to have localized Fermi levels (E_F) and thus predicted to have an unstable crystal structure.

Typically, the compounds ScAuSn and LuAuSn belonging to the half-Heusler class has VEN = 18, exhibiting closed-shell behaviour and are predicted to have filled bands at t_1, a_1, t_2 and e band symmetry, thereby showing either semi-metal or semiconducting properties. They are also predicted to exhibit a narrow energy gap near the Fermi level (E_F) in the band structure with zero magnetic moments, which is a key feature to achieve excellent transport properties for a compound [6]. The compound ScAuSn agrees well with the Slater-Pauling rule and exhibits a narrow band gap near E_F. The electronic states involved in causing the gap determines the properties like electronic conductivity, the optical response of the system and transport behaviour. LuAuSn half-Heusler alloy which also bears a VEN of 18 can also, therefore, be predicted to exhibit semi-metallic or semiconducting properties with zero magnetic moments like other 18 valence electron half-Heusler alloys. However, the zero magnetic moment criterion is satisfied in accordance with the Slater-Pauling rule but a gapless feature remains prominent in its electronic energy band. The gapless feature was observed on treating the exchange and correlation energies with Generalized Gradient Approximation (GGA), which is usually considered to be more accurate compared to treating the exchange and correlation energies with Local Density Approximation (LDA). Half-Heusler compounds with lanthanide or heavy elements like Bi, Pb, Pt or Au are known to exhibit gapless states. They are generally known as gapless semiconductors, due to the valence and conduction band touches at E_F but however, no crossing on it occurs. The zero band gap feature may have direct or indirect nature and the available electronic structure studies of such compounds [7] refer to direct gapless states as topological insulators. Surprisingly, on the inclusion of spin-orbit interaction in addition to GGA, LuAuSn shows semi-metallic property, which is highly unexpected, especially given the non-magnetic nature of the compound. Additionally, on treating the exchange and correlation with Trans and Blaha modified Becke-Johnson potential (TB-mBJ), the electronic structure is further modified and the compound shows a semiconducting property. A narrow indirect band gap near E_F now dominates the observed electronic structure. Neither the semi-metallic nor the semi conducting nature of the compound is likely, as the material property greatly depends on the treatment of the exchange and correlation energy. Interestingly, the mBJ potential which is a semilocal exchange potential with the ability to accurately mimic the orbital dependent behaviour, is a correction over GGA to yield band gaps with close accuracy compared to expensive approaches like GW or hybrid functionals and therefore the opening of a band gap with its employment for gapless states is extremely unlikely. Thus, the determination of the material property of LuAuSn using the first-principles method becomes extremely challenging. We, in this work, have tried to resolve the identity crisis faced by LuAuSn, employing up-to-date computational techniques. In addition, considering the close structural resemblance with ScAuSn and their isoelectronic characteristic, we have created a ScAuSn/LuAuSn superlattice to accurately understand the LuAuSn material property. Further, if the gapless state in LuAuSn corresponds to topological insulator behaviour, then the heterojunction formed by non-topological material (ScAuSn) and topological material (LuAuSn) should provide a means to tune the band inversion. Also, we provide the hybridization scheme that governs the band gap formation in these compounds. The different electronic states involved in the band gap formed employing mBJ potential in LuAuSn and other electronic structure properties are discussed in detail under the results section which follows later in the chapter.

METHODOLOGY OF STUDY

With the rapid advancement of reliable computational methods based on Density Functional Theory (DFT), a broad way has been paved towards increasing the importance of so-called "theoretical experiments". Such "theoretical experiments" with comprehensive calculations, forego or replace real experiments and potentially can predict unidentified materials as well as their unknown properties.

These "theoretical experiments" are comparatively low costs when compared to the actual laboratory activities. Further, the outcome of such "theoretical experiments" serves as extremely useful guides for setting up a proper direction, while searching for new efficient materials. Therefore, given the ever-growing significance of computational methods for physics and materials science, it is suitable to make these DFT-based "theoretical experiments" to be the backbone of the present study. Density functional theory (DFT) is among the most popularly used techniques of condensed matter physics, originally developed by Kohn and Sham [8]. It provides a modernized tool to examine the ground state properties of atoms, molecules, and solids. In principle, the ground-state properties of the many-body interacting system are reduced to one body non-interacting by an electron density. Thus, within the framework of Kohn and Sham, intractable electrons in a static external potential are reduced to tractable electrons moving in an effective potential. The Coulomb interaction between the electrons exchange and correlation along with the external potential is contained in the effective potential. In short, the Kohn and Sham formulation states that the electronic ground state density contains all the information available in the electronic wave function. To obtain a general potential, the Kohn-Sham equations are solved independently of shape approximation [8]. The local density approximation (LDA) [9] along with general gradient approximation (PBE-GGA) [10] are tested. Additionally, we will employ a precise method based on the full-potential linearized augmented-plane wave technique (FP-LAPW) for electronic structure calculations. In this technique, the atomic space is sub-divided into two regions namely, the non-overlapping atomic sphere region (region I) also known as the muffin-tin sphere region and the interstitial region (region II), *i.e* the space between two spheres (Fig. **1** below shows the two regions). In both regions, diverse basis sets are applied. For region I, the energy-independent Bloch function is used, which gives some modifications to the two regions mentioned above. In region, I, with the help of variational principle linear equations are constructed from secular equations which then reduces to the generalised eigenvalue equations that are solved through diagonalization. In region II, as charge density varies much gradually, the basis functions here are the solutions of kinetic energy dependent Helmholtz spherical equation. The so formed Fourier series corresponding to region II, outside the atomic sphere is the pseudo-wave function, whose shape accuracy within the muffin-tin region has no importance for continuous and differentiable boundary conditions with zero slopes at the origin of the sphere. Inside the atomic sphere **t** of radius $\mathbf{R_t}$, the basis functions are spherical harmonics times radial functions for a Bloch function

$$\phi_{kn} = \sum_{lm} \left[A_{lm} u_l \left(\vec{r}, E_l \right) + B_{lm} \dot{u}_l \left(\vec{r}, E_l \right) \right] Y_{lm} (\vec{r}) \tag{1}$$

where is the spherical potential inside the sphere with radial part of Schrodinger equation having energy E_l, is the energy derivative of for similar energy.

Region II which is the interstitial region has a plane wave extension applied as

$$\phi_{kn} = \frac{1}{\sqrt{w}} e^{ik_n r} \tag{2}$$

where k represents the wave vector in the 1st Brillouin zone.

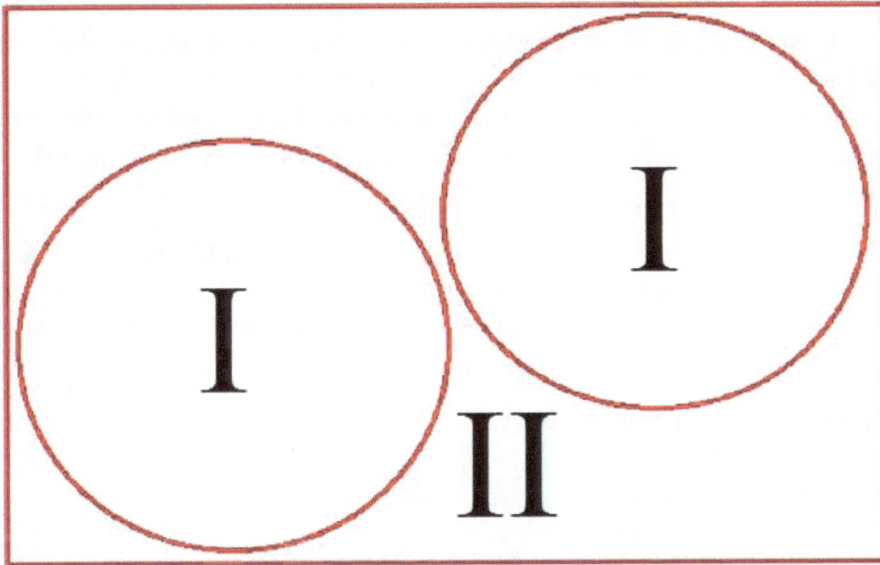

Fig. (1). Unit cell partitioning into atomic spheres (I) and an interstitial region (II).

Therefore, within the FP-LAPW technique, the potentials are expressed in general form within region I and region II as

$$V(\vec{r}) = \begin{cases} \sum_{lm} V_{lm}(\vec{r}) Y_{lm}(\hat{r}) & inside\,sphere \\ \sum_{K} V_K e^{ikr} & outside\,sphere \end{cases} \tag{3}$$

The crystal wave functions are toned to interstitial crystal waves at boundaries to ensure the continuity and differentiability of the crystal wave function. Potential obtained *via* charge density through Poisson's equation is utilized to solve the radial Schrodinger equation. Overlapping atomic charge density is taken as the density for the first iteration. A new solution is obtained by constructing a fresh charge density derived from eigenvectors calculated variationally. This is repeated until the self-consistency criterion is fulfilled. Taking these into consideration, Blaha *et al.* [11] developed a code, called WIEN2k, which will be the backbone of our computational methods.

The mBJ approach which is employed to determine the accuracy of band gap values is a correction over the self-interaction inconsistency contained in the LDA and GGA approach. The method is relatively cheap computation wise compared to the LDA+DMFT approach. Within DFT a reliable approach is the GGA+U method but is limited only to highly correlated electrons. Also, the calculation of U value required for the method in absence of experimental data shows no significant improvement over the originally achieved values within GGA and LDA approaches. Additionally, in the electronic structure calculation, the following basis sets were taken as valence states: Sc; $3d^1 4s^2$, Au; $4f^{14} 5d^{10} 6s^1$, Lu; $4f^{14}\ 5d^1\ 6s^2$; Sn; $5s^2 4d^{10} 5p^2$. Optimized k-mesh of $20 \times 20 \times 20$ over the first Brillouin-zone (IBZ) was integrated by the Monkhorst-Pack scheme. Experimental verifications have shown that both ScAuSn and LuAuSn show no magnetic ordering, therefore to make our calculations consistent with the experimental scenario, constrained magnetic calculations were performed.

RESULTS AND DISCUSSIONS

Crystal Structure

The compounds XAuSn (X= Sc, Lu) were synthesized by Sebastian *et al.* [4] and the authors have made a detailed study on the crystal structure of the materials. XAuSn follows the MgAgAs cubic crystal structure prototype and occupies the following Wyckoff's position-

Sc: 4b (1/2, 1/2, 1/2), Au: 4d (3/4, 3/4, 3/4) and Sn: 4a (0, 0, 0)
Lu: 4b (1/2, 1/2, 1/2), Au: 4c (1/4, 1/4, 1/4) and Sn: 4a (0, 0, 0)

The *4c* position is vacant in the case of ScAuSn and the *4d* position remains vacant in the case of LuAuSn (Fig. **2a**). The calculated lattice constants of the compounds are in close agreement with the available experimental results and are presented in Table **1**. For constructing the ScAuSn/LuAuSn superlattice, $2 \times 2 \times 1$ supercells of LuAuSn and ScAuSn were considered as shown in Fig. (**2b**). Then a $2 \times 2 \times 2$ superlattices was formed by combining the above supercells along the (001) direction, Fig. (**2c**). However, the creation of supercells drastically increases the number of basis atoms makes the calculation computationally expensive.

To reduce the high computing cost, a unit cell of the superlattice was considered. Fig. (**2d**) shows the top view of the 2 × 2 × 2 superlattices and the blue-lined box shows the unit cell of the superlattice for which the origin was moved by 0.75, 0.75, and 0.75 of a (lattice parameter) along the three principal axis. Finally, a tetragonal unit cell was taken for calculations.

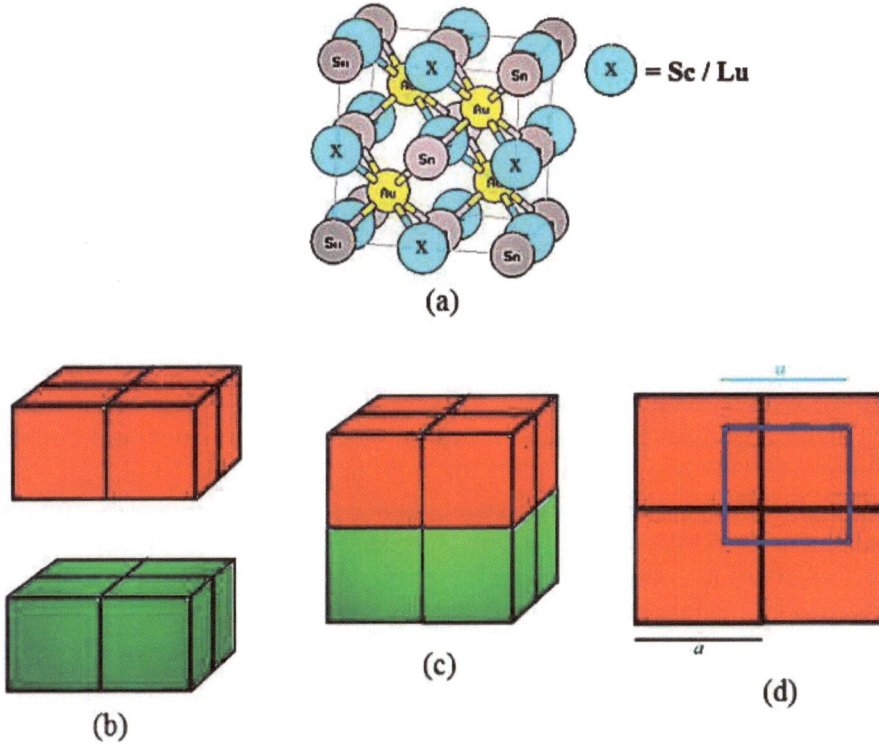

Fig. (2). (a) Unit cell crystal structure of XAuSn (X = Sc, Lu). (b) Schematic diagram of 2 × 2 × 1 supercells of LuAuSn (red colour) and ScAuSn (green colour), (c) Stacking of supercells along (100) direction to form a 2 × 2 × 2 superlattice of LuAuSn/ScAuSn, and (d) Top view of 2 × 2 × 2 supercells and a tetragonal unit cell of superlattice (blue box).

Table 1. Lattice constants (*a*), Bulk modules (*B*) and pressure derivative of bulk modulus (*B'*) obtained using Murnaghan's equation of state for XAuSn (X=Sc, Lu) and their superlattice. The lattice constants values are compared with the experimental results of Sebastian *et al.*, 2006 [4].

Compounds	Lattice Constant (*a*) Å		B (GPa)	B'	Difference in *a* (Δa)
	Experimental result	Calculated result			
ScAuSn	6.4194	6.5134	92.3875	5.2478	0.094
LuAuSn	6.5652	6.6501	84.8223	3.6871	0.0849
ScAuSn/LuAuSn Superlattice	-	a = b = 6.5652 c = 13.1304	-	-	-

Electronic Structure

The electronic structure of the studied materials has been investigated using three different approaches to get a better understanding and wider possibilities of accurately predicting their electronic properties. The presented results are based upon the GGA, mBJ and GGA+SOC methods. The GGA approximation is found to describe the electronic states with reliable accuracy [12] and has hence been employed. The mBJ approximation has the ability to accurately predict the energy band gaps saving computation time and the Spin-Orbit Coupling (SOC) incorporated with GGA is to account for additional states which remain hidden in absence of a non-relativistic approach. Fig. (**3a-c**) shows the total density of states (DOS) along with the partial density of states (PDOS), compared alongside the three methods employed in the study. The overall nature of DOS does not change in moving from GGA to GGA+SOC to mBJ in the case of ScAuSn except that for GGA+SOC the valence region of the DOS and the conduction region toches each other at 0 eV energy. However, in the case of LuAuSn, significant changes can be observed from the figure. Firstly, a change in DOS states/eV in both the conduction and valence region can be seen. The DOS states at 0 eV which represents the Fermi level crosses on either side of the valence and conduction region with GGA+SOC. Also, within the mBJ approach, the DOS states at 0 eV drop to 0 states/eV creating a small vacant space between the conduction and the valence region. A similar characteristic is observed for the superlattice structure. All these observed features will show a significant characteristic in the electronic band structure of the compound, which is discussed later in the section.

The partial DOS plots show the contribution of different atoms and their electronic states to the states available for occupation. In the case of ScAuSn, it can be seen that the Sc atoms dominate the entire conduction region and the valence region close to the Fermi level. A considerable contribution to the total DOS comes from the Au atoms in the conduction region, however, the Sn atom contribution dominates over Au in the valence region. For LuAuSn, the entire conduction region is dominated by Lu, and the valence region by Sn. In the case of super lattice, The most prominent contribution to the total DOS arises from Sc atom with significant contributions from the Sn atoms.

Band Formation

To understand the mechanism of band (or DOS) formation in the studied materials, the partial density of states (PDOS) of XAuSn (X=Sc, Lu) and their superlattice is referred from (Fig. **3**). The band formation is similar for the compounds under study and we discuss here the special case in LuAuSn which under mBJ opens up a gap. The Lu electron states mainly contribute above the

Fermi energy level (E_F) and are mostly unoccupied. The Au electron states on the other hand contribute predominantly around -6 eV below the E_F and are mostly occupied, while the Sn electron states are almost equally occupied and unoccupied.

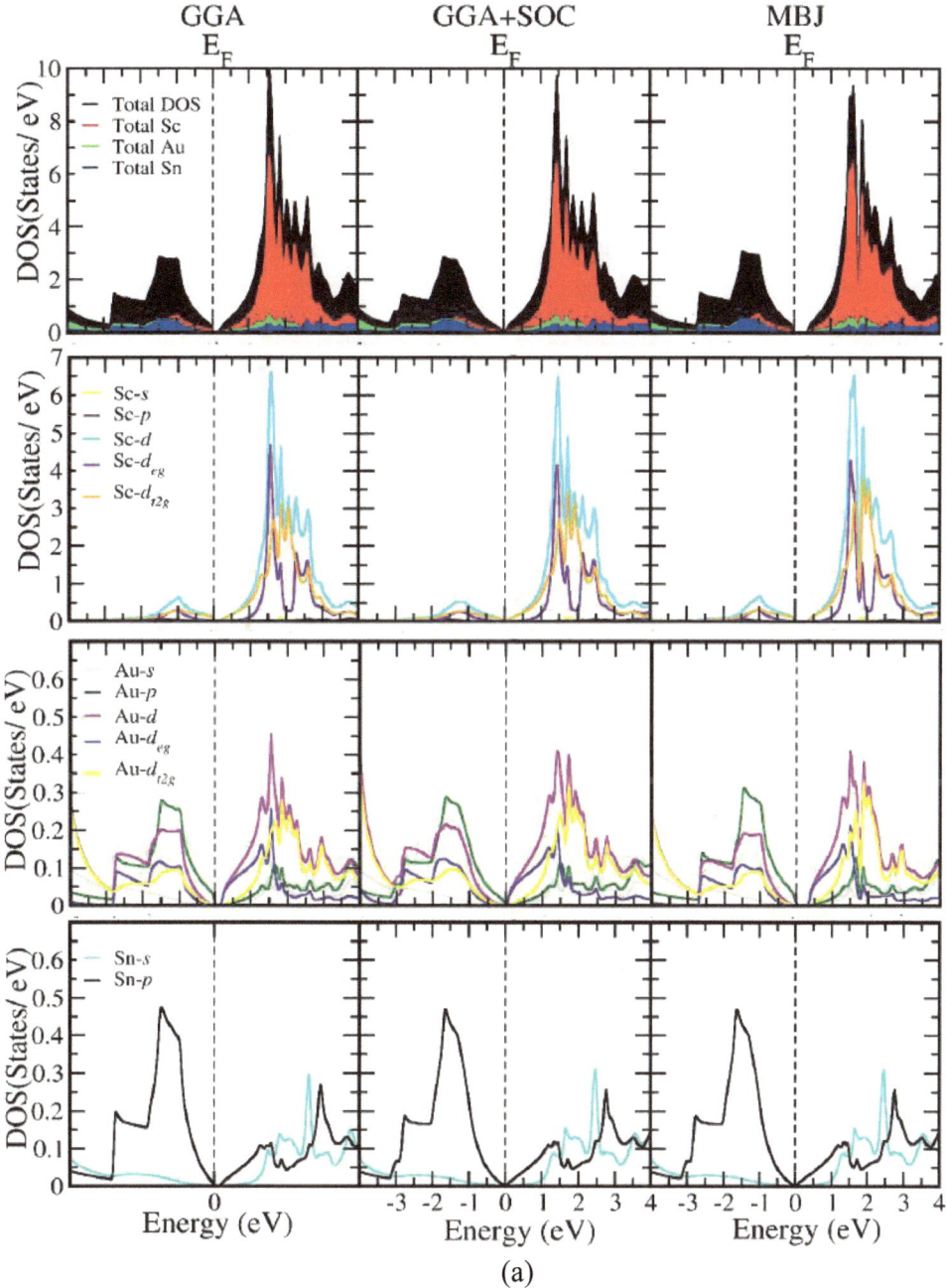

(a)

(Fig. 3) contd.....

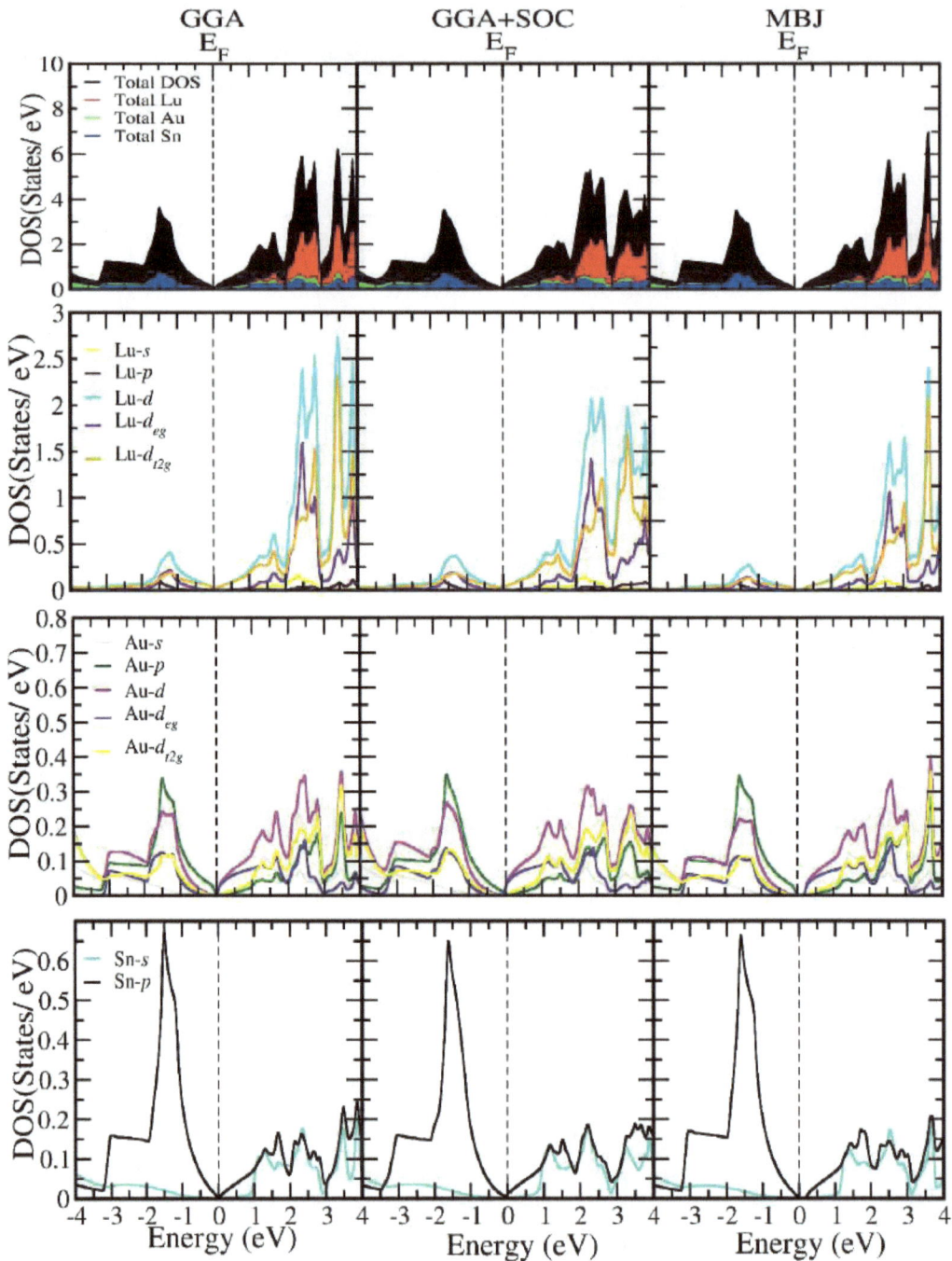

(b)

(Fig. 3) contd.....

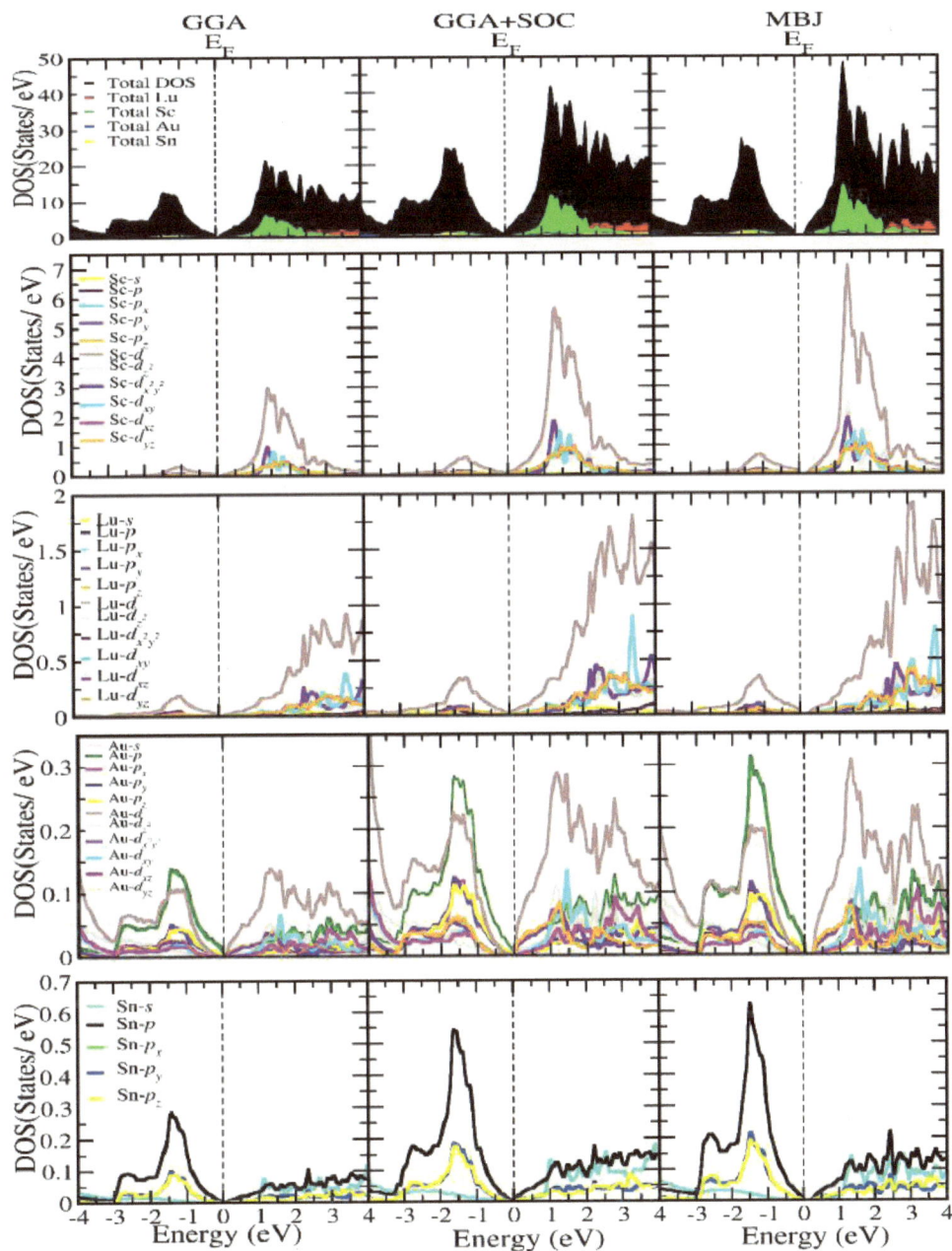

Fig. (3). Total and partial density of states for **(a)** ScAuSn **(b)** LuAuSn and **(c)** ScAuSn/LuAuSn superlattice plotted alongside with GGA, mBJ and GGA+SOC method.

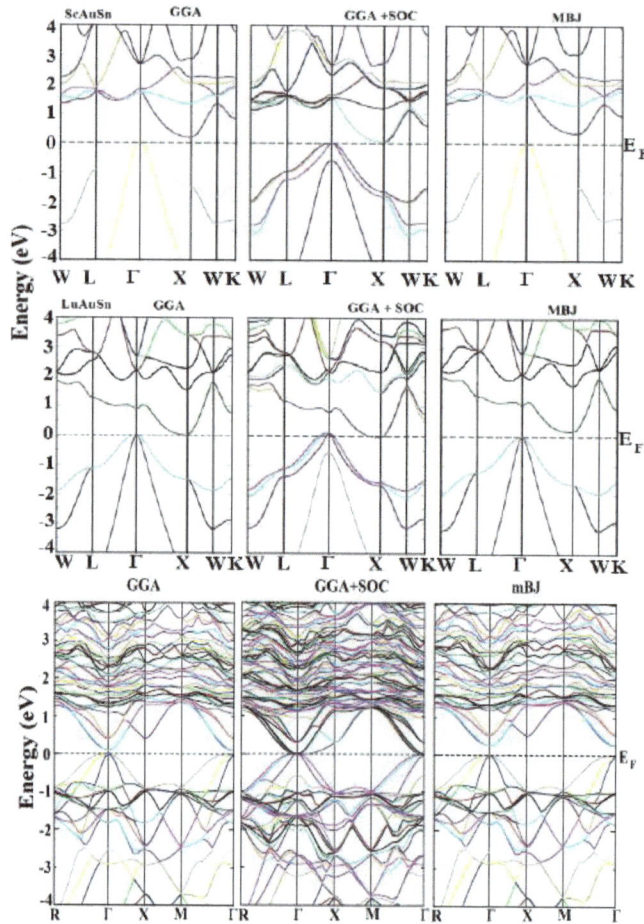

Fig. (4). The electronic band structure for XAuSn (X=Sc, Lu) and their superlattice obtained using GGA, GGA+SOC and mBJ.

The peaking of Au and Sn states around -6 eV, and that of Lu and Sn above E_F shows some kind of hybridization between their corresponding electronic states. A schematic representation of this hybridization is given in Fig. (**5**). The LuAuSn structure can be considered as a tetrahedral ZnS type structure formed by AuSn and the Lu filling the octahedral sites [13]. The LuSn on the other hand forms a NaCl sublattice. Hence a mixture of covalent and ionic characters is expected behind the band formation. The Lu being the most electropositive loses its all valence electron and exist as Lu^{3+}. The Au has the valence electron configuration of $4f^{14}5d^{10}6s^1$ and thus exist as Au^{1+} and Sn accepts all these electrons to fill all of its states and exist as Sn^{4-} [14]. In the first step (Fig. **5a**), the p, s, states of Au^+ and Sn^- hybridize to give bonding a_1 and t_2 states, and antibonding a_1^* and t_2^* states. The d states of Au^+ which are split by the crystal field remains as non-bonding states e and t_2 [15]. These electron states of $[AuSn]^{3-}$ now combine with Lu^{3+} electron states in the second step (Fig. **5b**) to finally give a_1, e, t_2 bonding states, and e and t_2 anti-bonding states along with non-bonding a_1^* and t_2^* states. The E_F in this case, lies between bonding t_2 and non-bonding a_1^* states. Thus there are exactly 9 states below the E_F occupying 18 valence electrons of the system. The scheme is similar for ScAuSn and the superlattice structure in which the d states further split into d_{xy}, d_{xz}, d_{yz}, and due to reduction in crystal symmetry of the superlattice structure.

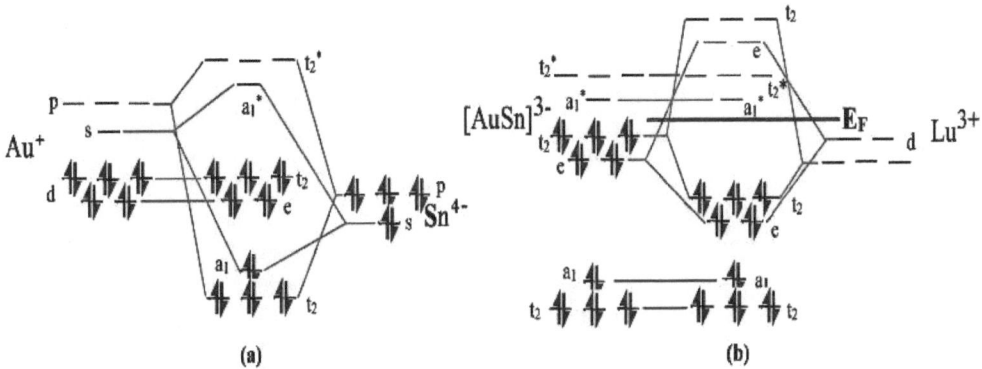

Fig. (5). **(a)** A schematic representation of the hybridization scheme in XAuSn and their superlattiice **(b)** The crystal field splitting in XAuSn and their superlattiice.

CONCLUSION

A first principle density functional theory calculation within FP-LAPW method was performed in LuAuSn and ScAuSn along with their superlattice. For better estimations of their electronic properties, three different approaches for exchange-correlation functional were employed namely GGA, mBJ and SOC. LuAuSn revealed an indirect spin-gapless semiconductor, while an indirect bandgap semiconductor was found with GGA. Whereas, their superlattice was found to tune the semiconducting bandgap from an indirect to a direct one. The mBJ was found to enhance the bandgap of these compounds without affecting their nature while the SOC predict all the studied compounds to be semi metal in nature. The analysis of DOS and PDOS showed the predominance of Sc and Lu states around the Fermi energy level and above in ScAuSn and LuAuSn, while the superlattice showed the dominance of Sc and Sn atom around the Fermi energy level.

CONSENT FOR PUBLICATION

Not applicable.

CONFLICT OF INTEREST

The authors declare no conflict of interest, financial or otherwise.

ACKNOWLEDGEMENTS

The authors acknowledge Amit Shankar for providing the computing facility.

REFERENCES

[1]   Felser, C.; Fecher, G.H.; Balke, B. Spintronics: A Challenge for Materials Science and Solid-State Chemistry. *Angew. Chem. Int. Ed.,* **2007**, *46*(5), 668-699.
        [http://dx.doi.org/10.1002/anie.200601815]

[2]   Snyder, G.J.; Toberer, E.S. Complex thermoelectric materials. *Nat. Mater.,* **2008**, *7*(2), 105-114.
        [http://dx.doi.org/10.1038/nmat2090] [PMID: 18219332]

[3]   Felser, C.; Hirohata, A., Eds. *Heusler Alloys: Properties, Growth, Applications*; Springer: Cham, **2016**.
        [http://dx.doi.org/10.1007/978-3-319-21449-8]

[4]   Sebastian, C.P.; Eckert, H.; Rayaprol, S.; Hoffmann, R-D.; Pöttgen, R. Crystal chemistry and spectroscopic properties of ScAuSn, YAuSn, and LuAuSn. *Solid State Sci.,* **2006**, *8*(5), 560-566.
        [http://dx.doi.org/10.1016/j.solidstatesciences.2006.01.005]

[5]   Galanakis, I.; Dederichs, P.H.; Papanikolaou, N. Slater-Pauling behavior and origin of the half-metallicity of the full-Heusler alloys. *Phys. Rev. B Condens. Matter,* **2002**, *66*(17), 174429.
        [http://dx.doi.org/10.1103/PhysRevB.66.174429]

[6]   Osterhage, H.; Gooth, J.; Hamdou, B.; Gwozdz, P.; Zierold, R.; Nielsch, K. Thermoelectric properties of topological insulator $Bi_2 Te_3$, $Sb_2 Te_3$, and $Bi_2 Se_3$ thin film quantum wells. *Appl. Phys. Lett.,* **2014**, *105*(12), 123117.
        [http://dx.doi.org/10.1063/1.4896680]

[7]   Al-Sawai, W.; Lin, H.; Markiewicz, R.S.; Wray, L.A.; Xia, Y.; Xu, S.Y.; Hasan, M.Z.; Bansil, A. Topological electronic structure in half-Heusler topological insulators. *Phys. Rev. B Condens. Matter Mater. Phys.,* **2010**, *82*(12), 125208.
        [http://dx.doi.org/10.1103/PhysRevB.82.125208]

[8]   Kohn, W.; Sham, L.J. Self-Consistent Equations Including Exchange and Correlation Effects. *Phys. Rev.,* **1965**, *140*(4A), A1133-A1138.
        [http://dx.doi.org/10.1103/PhysRev.140.A1133]

[9]   Von Barth, U.; Hedin, L. *J. Phys. C Solid State Phys.,* **1972**, *5*, 1629.
        [http://dx.doi.org/10.1088/0022-3719/5/13/012]

[10]   Perdew, J.P.; Burke, K.; Ernzerhof, M. Generalized Gradient Approximation Made Simple. *Phys. Rev. Lett.,* **1996**, *77*(18), 3865-3868.
        [http://dx.doi.org/10.1103/PhysRevLett.77.3865] [PMID: 10062328]

[11]   Blaha, P.; Schwarz, K.; Madsen, G.; Kvasnicka, D.; Luitz, J. *An Augmented Plane Wave Plus Local Orbitals Program for calculating Crystal Properties*; Tech. Universitat: Wien, Austria, **2014**.

[12]   Joshi, H.; Ram, M.; Limbu, N.; Rai, D.P.; Thapa, B.; Labar, K.; Laref, A.; Thapa, R.K.; Shankar, A. Modulation of optical absorption in m-$Fe_{1-x}Ru_xS_2$ and exploring stability in new m-RuS_2. *Sci. Rep.,* **2021**, *11*(1), 6601.
        [http://dx.doi.org/10.1038/s41598-021-86181-7] [PMID: 33758358]

[13]   Graf, T; Felser, C; Parkin, SS Simple rules for the understanding of Heusler compounds. **2011**, *Progress in solid state chemistry, 1;39*(1), 1-50.

[http://dx.doi.org/10.1016/j.progsolidstchem.2011.02.001]

[14] Graf, T; Felser, C; Parkin, SS Simple rules for the understanding of Heusler compounds. *Progress in solid state chemistry,* **2011**, *1;39*(1), 1-50.
[http://dx.doi.org/10.1016/j.progsolidstchem.2011.02.001]

[15] Toboła, J; Pierre, J Electronic phase diagram of the *XTZ (X= Fe, Co, Ni; T= Ti, V, Zr, Nb, Mn; Z= Sn, Sb)* semi-Heusler compounds *Journal of alloys and compounds,* **2000**, *10*, 243.

CHAPTER 4

Recent Trends in Nanosystems, Challenges and Opportunities

S. Kannadhasan[1,*] and **R. Nagarajan**[2]

[1] *Department of Electronics and Communication Engineering, Cheran College of Engineering, Tamilnadu, India*

[2] *Department of Electrical and Electronics Engineering, Gnanamani College of Technology, Tamilnadu, India*

Abstract: Nanotechnology is, in fact, a fairly extensive field of study and research at the moment. Physicists, chemists, biologists, material scientists, engineers, and computer scientists all contributed to its creation. In this paper, we examine how nanotechnology is evolving and growing, as well as how it may be utilised to construct a computer that is smaller, faster, and more trustworthy. In this work, we focus on the top-down and bottom-up manufacturing approaches to nanotechnology, which have a direct impact on current computer design and architecture. Researchers have been driven to enhance and produce a smaller, quicker, and more reliable computers due to the widespread usage of computers and their vast use in the contemporary world. Nanotechnology has the potential to achieve this goal. According to the definition of nanotechnology, it is the design, characterization, production, and application of structures, devices, and systems by controlled manipulation of size and shape at the nanometre scale atomic, molecular, and macromolecular scale, which results in structures, devices, and systems with at least one novel/superior characteristic or property. Passive nanostructures are the name given to this kind of structure.

Keywords: Nanobelt, Nanomaterials, Nanostructure and Applications, Properties.

INTRODUCTION

Nanobelt is a quasi-one-dimensional nanomaterial having a well-defined chemical composition, crystallographic structure, and surfaces, such as growth direction, top/bottom surface, and side surfaces.

This article discusses the nanobelt family of functional oxides, which includes ZnO, SnO_2, In_2O_3, Ga_2O_3, CdO, and PbO_2, as well as the hierarchical and sophisticated nanorods and nanowires made using a solid-vapor process. The nan-

---

[*] **Corresponding author S. Kannadhasan:** Department of Electronics and Communication Engineering, Cheran College of Engineering, Karur, Tamilnadu, India; Tel:+919677565511; E-mail: kannadhasan.ece@gmail.com

Dibya Prakash Rai (Ed.)

obelts have atomically flat surfaces and are single crystalline and dislocation-free. The oxides are semiconductors used to construct nanoscale functional devices such as field-effect transistors, gas sensors, nanoresonators, and nanocantilevers, which are significant in nanosystems and biotechnology. The oxide nanobelts were created using a solid-vapour technique. Thermal evaporation is a simple process in which a condensed or powdered source material is vaporised at a high temperature, and the resultant vapour phase condenses under particular circumstances like temperature, pressure, atmosphere, substrate, and so on to generate the required output [1 - 5].

We can develop electrical components and devices that can be used to build computers that are smaller, faster, and more reliable. Nanotechnology refers to the manipulation of matter on an atomic and molecular level. Two examples of applications are dispersed and contact nanostructures. Example Nanoparticle reinforced composites include aerosols, colloids, coatings, nanoparticle reinforced composites, nanostructured metals, polymers, and ceramics. It is a word that refers to active nanostructures. Some of the applications include robotics, guided assembly, 3D networking, and unique hierarchical architectures. Molecular nanosystems are the term for it. Only a few examples include molecular devices by design, atomic design, and emergent functionality. The fourth generation of nanotechnology is largely focused on the design and development of nanocomputers [6 - 10].

The topic was briefly examined in the paper, as well as its broad application. The purpose of this research is to show how nanotechnology is used in the creation of a clever little computer. Furthermore, the research seeks to explain why a certain field or topic of nanotechnology is closely linked to the development of a future sophisticated computer. Current research on bio-systems at the nano-scale and nanotechnology has developed one of the most active scientific and technological areas at the confluence of physical sciences, molecular engineering, biology, biotechnology, and medicine. This subject includes a better understanding of living and thinking systems, revolutionary biotechnology processes, new drug synthesis and targeted delivery, regenerative medicine, neuromorphic engineering, and building a sustainable environment. Nanobiosystems research is being prioritised by several countries, and its prominence within nanotechnology is expected to rise in the future. When pharmaceutical particles are decreased to the sub-micron range, the dissolving rate is greatly boosted, which enhances bioavailability. The use of nanoparticles (NPs) as carriers for small and large molecules to deliver medications has piqued scientists' curiosity. Targeting drug administration to ill regions is one of the most important aspects of a drug delivery system. They have been used in vivo to protect the drug entity in the systemic circulation, restrict drug access to the targeted locations, and deliver the medication to the target region at a controlled and sustained pace. Several polymers have been used in the creation of nanoparticles for drug delivery studies [11 - 16] to maximise therapeutic efficacy while reducing side effects. This review article highlights the most notable achievements in the field of nanotechnology as a drug delivery system. Pharmaceutical nanotechnology systems, preparation procedures, applications, advantages, and disadvantages. Throughout the past few decades, there has been a lot of interest in developing nanotechnology by using nanoparticles as transporters for small and large molecules.

NANOSYSTEMS

A number of polymers have been used to make nanoparticles. The most notable accomplishments in the field of nanotechnology are highlighted in this paper. The name 'Nano' is derived from the Latin word 'dwarf.' One nanometer is one-billionth of a metre (1n=109m). The term "nano" refers to a thousand millionth of a unit. Nanotechnology has been extensively employed in scientific areas such as electrical, physics, and engineering over many decades. On the other hand, biomedical and pharmaceutical fields have yet to be studied. Nanotechnology is an interdisciplinary field that combines basic and applied sciences such as biophysics, molecular biology, and bioengineering. Size reduction is a fundamental unit action in pharmacy that has a broad variety of applications. Some of the most major advantages of nanosizing are as follows: The surface area has been increased. An increase in the solubility of a substance, In the mouth, increases the rate of dissolution and bioavailability. The impact starts quickly. Less dose is required in the pharmacy business. These materials and technologies might be designed to interact with a high degree of functional specificity for medical and physiologic applications, allowing for a level of interaction between technology and biological systems that have never been feasible before. It is important to remember that nanotechnology isn't a single emerging scientific discipline in and of itself, but rather a fusion of established disciplines such as chemistry, physics, material science, and biology that brings together the necessary collective skills to build technologies (see Fig. **1**).

These are hexagonal carbon atom networks. These tubes have a diameter of 1-100nm and a length of 1nm. The two kinds of nanotubes are single walled nanotubes (SWNTS) and multi walled nanotubes (MWNTS) (MWNTS). The size, structure, and physical features of these small macromolecules are all unique. These semiconducting materials feature a semi-conductor core that is surrounded by a shell for improved optical properties. Their qualities are determined by their physical size, which ranges from 10-100A0 in radius. Imaging, biomolecular detection and analysis in vitro and in vivo, immunoassay, and DNA hybridization, as well as non-viral vectors for gene therapy, are all affected by these factors. Its principal function is to identify cells and therapeutic tools used in cancer therapy.

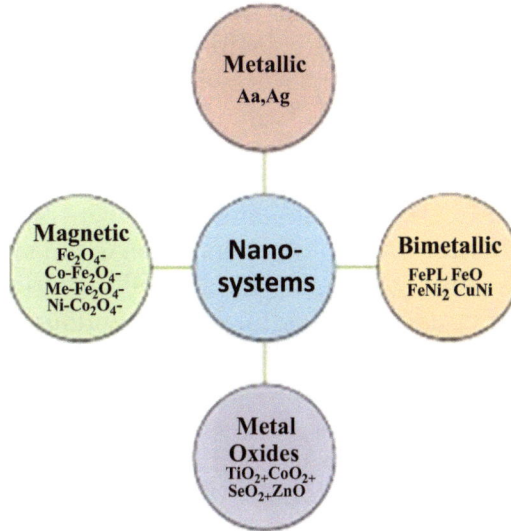

Fig. (1). Nanosystems.

Chemical polymer compartmentalization is used to create these enormously branched, tree-like structures. The core, branches, and surface are the three parts. The core is the central part, surrounded by branches that form an internal hollow and a sphere of groups. To match your demands, the branches may be modified or altered. Dendrimers may be transformed into more biocompatible molecules with low cytotoxicity and good bio-permeability, depending on the demands. Bioactives such as medications, vaccines, materials, and DNA may be transported to specified places using these. Amphiphilic block copolymers generate nanoscopic supramolecular core-shell structures called polymeric micelles. Their hydrophilic surface protects them from nonspecific reticuloendothelial absorption, and they are typically less than 100nm in size. Micelles are spherical aggregates formed by arranging component molecules in a spherical shape with a hydrophobic core sheltered from water by a mantle of hydrophilic groups in a solution. These are used to transport drugs that aren't water-soluble throughout the body. Drugs may be physically encased inside the hydrophobic cores or covalently bonded to micelle component molecules. High loading capacity, physiological stability, a slower rate of dissolution, high drug accumulation at the target region, and the ability to functionalize the end group for targeted ligand conjugation are all features of these. These are used in transfection procedures and develop naturally between nucleic acids and polycations or cationic ligands or hydrophilic polymers. The form, size distribution, and transfection capabilities of these complexes are determined by the quantity and charge ratio of nucleic acid to cationic lipid or polymer. The pharmacokinetic distribution of low molecular weight medications throughout the body and at the cellular level varies substantially when they are conjugated with polymer. As a consequence, they are designed to increase overall molecular weight, allowing for improved retention in cancer cells when using a passive delivery approach thanks to the EPR effect.

In the non-solvent phase, the monomer is emulsified with the help of surfactant molecules. As a consequence, monomers are generated, including inflated micelles and stable monomer droplets. Nanometer-sized swollen micelles have a greater surface area than monomer droplets. Polymerization occurs when a chemical or physical initiator is present. In the continuous phase, the initiator's energy creates free reactive monomers, which collide with neighbouring non-reactive monomers to start the polymerization chain reaction. Continuous phase diffusion from monomer droplets allows monomer molecules to diffuse into micelles, allowing polymerization to continue within the micelles. In this case, monomer droplets act as monomer reservoirs. The monomers are slightly soluble in the continuous phase. These techniques are utilised to detect a wide range of pathogenic proteins as well as physiological-biochemical markers connected to disease or changed metabolic conditions in the body.

A biosensor consists of a probe containing a sensitive biological recognition element or bio-receptor, a physiochemical detector component, and a transducer that amplifies and transforms these signals into a measurable form. A nano biosensor, also known as a nanosensor, is a biosensor with dimensions on the nanoscale scale. Biosensors are used for target identification, validation, assay development, ADME, and toxicity assessment. Nanoparticles may be used as delivery methods because of their tiny, controlled, and virtually zero-order kinetics; alternatively, they may cause toxicity when compared to I.V. Carrier molecules include liposomes, ethosomes, and transferosomes. By lowering peak plasma levels, allowing for more predictable and longer durations of action, lowering the need for re-dosing, and enhancing patient tolerance and compliance, they help to reduce the likelihood of adverse reactions. In both passive and active modalities, liposomes, polymeric micelles, dendrimers, iron oxide, and proteins are employed to regulate medication absorption. The enhanced permeability and retention (EPR) effect of nanoparticles is used to deliver drugs to tumours through passive delivery, which makes use of nanoparticles and the tumour's leaky vasculature. Improved site-specificity was achieved by employing site-specific ligands on the surface, either by covalent binding or adsorption using a carrier system. Drug absorption, dose frequency, and adherence to TB therapy have all been reported to improve with active distribution to lung cells.

Nanotechnology has aroused considerable attention and has been termed the twenty-first century's most important technology. The engineering of working systems at the molecular level is known as nanotechnology. It's all about controlling and using matter's structure on a massive scale below 100 nanometers. Nanotechnology will be the future path of improved advancement. It is now found in everything, from clothing to food, and it pervades every aspect of life. We should support it more for the benefit of our children's futures and current advances. In this essay, we looked at the concept of nanotechnology, as well as its history, applications, risks, and developments in India. The history of nanotechnology traces the growth of ideas and experimental activities that fall under the broad category of nanotechnology. The emergence of nanotechnology in the 1980s was fueled by a convergence of experimental breakthroughs such as the invention of the scanning tunnelling microscope in 1981 and the discovery of fullerenes in 1985. In the early 2000s, commercial nanotechnology applications became more prominent. Because of the advent of new techniques that allow scientists to inspect and manipulate materials at the nanoscale, nanotechnology seems to have taken off in the last ten years. Scanning tunnelling microscopy, magnetic force microscopy, and electron microscopy are just a few examples of technologies that allow scientists to witness events at the atomic level. Simultaneously, the electronics industry's economic needs have prompted the development of new lithographic technologies that will continue to decrease feature size and cost. Until recently, scientists were unable to learn more about the nanoscale due to a lack of high-quality equipment, just as Galileo's comprehension was limited by his time's technology.

Better equipment for viewing, controlling, and measuring occurrences at this scale will lead to further advances in our knowledge and capabilities. Nanowires and carbon nanotubes are two nanoscale structures that scientists are especially interested in right now. Nanowires are very small wires that may be as small as 1 nanometer in diameter. According to experts, they might be utilised to produce miniature transistors for computer chips and other electrical devices. In recent years, carbon nanotubes have eclipsed nanowires in popularity. We still have a lot to learn about these structures, but what we've found so far is interesting. A carbon nanotube is a cylinder made up of carbon atoms that is nanoscale in size. Consider a sheet of carbon atoms in the shape of a hexagon. By rolling the sheet into a tube, you can create a carbon nanotube. The way the sheet is rolled determines the features of carbon nanotubes. To put it another way, even though all carbon nanotubes are made of carbon, how the individual atoms are oriented may have a significant impact on how they appear. During the first phase, products will take advantage of the passive properties of nanomaterials like nanotubes and nanolayers. Because it absorbs and reflects UV rays, titanium dioxide is often used in sunscreens. When broken down into nanoparticles, it becomes transparent to visible light, minimising the white cream appearance associated with traditional sunscreens. Carbon nanotubes are stronger than steel while being much lighter. They are supposed to give improved stiffness without adding weight to tennis rackets that use them. Stain-resistant clothing, for example, might be manufactured from yarn covered with a nanolayer of material. When made at the nanoscale, each of these products takes advantage of a material's unique property. In each case, however, once the nanomaterial is encased within the product, it remains static. During use, active nanostructures change their state and respond to the environment in predictable ways. Before releasing a drug linked to cancer cells, nanoparticles may hunt for them. When a nanoelectromechanical device embedded in construction material senses strain, it releases an epoxy that cures the rupture. Alternatively, when a coating of nanomaterial is exposed to sunlight, it may produce an electrical charge that may be utilised to power a device. In this phase, goods must have a greater understanding of how the structure of a nonmaterial affects its properties, as well as the ability to create new materials. They also provide new challenges in complex production and deployment. At this phase, nanotool assemblies are working together to achieve a single goal. Getting the main components to operate together in a network, and maybe exchanging information is a major challenge. Proteins or viruses may construct little batteries. Self-assembling nanostructures might provide a lattice for bone or other tissues to grow on. Smart dust strewn around a room might detect human presence and convey their whereabouts. Small nanoelectromechanical devices may be able to identify cancer cells and prevent them from reproducing. At this moment, significant advancements in robotics, biotechnology, and next-generation information technology will begin to manifest themselves in things. Innovative pharmaceutical delivery systems based on nanotechnology technologies are being investigated for ailments such as cancer, diabetes, fungal infections, viral infections, and gene therapy. Medication targeting and an enhanced safety profile are two of the main advantages of this therapeutic strategy. In diagnostic medicine, contrast agents, fluorescent dyes, and magnetic nanoparticles are all examples of nanotechnology.

NANOTECHNOLOGY

Carbon nanotubes and other carbon-based nanomaterials are a kind of carbon-based nanomaterial. Carbon nanotubes are long, straight molecules that are completely made up of carbon atoms. Their ability to lengthen or contract in suitable electrolytes at very low voltages is now being investigated, and this property might make them incredibly useful as actuators or sensors in a variety of medical devices. Another potentially relevant aspect is their potential use as sensors, such as for CO_2 monitoring in anaesthesia. The lack of an internal hollow distinguishes nanowires from nanotubes. In the detection of viruses in solution, semiconducting silicon-based nanowires are showing promise, and their capabilities may exceed present technologies. Nanoporous materials are porous materials having nanoporous pores. Carbon, silicon, ceramic, or polymer-based nanoporous materials with holes in the range of 100nm have a greatly increased surface area and may have especially relevant catalytic, adsorbent, and absorbent characteristics. For example, this might be beneficial in implant technology or drug delivery. Surgical blades that have been nanocoated. Nanoparticulate coatings on precisely prepared hard metal substrates, such as plasma polished diamond nanolayers, may be used to make surgical blades with exceptional sharpness and minimal friction, which are particularly suited to optical and neurosurgery. Needles. Nanocoated needles are now available for exceptionally sensitive suturing in demanding applications. The ductility, strength, and corrosion resistance of these needles are exceptional.

In minimally invasive surgeries, catheters are utilised. Carbon nanotubes, for example, have been successfully used in minimally invasive surgical catheters to increase their strength and flexibility while minimising their thrombogenic impact. The phrase "in-vitro diagnostics" refers to tests that are carried out in a laboratory. In the realm of in-vitro diagnostic medical equipment, nanotechnology has a lot of potentials. The development of microfluidic and nanofluidic systems has enabled the use of tiny analyte mounts, and the degree of miniaturisation that can be achieved will allow the production of true "lab-on-a-chip" devices capable of completing dozens, if not hundreds, of analyses in real-time. When linked to other devices, this allows for continuous monitoring of the patient's condition as well as changes in treatment, such as medication delivery, to better suit the patient's needs.

Paints and coatings, textiles and apparel, cosmetics, food science, catalysis, and other industries might benefit from nanotechnology. Furthermore, nanotechnology opens us with new possibilities for improving how we measure, monitor, and manage. Nanotechnology has emerged as a fast-evolving and developing area. New nanomaterial generations will emerge, bringing with them new and potentially unanticipated challenges. Nanotechnology is the way advanced progress will be in the future. It is everything nowadays, from garments to meals, and it covers every area. We should encourage it more for the sake of our future and for further advancements in our present lives.

Nanotechnology is a rapidly expanding industry with substantial global economic ramifications. Nanoscale materials' features have spawned new technologies with applications in healthcare, industry, and the environment. The potential for nano-sized particles to have detrimental effects on the environment is the environmental impact of nanotechnology. Nanoparticles' impact on environmental variables such as soil, water, air, bacteria, plants, and animals are currently being raised as a result of nanotechnology's fast progress. Given the lack of nanotoxicology data, any scientific input on nanoparticle environmental risks could contribute in the regulation of artificial nanomaterials' usage and manufacture. Examining the detrimental impacts of nanoparticles and their likely accumulation in plants or microorganisms are two current study fields that investigate the impact of nanoparticles on the environment.

According to certain studies, metal oxide nanomaterials may be harmful to plants, underlining the need for ecologically responsible metal oxide nanoparticle waste management and advocating for further experimental study into the possible effects of generated nanoparticles on environmental systems. Various nanoparticle types have different effects on plant growth during early ontogenetic stages. Inorganic nanosystems are nanoscale chemical objects with entirely inorganic components that exhibit remarkable behaviours as a result of quantum size and geometry effects. The so-called "top-down" and "bottom-up" approaches of creating nano-objects are two major synthetic routes. Chemistry plays a key role in the assembly and creation of larger nanometric units from smaller ones, particularly in the latter approach. Nanosystems may be defined and classified using the hierarchical order of dimensionality. Zero-dimensional systems include pseudospherical objects such as nanoclusters and nanoparticles supported on inorganic bulk supports as well as colloidal fluids or ceramic nanopowders. One-dimensional systems include carbon-based, metal-based, or even oxide-based systems in which one dimension's expansion dominates the other two. Solid nanofibers, nanowires, nanorods, and hollow nanotubes are examples of one-dimensional systems two-dimensional nanosystems are crystalline flat nanometric materials, such as nanodiscs or nanoprisms, as well as amorphous nanofilms and nanomembranes. Then there are three-dimensional nanosystems, which are composed of crystalline and amorphous nanostructures such as nanocrystals and a variety of ordered nanoarranged porous materials. Simpler components, such as nanoparticles or nanorods, may be employed to create three-dimensional arrangements, resulting in better superstructures or superlattices. Atoms and molecules are the building blocks of everything. Understanding the characteristics of these essential components, as well as their reciprocal relationships, requires a knowledge of how they are constructed. Effective control of the synthetic pathways is crucial during the creation of nano building blocks of varied sizes and shapes that may lead to the manufacture of new gadgets and technologies with improved performance. To achieve this, two opposed but complementary strategies are required. The top-down miniaturisation of existing components and materials is one method, while the bottom-up development of ever-more-complex molecular structures, atom by atom or molecule by molecule, is another. These two methods emphasise nanosystems' organisational level as the connection between molecular objects and bulk materials.

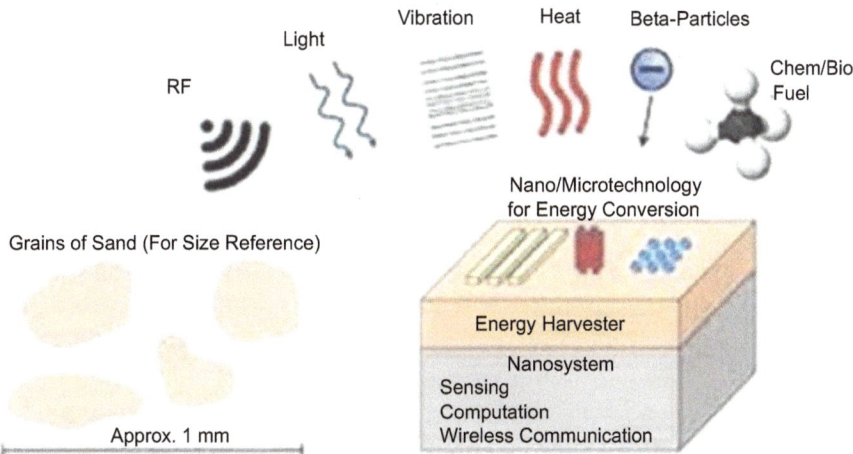

Fig. (2). Nanotechnology.

In many fields of contemporary science and technology, micro- and nanotechnologies have emerged as the driving force. Nanotechnology focuses on bottom-up synthesis and self-assembly, while microtechnology is mostly centred on top-down machining. Nanosystems are systems having components that are less than 100 nanometers in size. Microsystems are systems having structures that are smaller than one millimetre in size. Because objects operate differently than predicted with our intuition for the human scale of metres, the capacity to purposefully develop structures in this length scale offers up new phenomena and functionality. Surface-based phenomena become more prominent than volume-based phenomena as the size decreases, according to the fundamental scaling rule, the square-cube law. The following are some biological examples that demonstrate the various size scales. Proteins are complex molecules with sizes ranging from a few to 10 nanometers. The size of a more complicated system, such as a virus, is roughly 100 nanometers. Bacteria, for example, have sizes ranging from a hundred nanometers to a micrometre. The building blocks of sophisticated creatures in the plant and animal worlds are cells, which are a few micrometres in size. Surprisingly, biological systems have size scales that are similar to micro-and nanosystems. As a result, micro-and nanotechnology may be used to develop tools for building and manipulating biological systems at the molecular and cellular levels is shown in Fig. (**2**). Nanomedicine allows for early detection and prevention, as well as accurate diagnosis, treatment, and follow-up of disorders. Thanks to the use of nanoscale particles as tags and labels, biological testing has become more sensitive and adaptable. Gene sequencing has been increasingly effective with the development of nanodevices such as gold nanoparticles. When tagged with small pieces of DNA, these gold nanoparticles may be utilised to detect genetic sequences in a sample.

Nanotechnology may be used to duplicate or mend damaged tissue. Tissue engineering uses artificially stimulated cells and has the potential to revolutionise organ transplants and artificial implant placement. Carbon nanotubes have the potential to be employed to make improved biosensors with novel characteristics. These biosensors might be used in astrobiology and might provide information about the origins of life. This method is also being used in the development of cancer detection sensors. Despite its inert nature, CNT may be functionalized by adding a probe molecule to the tip. In their study, AFM is employed as an experimental platform. The environment must be safeguarded, and sufficient energy must be given for a growing world. Energy storage, conversion to multiple forms, environmentally acceptable material creation, and better renewable energy sources may all benefit from nanotechnology's advanced processes.

Nanotechnology may be used for less expensive energy production and renewable energies in solar technology, nano-catalysis, fuel cells, and hydrogen technology. Fuel cells made of carbon nanotubes are used to store hydrogen and have potential in electric cars. Nanotechnology is used in photovoltaics to make them cheaper, lighter, and more efficient, as well as catalytic converters made of nanoscale noble metal particles, catalytic coatings on cylinder walls, and catalytic nanoparticles as fuel additives, which can reduce engine pollutants combustion by nanoporous filters and clean the exhaust mechanically with the help of catalytic converters.

Nanotechnology may contribute to the creation of new pollution-reducing ecologically friendly and green technologies. The use of solid-state lighting has the potential to reduce total power consumption. Nanotechnological advancements might lead to a large reduction in lighting energy use. Nanomedicine is a very new field of science and technology. By interacting with biological molecules at the nanoscale, nanotechnology broadens the scope of research and application. The interactions of nanodevices with biomolecules may be examined both outside and within human cells. Physical properties such as the volume/surface ratio, which are not observable at the microscale, may be utilised at the nanoscale.

Two examples of nanomedicine that have been tested in mice and are awaiting human trials include the use of gold nanoshells to help diagnose and treat cancer and the use of liposomes as vaccine adjuvants and drug delivery vehicles. Another use of nanomedicine that has been successfully tried in rats is drug detoxification. Medical technology now allows for the use of smaller, less invasive devices that may be implanted inside the body, as well as substantially quicker biochemical reaction times. Traditional pharmaceutical delivery techniques are slower and less sensitive than nanotechnologies.

In nanotechnology, nanoparticles are used to deliver medications to specific places. This treatment uses the required drug amount, and side effects are much decreased since the active ingredient is only deposited in the diseased region. Patients may save money and have less agony as a result of this highly focused procedure. As a consequence, a variety of nanoparticles are employed, including dendrimers and nanoporous materials. Micelles produced from block co-polymers are used to encapsulate medications. They convey small pharmaceutical molecules to where they're supposed to go. Nanoelectromechanical devices are also employed for the active release of medications, as seen in Fig. (**3**). In cancer treatment, the use of iron nanoparticles or gold shells is gaining favour. Targeted therapy reduces pharmaceutical intake and treatment expenses, lowering the cost of patient care.

Fig. (3). Nanoelectromechanical Devices.

For drug delivery, nanomedicines, which are nanoscale particles or substances that may boost medication bioavailability, are used. To maximise bioavailability both at specific places in the body and over time, molecular targeting is done utilising nano-designed devices such as nanorobots. The molecules are carefully targeted, and the drug is delivered with cell-level precision. Another area where nanotechnology is being investigated is nanotools and devices for in vivo imaging. In nanoparticle images, such as ultrasound and MRI, nanoparticles are used as contrast.

CONCLUSION

Nanoengineered materials are being developed to better effectively cure disorders and illnesses such as cancer. Because of breakthroughs in nanotechnology, self-assembled biocompatible nanodevices that detect cancerous cells and autonomously analyse, treat, and write reports are now conceivable. Nanomaterials are useful for drug and gene delivery, biomedical imaging, and diagnostic biosensors because they have a greater surface area and show nanoscale effects. Nanomaterials have unique physicochemical and biological properties when compared to their larger counterparts. The size, shape, chemical composition, surface structure, charge, solubility, and aggregation of nanomaterials may have a significant influence on their interactions with biomolecules and cells. Nanoparticles, for example, have been used as high-efficiency biomolecule delivery transporters into cells while single-walled carbon nanotubes have been used as high-efficiency biomolecule delivery transporters into cells. Because of its fusion with other technologies and the emergence of complex and new hybrid technologies, nanotechnology has a bright future. Nanotechnology and biology-based technologies are closely intertwined; nanotechnology is already utilised to manipulate genetic material, and nanomaterials are made using biological components. The ability of nanotechnology to design matter at the atomic level is revolutionising industries such as information technology, cognitive science, and biology, as well as spawning new and interlinked sciences. With further research, nanotechnology has the potential to be useful in every aspect of human life.

CONSENT FOR PUBLICATION

Not applicable.

CONFLICT OF INTEREST

The authors declare no conflict of interest, financial or otherwise.

ACKNOWLEDGEMENTS

Declared none.

REFERENCES

[1]   Huang, Y.; Duan, X.; Cui, Y.; Lauhon, L.J.; Kim, K.H.; Lieber, C.M. Logic gates and computation from assembled nanowire building blocks. *Science,* **2001**, *294*(5545), 1313-1317.

[http://dx.doi.org/10.1126/science.1066192] [PMID: 11701922]

[2]   Lu, W.; Lieber, C.M. Semiconductor nanowires. *J. Phys. D Appl. Phys.,* **2006**, *39*(21), R387-R406.
      [http://dx.doi.org/10.1088/0022-3727/39/21/R01]

[3]   Iijima, S.; Ichihashi, T. Single-shell carbon nanotubes of 1-nm diameter. *Nature,* **1993**, *363*(6430), 603-605.
      [http://dx.doi.org/10.1038/363603a0]

[4]   Martel, R.; Schmidt, T.; Shea, H.R.; Hertel, T.; Avouris, P. Single- and multi-wall carbon nanotube field-effect transistors. *Appl. Phys. Lett.,* **1998**, *73*(17), 2447-2449.
      [http://dx.doi.org/10.1063/1.122477]

[5]   Huard, B.; Sulpizio, J.A.; Stander, N.; Todd, K.; Yang, B.; Goldhaber-Gordon, D. Transport measurements across a tunable potential barrier in graphene. *Phys. Rev. Lett.,* **2007**, *98*(23), 236803.
      [http://dx.doi.org/10.1103/PhysRevLett.98.236803] [PMID: 17677928]

[6]   Boisseau, P.; Loubaton, B. Nanomedicine, nanotechnology in medicine. *C. R. Phys.,* **2011**, *12*(7), 620-636.
      [http://dx.doi.org/10.1016/j.crhy.2011.06.001]

[7]   LaVan, D.A.; McGuire, T.; Langer, R. Small-scale systems for in vivo drug delivery. *Nat. Biotechnol.,* **2003**, *21*(10), 1184-1191.
      [http://dx.doi.org/10.1038/nbt876] [PMID: 14520404]

[8]   Cavalcanti, A.; Shirinzadeh, B.; Freitas, R.A., Jr; Hogg, T. Nanorobot architecture for medical target identification. *Nanotechnology,* **2008**, *19*(1), 015103.
      [http://dx.doi.org/10.1088/0957-4484/19/01/015103]

[9]   Allen, T.M.; Cullis, P.R. Drug delivery systems: entering the mainstream. *Science,* **2004**, *303*(5665), 1818-1822.
      [http://dx.doi.org/10.1126/science.1095833] [PMID: 15031496]

[10] Ullah, Zobair Nanotechnology and Its Impact on Modern Computer *Global Journal of Researches in Engineering General Engineering,* **2012**, *12*(4), Version 1.0.

[11] Boonserm, K.; Peter, J.B. Computer Science for Nanotechnology: Needs and Opportunities In: *Department of Computer Science, University College London*;

[12] Srivastava, D.; Atluri, S.N. Computational Nanotechnology: A Current Perspective. *CMES,* **2002**, *3*(5), 531-538.

[13] Rambidi, N.G. Computer Engineering and Nanotechnology *Molecular Computing,*
      [http://dx.doi.org/10.1007/978-3-211-99699-7_2]

[14] Kumar, S.; Pant, G.; Sharma, V.; Pooja, B. Nanotechnology in Computers *International Journal of Information & Computation Technology,* **2014**, *4*(15), 1597-1603.

[15] Noriega-Luna, B.; Godínez, L.A.; Rodríguez, F.J.; Rodríguez, A.; Zaldívar-Lelo de Larrea, G.; Sosa-Ferreyra, C.F.; Mercado-Curiel, R.F.; Manríquez, J.; Bustos, E. Applications of Dendrimers in Drug Delivery Agents, Diagnosis, Therapy, and Detection. *J. Nanomater.,* **2014**, *2014*, 1-19.
      [http://dx.doi.org/10.1155/2014/507273]

[16] Kesavan, A.; Ilaiyaraja, P.; Sofi Beaula, W.; Veena Kumari, V.; Sugin Lal, J.; Arunkumar, C.; Anjana, G.; Srinivas, S.; Ramesh, A.; Rayala, S.K.; Ponraju, D.; Venkatraman, G. Tumor targeting using polyamidoamine dendrimer–cisplatin nanoparticles functionalized with diglycolamic acid and herceptin. *Eur. J. Pharm. Biopharm.,* **2015**, *96*, 255-263.
      [http://dx.doi.org/10.1016/j.ejpb.2015.08.001] [PMID: 26277659]

CHAPTER 5

Improvement of Performance of Single and Multicrystalline Silicon Solar Cell Using Low-temperature Surface Passivation Layer and Antireflection Coating

Tapati Jana[1,*] and **Romyani Goswami**[2]

[1] *Department of Physics. Sarojini Naidu College for Women, 30 Jessore Road, Kolkata 700 028, India*

[2] *Department of Physics. Surya Sen Mahavidyalaya, Surya Sen Colony, Siliguri 734004, West Bengal, India*

Abstract: In this work, amorphous silicon oxide (a-SiO$_x$:H) and silicon nitride (a-SiN$_x$:H) layers are deposited at a very low substrate temperature of 250°C -300°C by the chemical Vapour deposition technique. Interface charge density (D_{it}) and fixed charge density (Q_f) have been estimated by high frequency (1 MHz) capacitance-voltage measurement on Metal-Insulator–Silicon structure (CV-MIS). The low interface charge density (D$_{it}$) reduces the surface recombination velocity. Fixed positive charges (D$_f$) stored in SiO$_x$:H/a-SiN$_x$:H layer forms negative charges at silicon film. The band bending due to negative charges provides a very effective field-induced surface passivation. A significant improvement in efficiency and short circuit current has been observed using developed a-SiO$_x$:H and a-SiN$_x$:H on the front surface of c-Si solar cells. As the refractive index of the films is close to silicon, hence it also acts as an anti-reflection coating (ARC) to reduce optical losses in silicon solar cells.

Keywords: A passivation layer, Antireflection coating, Chemical Vapour deposition, Fixed charge density, Interface charge density.

INTRODUCTION

The crystalline silicon is an indirect bandgap semiconductor and a large density of defect states due to dangling bonds (non-saturated bonds) exists at the surface of the crystalline silicon. The recombination losses *via* defect states within the bandgap reduces the cell efficiency, short circuit current. The conversion efficiency of solar cell may be reduced due to large surface recombination losses.

---

* **Corresponding author Tapati Jana:** Department of Physics. Sarojini Naidu College for Women, 30 Jessore Road, Kolkata 700 028, India; Tel: +919051080671; E-mail: tapati@sncwgs.ac.in

Dibya Prakash Rai (Ed.)

The minimization of surface recombination losses becomes very much important for further improvements of single and multicrystalline Si solar cells. Atomic hydrogen is not efficient in passivating the surface of the multicrystalline silicon solar cells [1, 2]. Low interface state density (D_{it}) and a large number of fixed charges (Q_f) at the interface (Si/SiN$_x$: H or Si/SiO$_x$: H) cause a strong band bending in the Si, reducing the surface recombination rate at the interface. The reflection losses in polished bare silicon are quite high, with about 35% of the incident radiation between 300 and 1200 nm wavelength [3]. The surface texturing and coating of the surface of the cell with one or more dielectric layers of a slightly lower refractive index than the cell itself can minimize the reflection losses [4, 5].

Thermally grown silicon oxide is widely used in different devices but its processing temperature is high (900 – 1000°C) [6]. The stress in thermal SiO_2 is expected due to mismatch in thermal expansion coefficients between the silicon and silicon oxide. Moreover, high oxidation temperature reduces the bulk minority carrier lifetime. The metal contacts cannot withstand high oxidation temperatures. Therefore, the low-temperature passivation layers would avoid some drawbacks associated with the high-temperature process.

In recent years, several authors have deposited silicon oxide or nitride layers for solar cell passivation and antireflection coating using different methods such as hot-wire CVD, catalytic-CVD, sol-gel process, rf Magnetron sputtering, *etc.* [7 - 10]. Each of these techniques has its own problems. Hot-wire CVD and catalytic-CVD are high-temperature deposition methods. With rf magnetron sputtering, defect densities in the films are high and the films are also nonuniform in thickness.

In this work, we have developed insulating silicon oxide films at low temperatures (~250° C), using silane (SiH_4) and nitrous oxide (N_2O) as the source gas by ion-damage-free Hg-sensitized photo-CVD technique. In this process, gas molecules are dissociated by photon absorption and no ions are formed. The silicon nitride films are also prepared at a sufficiently low substrate temperature of 300°C by rf PECVD (Anelva) technique, decomposing a mixture of silane (SiH_4), ammonia (NH_3) and H_2 gases.

EXPERIMENTAL DETAILS

Insulating silicon oxide films (a-SiOx:H) are deposited using Hg-sensitized photo-CVD system (SAMCO, Japan) decomposing a gas mixture of SiH_4, N_2O, and H_2 at 250°C substrate temperature. Hydrogen gas has been passed through a heated pot of mercury (70°C) before entering the chamber. The Hg atoms resonantly absorb UV radiation of wavelength 254 nm and 185 nm and are raised to an excited state, catalytically transferring energy to the source gases to produce the precursors.

$$Hg + h\nu \ (254 \ nm) \rightarrow Hg^*$$

In Hg-sensitized photo-CVD, N_2O is typically the oxygen source because it readily reacts with excited Hg atoms and yields neutral atomic oxygen in the ground state, SiH_4 reacts with atomic oxygen to form silicon monoxide or dioxide. SiH_4 can also react with excited Hg atoms to form reactive SiH_x and H radicals as a by-product. The main film growing precursors for the SiH network are SiH_3 and SiH_2.

The silicon nitride films are prepared to decompose a mixture of silane (SiH_4), ammonia (NH_3) and H_2 gases by rf PECVD (Anelva) technique at 300°C. The effect of hydrogen dilution ($R_H=H_2/(SiH_4+NH_3+H_2)$) and ammonia variation ($R_{NH3}=NH_3/(SiH_4+NH_3+H_2)$) on the film properties have been studied keeping the power density fixed at a very low value of 30 mW/cm^2.

The silicon oxide and nitride films are deposited on 7059 corning glass substrate for optoelectronic properties and c-Si wafer for FTIR studies. The optical absorption of the samples is determined using transmission and reflection measurements in the UV-VIS region by a double-beam spectrophotometer (Hitachi 330, Japan) after measuring the thickness by a stylus type instrument. The refractive index of the films are measured by ellipsometry technique. The dark and photoconductivities (under 50 mW/ cm^2 white light insolation) of the films are measured in vacuum (~10^{-6} Torr) after annealing the samples at 150°C for 1 hr. The dark conductivity (σ_D) activation energy is measured from the temperature dependence of dark conductivity values (σ_D vs 1000/T). The bonding configuration of the a-SiO: H films is investigated using a Fourier Transform Infrared Spectrophotometer (Perkin Elmer 1700, UK). The amount of Hg incorporation in the films has been estimated from the sputtered neutral mass spectroscopy (SNMS) measurement. High frequency (1 MHz) Capacitance–Voltage measurement on Metal–Insulator–Semiconductor structure (CV–MIS) is used to estimate the interface charge density (D_{it}) and fixed charge density (Q_f). The developed insulating silicon oxide films have been applied on c-Si solar cells and their effect on the performance of the solar cells have been studied.

RESULTS AND DISCUSSIONS

Silicon Oxide Films Developed by Photo-CVD Technique

Hydrogenated amorphous silicon-oxygen alloy films (a-SiO$_x$:H) are prepared by varying N_2O to SiH_4 ratio (R) from 7.1 to 42.6, while SiH_4 and H_2 flow were kept constant. The deposition rate is found to decrease from 40 Å min^{-1} to 25 Å min^{-1} with the increase in R from 7.1 to 42.6. The variation of optical gap (E_g) and refractive index (n_f) as a function of *R is* shown in Fig. (**1**). The optical gaps (Eg) of the films are determined from Tauc's plot and it increases monotonically from 2.17 eV to 2.60 eV with the variation of *R* from 7.1 to 42.6. This is due to increased oxygen incorporation in the film matrix with the increase of *R*, measured from Infrared studies. A similar variation of optical gap (Eg) and refractive index (n_f) are also observed in a-SiOx:H films prepared by the photo-CVD technique using CO_2 gas as a source gas of oxygen [11]. The stronger bond energy of the Si–O bond (8.4 eV) compared to those of Si–Si (2.4 eV) and Si–H (3.0 eV) increases the optical gap (E_g) with the increase of Si-O bonding. On the other hand, an increase in the optical gap is considered to be partly due to an increase in the Si-Si bond energy accompanied by the strong electronegative O atoms back bonded to Si-Si bonds which may arise due to either a decrease in the bandwidth or an increase in the energy difference between bonding and antibonding splitting of Si-Si bonds [12]. Refractive index (n_f) decreases rapidly from 2.43 to 2.06 with an increase of R from 7.1 to 14.2, then decreases slowly to 1.96 with a further increase of R from 14.2 to 42.6. The decrease in n_f is considered to be due to increased oxygen content in the film with an increase of R [13, 14]. An increase-maximum-decrease pattern of deposition rate with increasing silane partial pressure ratio is also observed by Lan *et al.* [15]. Although stoichiometric PECVD-deposited SiO_2 can be suitable as a passivation layer of the silicon surface, it is less suitable as an antireflection coating of silicon solar cells [16]. The optimum value of the refractive index is 2.3 for a single layer antireflection coating of a silicon solar cell underneath a glass encapsulation. The developed *a*-SiO$_x$:H films with $n_f \sim 2.00$ is more suitable for antireflection (AR) coating for silicon solar cells than thermal silicon oxide (SiO_2) with refractive index 1.46. The decrease of deposition rate with the increase of R is due to less amount of SiH$_x$ radicals adsorption on the surface. A similar trend of deposition rate has been obtained by Cobianu *et al.* [17].

Fig. (**2**) depicts the variation of dark (o_D) and photoconductivity (o_{ph}) and activation energy (E_a) as a function of R. The σ_D and σ_{ph} of SiOx:H films are $\sim 10^{-14}$ S cm^{-1} and $\sim 10^{-9}$ S cm^{-1}, respectively with activation energy (Ea) of 1.10 eV. These results indicate the amorphous nature of the film. Further increase of R decreases both σ_D and o_{ph} but Ea shows the reverse replica of the σ_D, σ_{ph} curve. An increase of N_2O to SiH_4 ratio (R) leads to more oxygen incorporation in the film matrix, which causes an increase of an optical gap and reduction in the dark- and photoconductivity. The observed deterioration in the photoconductivity with the incorporation of oxygen may be due to an increase in electronic and structural disorders. The Sputtered Neutral Mass Spectrometry (SNMS) mass spectra do not show any significant mass peaks corresponding to Hg.

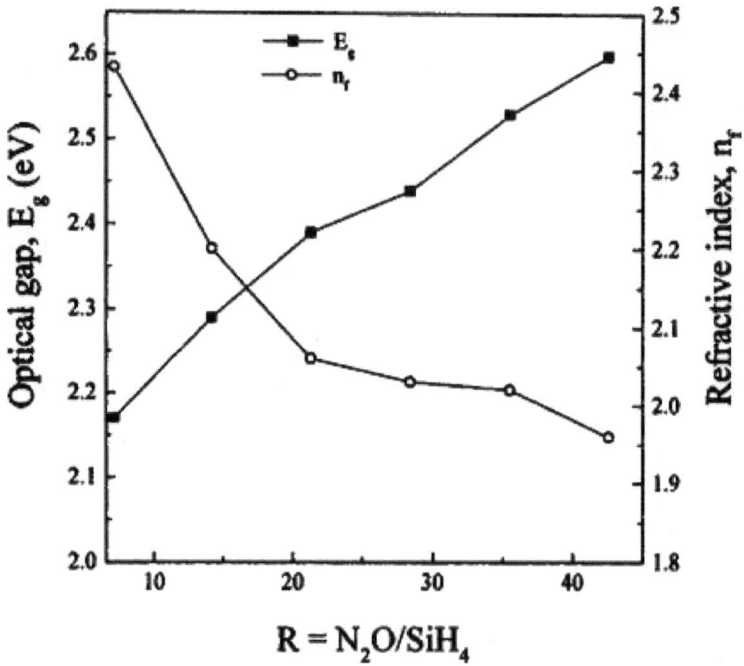

Fig. (1). Optical gap (E_g) and refractive index (n_r) of silicon oxide films with variation of R (N_2O/SiH_4).

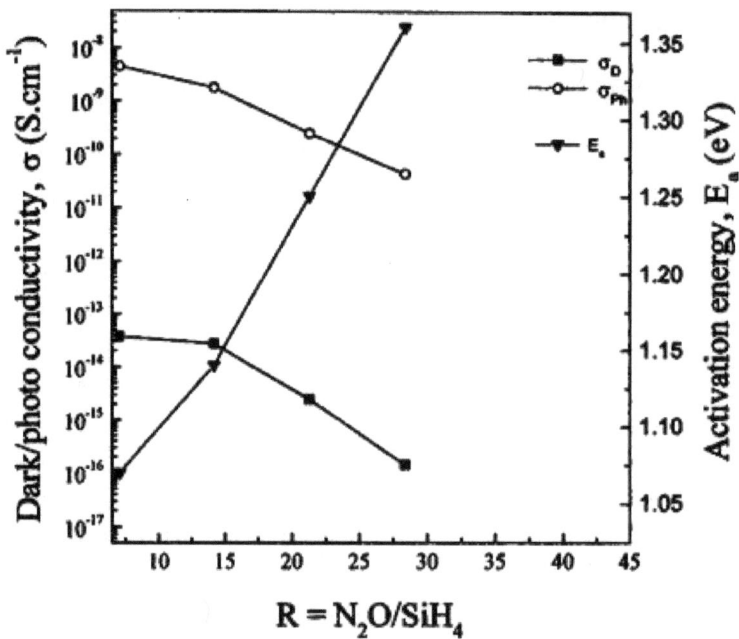

Fig. (2). Dark and photo conductivity and its activation energy (E_a) with variation of R (N_2O/SiH_4).

The IR absorption spectrum of SiOx:H films in the range 400 – 4000 cm^{-1} prepared at R=14.2 is shown in Fig. (3). The Si–H stretching and bending absorptions in a-SiOx: H films occur over the ranges 2000 – 2300 cm^{-1} and 600 – 900 cm^{-1}, respectively [18, 19]. On the other hand, SiO stretching, bending, and rocking absorptions occur at 950 – 1100 cm^{-1} 600 – 800 cm^{-1}, and 500 cm^{-1}, respectively [19, 20]. Thus the IR spectra over the range 600 – 800 cm^{-1} are composed of both SiO and SiH bending absorption bands.

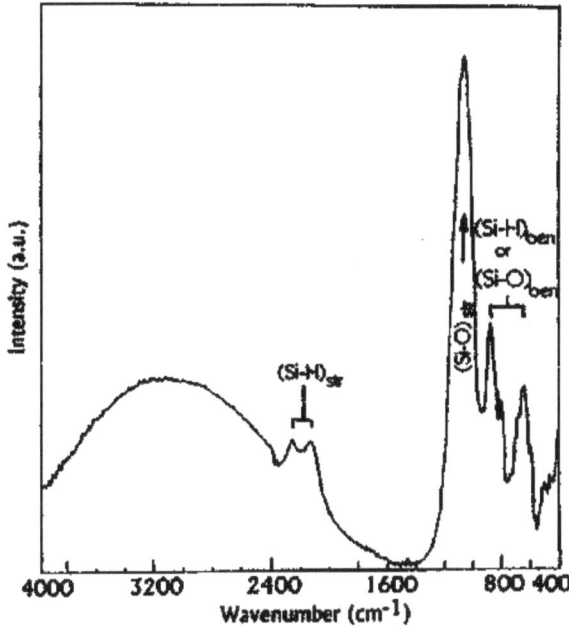

Fig. (3). Fourier Transform Infrared absorption spectra of silicon oxide film prepared at $R = 14.2$.

The basic configuration of SiO$_x$:H films can be described as H–Si(Si$_{3-n}$O$_n$) according to the Modified Random Bonding Model (RBM), where n=0,1,2,3,. The Si–H stretching absorptions peak positions of H–Si(Si$_3$), H–Si(Si$_2$O), H–Si(SiO$_2$), and H–Si(O$_3$) occurs at 2000, 2115, 2200, and 2260 cm^{-1} respectively [19]. The corresponding bending modes of the H–Si(Si$_2$O), H–Si(SiO$_2$), and H–Si(O$_3$) configuration appears at 790 cm^{-1}, 840 cm^{-1}, and 880 cm^{-1}, respectively [19]. In case of Si–O stretching absorptions, the peaks corresponding to H–Si(Si$_{3-n}$O$_n$), for n =1,2,3, are at 983 cm^{-1}, 1012 cm^{-1}, and 1034 cm^{-1} respectively [20]. Therefore, any change in the H–Si(Si$_{3-n}$O$_n$) configuration will be reflected both in the SiH and SiO stretching mode.

Fig. (**4a** and **4b**) represent the total integrated absorption intensities (I_v cm^{-2}) for the subbands (n =1–3) of the Si–H stretching absorption for H–Si(Si$_{3-n}$O$_n$)

configuration and their peak frequency (v_{SiH}) for each individual subband as a function of R, respectively. The shift of peak frequency towards the higher wavenumber side with an increase of R corresponding to each bond is consistent with the fact that higher oxygenation occurs at higher R. Integrated intensity corresponding to each band also increases with more oxygen incorporation in the film which indicates an increase of N_2O/SiH_4 ratio increases oxygen incorporation in film matrix. The absorption spectra of Si–O stretching mode also shows the same behaviour as the Si–H stretching mode of a-SiO_x:H films. The absorption intensities (I_v cm^{-2}) and the peak frequency (v_{SiO}) for each subband (n =1–3) in the Si–O stretching mode are shown in Fig. (**5a** and **5b**) respectively. The absorption intensity (I_v cm^{-2}) corresponding to each subband increases with the increase of R. The peak position of the band for n = 1 and 2 shifted to a higher wavenumber, whereas the peak for n = 3 remains almost constant. The peak frequencies of both SiO and SiH stretching bonds shift to higher frequencies with an increase of R $i.e$ with an increase of oxygenation in the network.

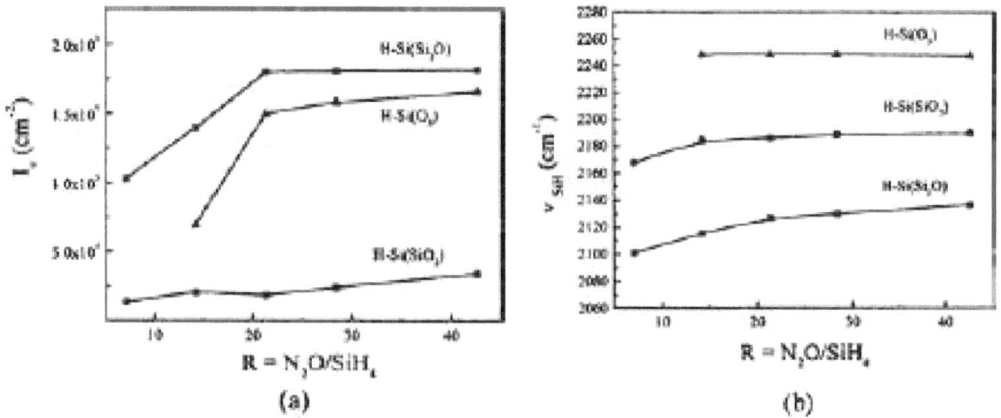

Fig. (4). (a) Total integrated absorption intensities (I_v cm^{-2}) and **(b)** their peak frequency (v_{SiH}) for each individual subbands of the Si–H stretching absorption as a function of R.

Fig. (**6**) shows the high frequency (1 MHz) C–V characteristics of an M–I–S structure using silicon oxide as the insulator layer with 2.99 eV optical gap. The calculated interface-trap density (D_{it}) and fixed charge density (Qf) are 2.47×10^{11} cm^{-2} eV^{-1} and of 3.51×10^{12} cm^{-2} [21]. This silicon oxide material with low interface-trap density and large fixed charge density will be very much effective for field-effect passivation of silicon solar cells. It can also be used as an effective antireflection coating for crystalline and multi-crystalline silicon solar cells as the refractive index of the silicon oxide material is ~2.00.

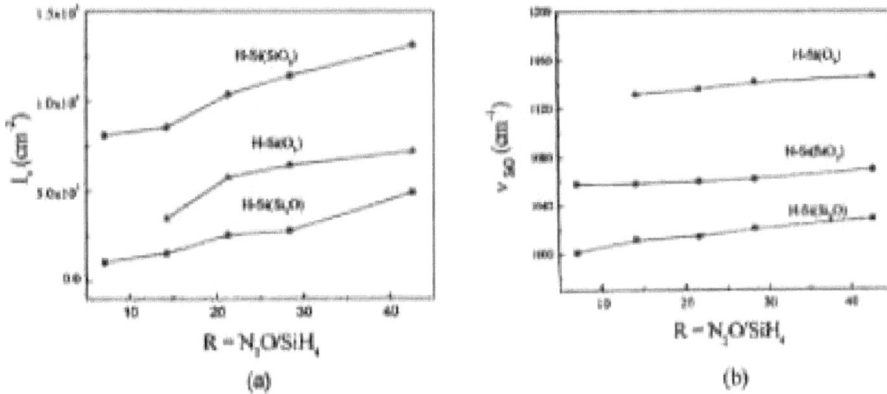

(a)   (b)

Fig. (5). (a) Total integrated absorption intensities (I_v cm^{-2}) and **(b)** their peak frequency (v_{SiH}) for each individual subbands of the Si–O stretching absorption as a function of R.

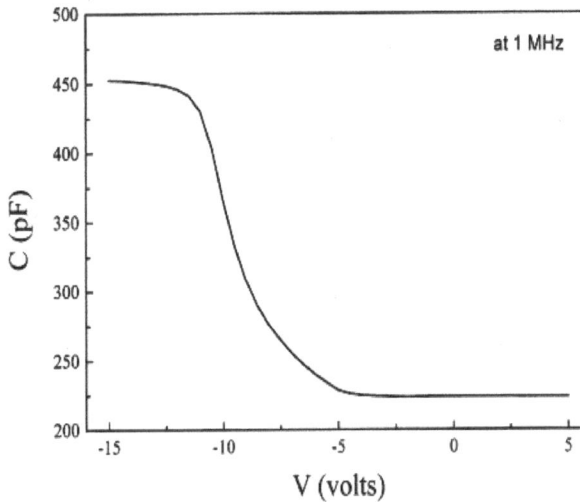

Fig. (6). High frequency (1 MHz) $C–V$ characteristics using photo-CVD silicon oxide MIS structure.

Silicon Nitride Films Developed by rf-PECVD Technique

The variation of optical gap (E_g) with the percentage change of ammonia in the gas mixture (R_{NH3}=NH$_3$/(SiH$_4$+NH$_3$+H$_2$)) is shown in Fig. (7). The change of R_{NH3} from 48.2% to 66.9% increases E_g linearly from 2.96 to 4.17 eV due to an increase of Si-N bonding in the film matrix, associated with the increase of the percentage of ammonia in the gas mixture (R_{NH3}). The optical gap increases sharply with the increase of the percentage of ammonia in the gas mixture (R_{NH3}) as the binding energy of Si–N (105 kcal/mol) is much greater than that of N–H (75 kcal/mol) and Si–H (71 kcal/mol).

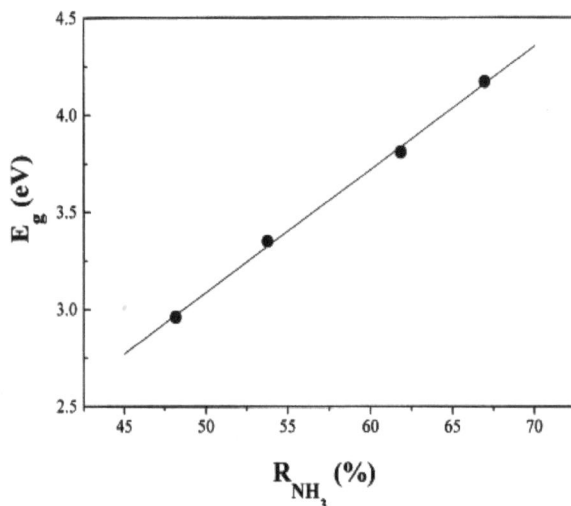

Fig. (7). Variation of optical gap (E_g) with R_{NH3} in silicon nitride films.

The effect of hydrogen dilution was investigated at the same time by keeping the R_{NH3} fixed at 61.8%. Table **1** shows the variation of optical gap (E_g), deposition rate (Å/min), hydrogen content (N_H) calculated from Si–H stretching mode and N–H stretching mode, with the variation of hydrogen dilution (R_H=H$_2$/(SiH$_4$+ NH$_3$+H$_2$)). Fig. (**8**) shows the IR spectra of the films deposited at hydrogen dilution of 21.6% and R_H=35.52%. It clearly shows an increase of R_H that enhances the intensity of Si–N stretching mode (750–1050 cm^{-1}) and decreases Si–H and N–H bonds. So it is concluded that hydrogen dilution in the gas mixture favours the formation of Si–N bonds compared to the Si–H, N–H bonds. The increase of nitrogen incorporation with an increase in hydrogen dilution is also reported by Giorgis *et al.* [22]. The hydrogen content has been calculated both from Si–H and N–H stretching mode using the formula [23].

Table 1. Optical gap (E_g), deposition rate and hydrogen content with the variation of hydrogen dilution R_H=H$_2$/(SiH$_4$+NH$_3$+H$_2$).

Sample No.	R_H (%)	(E_g) eV	Dep rate (Å/min)	N_H from Si–H Stretching Mode	N_H from N–H Stretching Mode
1	21.60	3.25	40.4	2.02×10^{22} (40.3 at%)	1.17×10^{22}
2	35.52	3.47	32.0	1.60×10^{22} (32.0 at%)	1.13×10^{22}
3	45.25	3.59	26.9	1.35×10^{22} (26.9 at%)	9.12×10^{21}

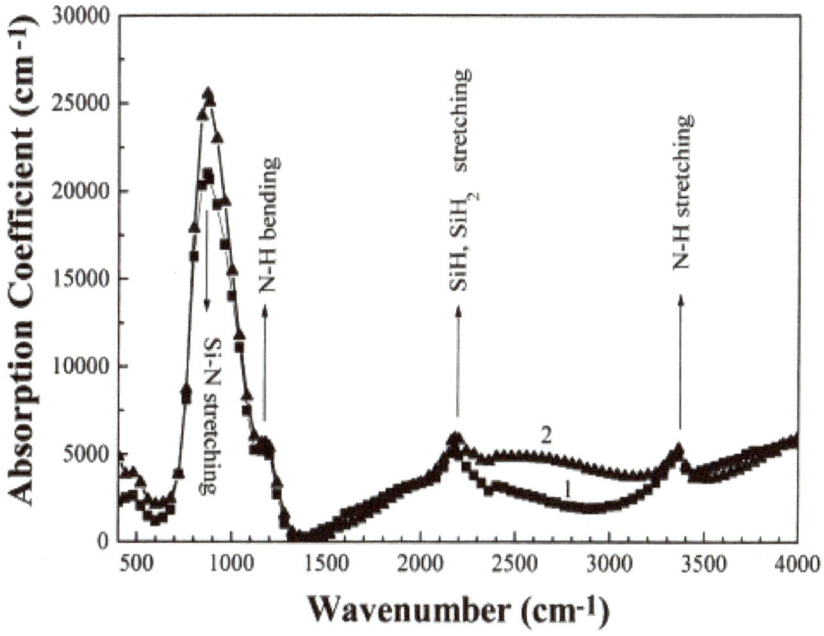

(a)

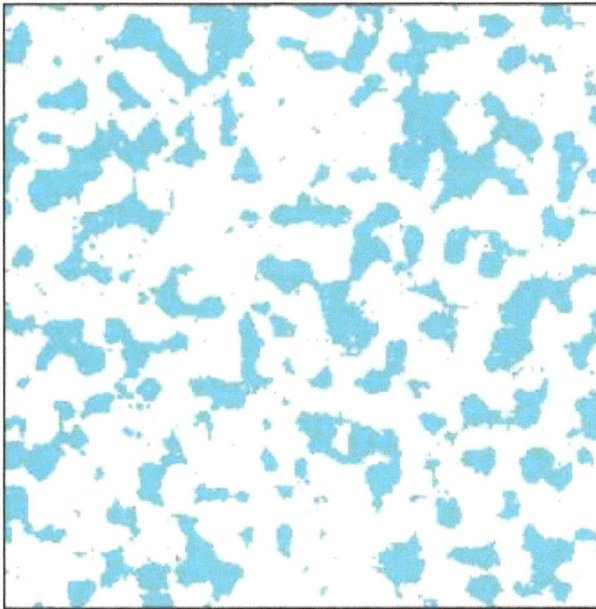

(Fig. 8) contd.....

(b)

(c)   Kn=10

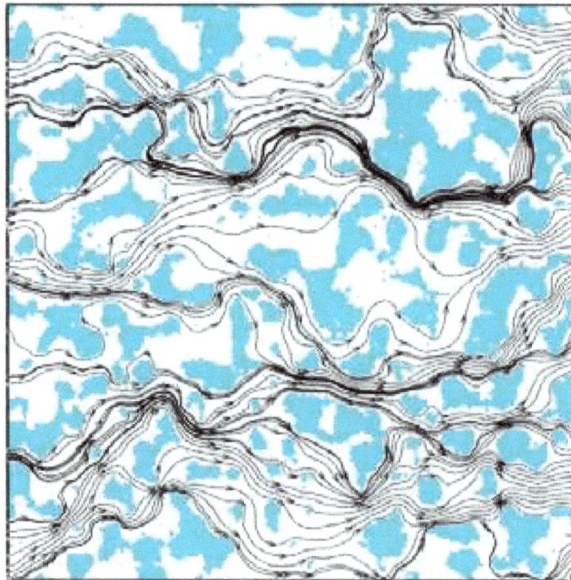

Fig. (8). Infrared absorption spectra of silicon nitride films prepared at different hydrogen dilution, curve (—■—) and (—▲—) are for R_H=21.60% and R_H=35.52% respectively.

$$NH = A\int \alpha(\omega)\omega_o d\omega$$

where N_H, α, ω, ω_o are the hydrogen content, the absorption coefficient, the wavenumber and the peak wave number, respectively. The magnitude of A is taken as 1.4×10^{20} cm^{-2} for Si–H stretching mode and 2.8×10^{20} cm^{-2} for N–H stretching mode.

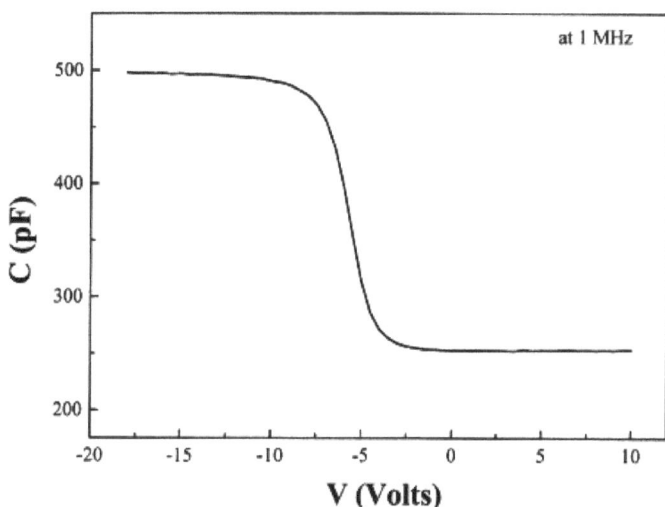

Fig. (9). *C–V* curve using PECVD silicon nitride MIS structure (1 MHz frequency).

The *C–V* characteristics of the MIS structure using insulating SiN$_x$ with 3.41 eV optical gap measured at 1 MHz frequency is shown in Fig. (9). A significantly low D_{it} value of 2.52×10^{11} cm^{-2} eV^{-1} and a sufficiently high fixed charge density of 1.34×10^{12} cm^{-2} at the interface have been obtained from the *C–V* characteristics [21]. On the other hand, the refractive index of this material deposited on a quartz substrate is 1.9, which is close to the optimum value for antireflection coating in Si solar cells. So this material will be very effective to reduce surface recombination velocity as well as antireflection coating on c-Si collar cells.

Application of silicon oxide and nitride in crystalline silicon solar cells

To investigate the effect of surface passivation as well as antireflection coating on the performance of c-Si solar cells, the developed silicon oxide and silicon nitride materials have been deposited on c-Si solar cells. The silicon oxide passivation layer having 2.55 eV optical gap and 2.02 refractive index is deposited on the front surface of the solar cell and the results are shown in Table **2**. The Improvement in I_{sc} and efficiency are 11.4% and 11.2% respectively, whereas V_{oc} remains unchanged.

Table 2. Improvement of the performance of c-Si solar cell using silicon oxide passivation layer.

C-Si solar cell condition	V_{oc} (Volts)	I_{sc} (mA)	FF	η (%)
Before passivation	0.57	173	0.625	13.87
After 800 Å silicon oxide deposition	0.57	192.7	0.630	15.43

The silicon nitride layer with 3.41 eV optical gap and 1.9 refractive indexes is deposited on the front surface of the solar cell. Table **3** represents the cell parameters before and after depositing the silicon nitride layer with 3.41 eV optical gap and 1.9 refractive indexes. The improvement of cell efficiency by 11.16% and I_{sc} by 10.2% have been achieved.

Table 3. Improvement of the performance of c-Si solar cell using silicon nitride passivation layer.

c-Si solar cell condition	V_{oc} (Volts)	I_{sc} (mA)	FF	η (%)
Before passivation	0.56	162.4	0.560	9.50
After 800 Å silicon nitride deposition	0.56	179.0	0.568	10.56

CONCLUDING REMARKS

The high bandgap insulating silicon oxide surface passivation, as well as antireflection coating, have been developed by ion-damage-free Hg-sensitized photo-CVD technique suitable for crystalline and microcrystalline silicon solar cells at low substrate temperature (250°C). The optical gap of the films is increased linearly with an increase in N_2O to SiH_4 ratio (R). The SiH and SiO stretching absorption profiles are characterized as a $H–Si(Si_{3-n}O_n)$ configuration, where n is a positive integer varying from 0 to 3. The total oxygen content, as well as each band of $H–Si(Si_{3-n}O_n)$ configuration (n=1–3), are increased almost linearly with an increase in R, as observed for both SiH and SiO stretching modes. The a-SiO_x:H films with 2.55 eV optical gap have been applied as passivation layers as well as antireflection coatings in c-Si solar cells. Efficiency (η) and short circuit current (I_{sc}) have been improved by 11.2% and 11.4%, respectively.

Insulating silicon nitride (SiN_x: H) films also have been developed as a passivation layer and antireflection coating for crystalline and multi crystalline silicon solar cells by plasma-enhanced chemical vapour deposition technique at a sufficiently low substrate temperature of 300°C. The optical gap increases both with the increase of the percentages of NH_3 and H_2. The improvement in I_{sc} by 10.2% and efficiency by 11.16% have been achieved using 3.41 eV optical gap and 1.9 refractive index silicon nitride material on the front surface of c-Si solar cells.

CONSENT FOR PUBLICATION

Not applicable.

CONFLICT OF INTEREST

The authors declare no conflict of interest, financial or otherwise.

ACKNOWLEDGEMENTS

Authors are highly obliged to Prof. Swati Ray for her enormous guidance and help. The authors are also grateful to the Indian Association for the Cultivation of Science, Jadavpur, Kolkata, India for giving us the opportunity to work in the laboratory.

REFERENCES

[1]   Ghannam, M.; Palmers, G.; Elgamel, H.E.; Nijs, J.; Mertens, R.; Peruzzi, R.; Margadonna, D.; Margadonna, D. Comparison between different schemes for passivation of multicrystalline silicon solar cells by means of hydrogen plasma and front side oxidation. *Appl. Phys. Lett.,* **1993**, *62*(11), 1280-1282.
        [http://dx.doi.org/10.1063/1.108707]

[2]   Elgamel, H.E.; Ghannam, M.Y.; Vinckier, C.; Nijs, J.; Mertens, R.; Van Overstraeten, R. Boosting the efficiency of solar cells fabricated on electromagnetic cold crucible cast multicrystalline silicon by means of hydrogen passivation. *Sol. Energy Mater. Sol. Cells,* **1994**, *34*(1-4), 237-241.
        [http://dx.doi.org/10.1016/0927-0248(94)90045-0]

[3]   Sexton, F.W. Plasma nitride AR coatings for silicon solar cells. *Sol. Energy Mater.,* **1982**, *7*(1), 1-14.
        [http://dx.doi.org/10.1016/0165-1633(82)90091-0]

[4]   Min, W.L.; Jiang, B.; Jiang, P. Bioinspired Self-Cleaning Antireflection Coatings. *Adv. Mater.,* **2008**, *20*(20), 3914-3918.
        [http://dx.doi.org/10.1002/adma.200800791]

[5]   Sharma, J.R.; Banerjee, P.; Mitra, S.; Ghosh, H.; Bose, S.; Das, G.; And, S. Mukhopadhyay, "Potential of zinc oxide nanowhiskers as antirefection coating in crystalline silicon solar cell for cost efectiveness Journal of Materials Science. *Materials in Electronics,* **2019**, *30*, 11017-11026.
        [http://dx.doi.org/10.1007/s10854-019-01443-5]

[6]   Aberle, A.G.; Glunz, S.; Warta, W. Impact of illumination level and oxide parameters on Shockley–Read–Hall recombination at the Si☐SiO$_2$ interface. *J. Appl. Phys.,* **1992**, *71*(9), 4422-4431.
        [http://dx.doi.org/10.1063/1.350782]

[7]   Holt, J.K.; Goodwin, D.G.; Gabor, A.M.; Jiang, F.; Stavola, M.; Atwater, H.A. Hot-wire chemical vapor deposition of high hydrogen content silicon nitride for solar cell passivation and anti-reflection coating applications. *Thin Solid Films,* **2003**, *430*(1-2), 37-40.
        [http://dx.doi.org/10.1016/S0040-6090(03)00131-7]

[8]   He, Y.; Huang, H.; Zhou, L.; Yue, Z.; Yuan, J.; Zhou, N.; Gao, C. a-SiOx:H passivation layers for Cz-Si wafer deposited by hot wire chemical vapor deposition. *Mater. Sci. Semicond. Process.,* **2017**, *61*, 1-4.
        [http://dx.doi.org/10.1016/j.mssp.2016.12.031]

[9]   Voz, C.; Martin, I.; Orpella, A.; Puigdollers, J.; Vetter, M.; Alcubilla, R.; Soler, D.; Fonrodona, M.; Bertomeu, J.; Andreu, J. Surface passivation of crystalline silicon by Cat-CVD amorphous and nanocrystalline thin silicon films. *Thin Solid Films,* **2003**, *430*(1-2), 270-273.

[http://dx.doi.org/10.1016/S0040-6090(03)00130-5]

[10]   Deligiannis, D.; Vliet, J. V.; Vasudevan, R.; Swaaij, R.; Zeman, M. Passivation mechanism in silicon heterojunction solar cells with intrinsichydrogenated amorphous silicon oxide layers. *J. Appl. Phys,* **2017**, *121*, 085306-1-085306-7.

[11]   Jana, T.; Ghosh, S.; Ray, S. Silicon oxide thin films prepared by a photo-chemical vapour deposition technique. *Journal of Materials Science,* **1997**, *32*, 4895-4900.

[12]   Umezu, I.; Miyamoto, K.; Sakamoto, N.; Maeda, K. Optical Bond Gap and Tauc Gap in a-SiO x:H and a-SiN x:H Films. *Jpn. J. Appl. Phys.,* **1995**, *34*(Part 1, No. 4A), 1753-1758.
       [http://dx.doi.org/10.1143/JJAP.34.1753]

[13]   Sassella, A.; Borghesi, A.; Corni, F.; Monelli, A.; Ottaviani, G.; Tonini, R.; Pivac, B.; Bacchetta, M.; Zanotti, L. Infrared study of Si-rich silicon oxide films deposited by plasma-enhanced chemical vapor deposition. *J. Vac. Sci. Technol. A,* **1997**, *15*(2), 377-389.
       [http://dx.doi.org/10.1116/1.580495]

[14]   Sassella, A. Tetrahedron model for the optical dielectric function of H-rich silicon oxynitride. *Phys. Rev. B Condens. Matter,* **1993**, *48*(19), 14208-14215.
       [http://dx.doi.org/10.1103/PhysRevB.48.14208] [PMID: 10007835]

[15]   Lan, W.H.; Lin, W.J.; Tu, S.L.; Yang, S.J.; Huang, K.F. Reaction Mechanism of Mercury-Sensitized Photochemical Vapor Deposited Silicon Oxide. *Jpn. J. Appl. Phys.,* **1993**, *32*(Part 1, No. 1A), 150-154.
       [http://dx.doi.org/10.1143/JJAP.32.150]

[16]   Leguijt, C.; Lölgen, P.; Eikelboom, J. A.; Weeber, A. W.; Schuurmans, F. M.; Sinke, W. C.; Alkemade, P. F. A.; Sarro, P. M.; Maree, C. H. M.; Verhoef, L. A. Low temperature surface passivation for silicon solar cells. *Sol. Energy Mater. Sol. Cells,* **1996**, *40*, 297-305.
       [http://dx.doi.org/10.1016/0927-0248(95)00155-7]

[17]   Cobianu, C.; Pavelescu, C. A Theoretical Study of the Low□Temperature Chemical Vapor Deposition of SiO2 Films. *J. Electrochem. Soc.,* **1983**, *130*(9), 1888-1893.
       [http://dx.doi.org/10.1149/1.2120118]

[18]   Lucovsky, G.; Yang, J.; Chao, S. S.; Tyler, J. E.; Czubatyj, W. Oxygen-bonding environments in glow-discharge-deposited amorphous silicon-hydrogen alloy films. *Phys. Rev. B,* **1983**, *28*, 3225-3238.
       [http://dx.doi.org/10.1103/PhysRevB.28.3225]

[19]   He, L.; Kurata, Y.; Inokuma, T.; Hasegawa, S. Analysis of SiH vibrational absorption in amorphous SiO_x:H ($0{\leq}x{\leq}2.0$) alloys in terms of a charge□transfer model. *Appl. Phys. Lett,* **1993**, *63*, 162-164.
       [http://dx.doi.org/10.1063/1.110386]

[20]   Morimoto, A.; Noriyama, H.; Shimizu, T. Structure and Defects in Amorphous Si–O Films. *Jpn. J. Appl. Phys.,* **1987**, *26*(Part 1, No. 1), 22-27.
       [http://dx.doi.org/10.1143/JJAP.26.22]

[21]   Sze, S.M. *Physics of Semiconductor Devices*; Wiley Eastern Limited: New Delhi, **1991**.

[22]   Giorgis, F.; Pirri, C.F.; Tresso, E. Structural properties of a-$Si_{1-x}N_x$:H films grown by plasma enhanced chemical vapour deposition by SiH_4 + NH_3 + H_2 gas mixtures. *Thin Solid Films,* **1997**, *307*(1-2), 298-305.
       [http://dx.doi.org/10.1016/S0040-6090(97)00272-1]

[23]   Morimoto, A.; Kobayashi, I.; Kumeda, M.; Shimizu, T. Hydrogen Evolution from Amorphous Si-N Films. *Jpn. J. Appl. Phys.,* **1986**, *25*(Part 2, No. 9), L752-L754.
       [http://dx.doi.org/10.1143/JJAP.25.L752]

CHAPTER 6

Advanced Materials and Nanosystems: Synthesis and Characterization

Rekha Garg Solanki[1,*]

[1] *Department of Physics, Dr. Harisingh Gour University, Sagar (M.P.), India*

Abstract: The nanomaterials are materials of dimensions 1-100 nm. Advanced materials and Nanosystems; synthesis and characterization is an emerging field of research. The day by day modification in synthesis approaches may generate a new synthetic approach. The gradual development in the synthesis techniques from the bulk to nanoparticles synthesis has been reported in this chapter. We are focused on all types of synthesis methods for advanced nanomaterials and the techniques used for nanomaterials characterization. For the Commercial production of nanomaterials, various bottom-up and top-down approaches have been developed during the last two decades. Here, we are trying to summarize the basic principle of solid phase, vapour phase and liquid phase synthetic techniques in detail with a schematic setup.

Keywords: Ball milling, Coprecipitation, Epitaxial growth, Evaporation, Green or biomimetic synthesis, Liquid and vapour phase synthesis, Micelle & inverse micelle, Polyol, Sol-gel, Solid, Spray pyrolysis, Sputtering.

INTRODUCTION

Advanced nanomaterials synthesis, characterization and applications are the new branches of research for the past few years. Various advanced materials like quantum clusters, nanocomposites, anisotropic nanocrystals, one-dimensional nanostructures such as nanowires and nanorods, nanoparticles, quantum dots, macromolecules, dendrimers and hybrid nanosystems [1, 2]. All the materials are known as advanced nanomaterials. Each of them is different from the others only in their shape and size. Shape and size play a major role in the modification of the properties of materials. Whenever the material composition remains the same, only changes in size; the material shows dramatically different properties and enhances applicability many times. The chapter has a brief description of all kinds of synthetic approaches employed for nanomaterials like quantum clusters, nanoparticles, quantum dots, nanowires, nanorods, *etc*.

---

[*] **Corresponding author Rekha Garg Solanki:** Department of Physics, Dr Harisingh Gour University, sagar (MP)-470003 India; Tel: 91-8770781460; E-mail: rgsolanki@dhsgsu.edu.in, sorekha49@gmail.com

Quantum clusters (QCs) refer to a new group of materials with core dimensions is of the order of 1-10 nm [1, 3]. Quantum clusters are totally different from metallic nanoparticles. They act as an intermediate chain between molecules and nanoparticles. Nanoparticles are three-dimensional arrangements of nano (10^{-9} m) sized particles in a micron regime. They are also known as superlattices (SLs) [1]. In general, the nanoparticles are assumed to be spherical in shape, which shows properties are the same in all directions and are called isotropic nanoparticles. Anisotropic nanoparticles (ANPs) are nanomaterials with direction-dependent properties. Anisotropic nanomaterials can exist in various forms such as nanowires, nanotubes, nanorods, nanoplates and nanostars.

IMPORTANCE OF NANOMATERIALS

Nanomaterials are materials of dimensions 1 to 100 nm [2, 3]. The materials can be spherical particles, wires, rods, films, sheets, flakes, flowers of nano dimensions. The materials with at least one dimension in the nano range show dramatically different properties than their bulk counterparts. Their applicability increases many times than bulk. Some important applications of nanomaterials are represented in Fig. (1).

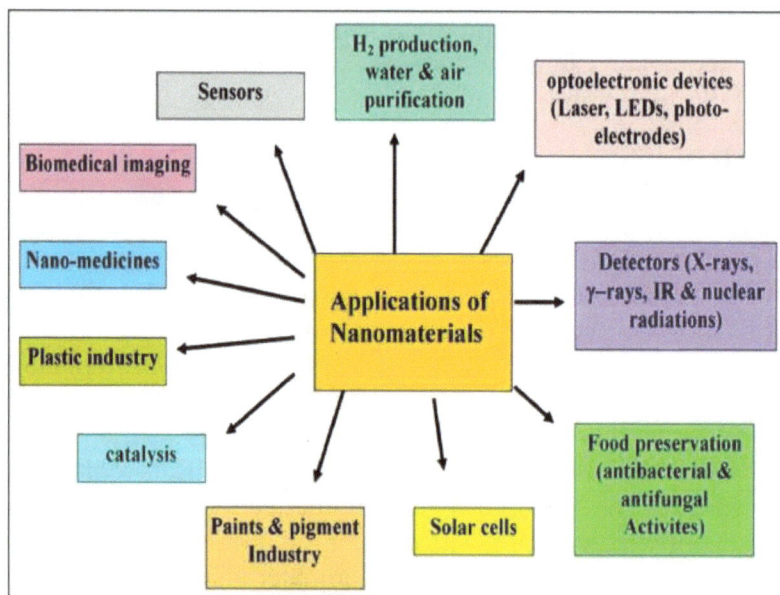

Fig. (1). Some of the important applications of nanomaterials.

UNIQUE PROPERTIES OF NANOMATERIALS

The nanomaterials are showing extremely different properties compared to their bulk counterparts. Size-dependent properties become more prominent at the nanoscale; hence the properties of materials can be tuned *via* tuning the size. The properties of materials such as electronic, mechanical, optical, *etc.* are substantially changed by reducing the materials into nano regime. The mechanical properties of nanomaterials are considerably improved compared to bulk form because of an increase in crystal perfection or maybe because of the reduction in crystallographic defects [1 - 3]. Among a sequence of unique properties, some of the key properties such as surface, mechanical, optical, anti-bacterial and anti-fungal are modified with the size and shape of nanomaterials.

Surface Properties

As the size of the materials decreases, the number of atoms on the surface will increase. The size reduction process increases the number of particles on the surface. For example, a large spherical particle of size 1 cm, its surface area is 4π cm^2 and volume is 4π cm^3. When the large particle is reduced to the small particles of size 1nm or 10^{-7}cm, the surface area is $4 \times 10^{-7}\pi$ cm^2 and the volume is $4 \times 10^{-21}\pi$ cm^3 [4, 5]. Now the volume ratio of large to small spherical particles and surface ratios will be;

$$\frac{V_{large}}{V_{small}} = \frac{4\pi \ cm^3}{4 \times 10^{-21} \ \pi \ cm^3} = 10^{21}$$

$$\frac{S_{large}}{S_{small}} = \frac{4\pi \ cm^2}{4 \times 10^{-14} \ \pi \ cm^2} = 10^{14}$$

It means by reducing the 1 cm particle to 1 nm; the volume occupied by small-sized particles and surfaces occupied by small particles. The surface to volume ratio increases times if a cm size particle is reduced to 1nm-sized small particles.

Mechanical Properties

Nanomaterials show improved mechanical properties which are absent in their bulk form. The increase in mechanical properties such as the strength of wearing force and corresponding strain *etc.* may be because of a reduction in size.

Catalytic Properties

Catalysis is an important chemical phenomenon in which by the presence of a foreign entity or substance the reaction rate enhances many times than the reaction rate in normal conditions [1 - 7]. When nanoparticles are used as a catalyst in any chemical reaction then because of the large surface to volume ratio; they reduced the reaction time many times than the bulk catalyst of the same material. The 2D sheets or films of various nanomaterials have provided more active channels for showing excellent catalytic activity and reduced the lifetime of the chemical reaction. Hence, the size from bulk to nano enhances the catalyst performance.

Optical Properties

Quantum effects are observed more prominently at the nanoscale regime. However, the size and nature of the semiconductor material will strongly be responsible for the demonstration of the quantum confinement effect [1, 3, 7]. The optical properties like emission, absorption, luminescence, transmittance, *etc.* of nanomaterials vary with their shape and size. For semiconductors, the exciton diameter is an important factor that actively modifies the optical properties. If the particle size of semiconducting material is less than or of the order of the exciton radius; showing a quantum confinement effect [1, 8]. It means the optical properties of the materials changes with change in size; with the reduction in size, the optical properties shifted towards a higher wavelength or lower energy region. Fig. (2) shows the change in optical properties with the size of nanoparticles. A photo-generated electron-hole pair is known as an exciton [6, 7]. Hence, the absorption and emission of light can be tuned by controlling the size of nanoparticles in order of exciton radius.

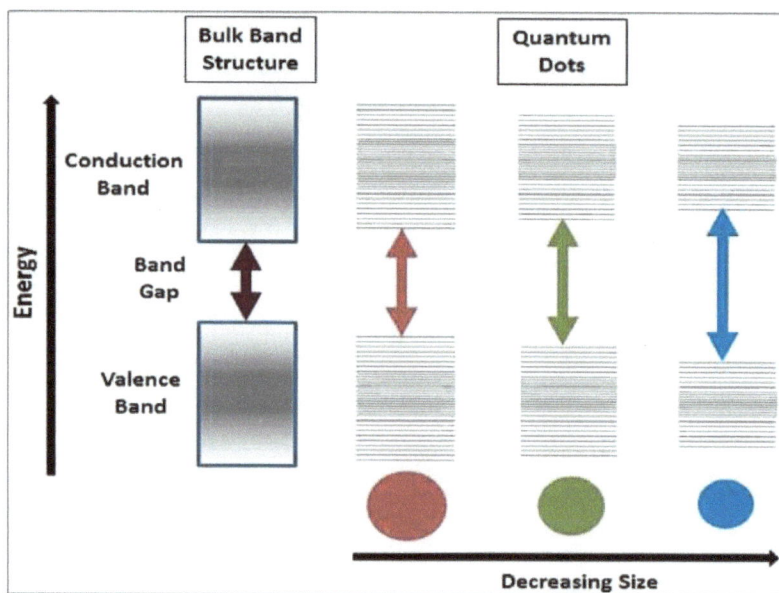

Fig. (2). Size-dependent optical gap and separation of energy states the figure is redrawn from reference [50].

Antibacterial and Antifungal Activity

Some nanomaterials demonstrate antibacterial, antifungal and antiviral, properties and have an excellent tendency to deal with pathogen-related diseases [1, 8]. The property has made some of the nanomaterials more prominent for enhancing the performance of various devices and materials in a wide variety of areas. The antifungal and antibacterial properties of nanomaterials make them a potential candidate for medical applications like medicines, pre and post-surgery apparatus and many more. Nanomaterials are applicable in food preservation and food packaging because of their antibacterial properties.

SYNTHESIS METHODS FOR NANOMATERIALS

Nanomaterials are synthesized by a diverse variety of synthesis methods. The performance of materials affects by their properties. The material properties depend on the adopted synthesis process and growth conditions [8]. The properties like crystal structure, composition, microstructural properties like stress and strain, defects, interfaces and other properties can be controlled by controlling the thermodynamics conditions and chemical kinetics of the reaction [3, 4]. The starting materials are the key factors for the synthesis of desired nanomaterials. The starting materials are either bulk or atomic materials. Fig. (3) shows the demonstration of top-down and bottom-up approaches for the synthesis of nanomaterials.

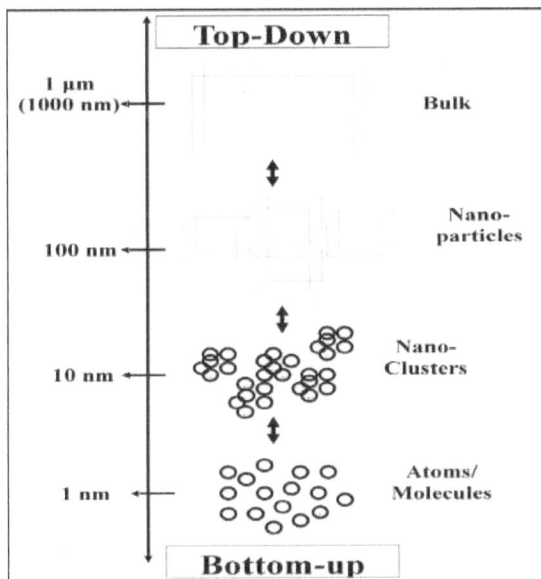

Fig. (3). Schematic demonstration of top-down and bottom-up techniques.

On the basis of starting materials the synthesis methods are classified as follows;

 i. Top-down
 ii. Bottom up

Top Down

In this way of synthesis, the bulk material is used as source material and breaking down it into nano-dimensional structures. The process basically is the miniaturization of bulk fabrication processes for producing the desired structure with nano-dimensional particles. The micropatterning techniques, such as electron beam lithography, high-energy wet ball milling, atomic force manipulation, *etc.* are in the category of top-down approaches. The main problem with the top-down approach is the imperfection of surface structures.

Bottom-Up

The bottom-up approach assigns to the buildup of material from atom-by-atom, molecule-by-molecule, or cluster-by-cluster. The process is basically the formation of material by combinations of basic units; atoms or molecules. The bottom-up techniques are also called chemical or liquid phase synthesis techniques. These techniques of preparation of nanomaterials have the potential of creating less waste, environment friendly and more economical than top-down approaches.

The liquid phase precursors are used in liquid & vapour phase synthetic approaches like sol-gel, solvothermal, hydrothermal, hot injection, electrochemical synthesis, evaporation, sputtering *etc.* Fig. (**4**) demonstrates all types of synthetic techniques of nanomaterials. The state of precursors and adopted instrumentation helps in choosing a suitable approach. Nanomaterials Synthetic techniques are further divided as:

- **Solid-phase synthesis techniques**: The solid-phase methods employed solid precursors and sophisticated instrumentation; some examples are ball milling and lithographic techniques.
- **Vapour phase synthesis techniques**: The synthetic approach employs the gaseous precursors; the approach is vapour-phase synthesis and some of the examples are physical vapour deposition techniques and sputtering techniques.
- **Liquid phase techniques:** The liquid phase techniques are under development for the nanomaterials produced at the commercial and industrial levels. Among all the above approaches the liquid phase approaches are quite simple, low cost

and easy techniques that can fulfil the criterion of industrial feasibility have convenient control of growth parameters and produced results.

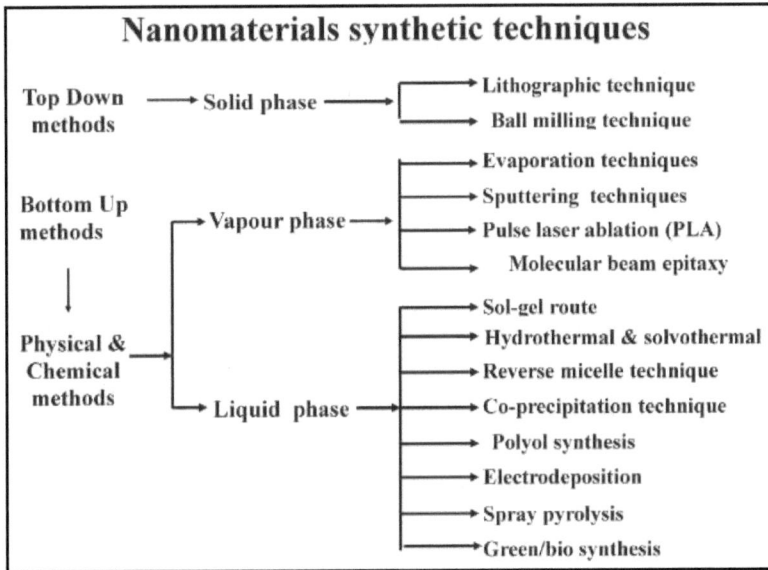

Fig. (4). Schematic diagram of nanomaterials synthesis approaches.

VAPOUR PHASE SYNTHESIS TECHNIQUES

The vapour phase synthesis techniques for nanomaterials are the most versatile among all other synthesis techniques, although the involvements of a large number of variables make these techniques relatively difficult. The requirement of sophisticated instrumentation; hence the techniques are very costly. Some of the important variables are density and nature of source materials, driving force or carrier gas, the stoichiometry of reactants and products, heating source *etc.* some of the important vapour phase techniques are as follows;

- Evaporation techniques
- Sputtering techniques
- Pulse laser ablation (PLA)
- Molecular beam epitaxy (MBE)

Evaporation Techniques

Evaporation techniques are the most well-known vapour phase synthesis technique among all other techniques. Generally, the materials synthesized by this technique are in the form of 2D-nanomaterials or films. In this technique, the material is converted into vapours using resistive/RF heating and transported through a vacuum to get deposited onto the substrate [9]. The mean free path between collisions become large enough so that the vapour beam arrives at the substrate without scattering only when the pressure inside the chamber is less or equal to 10^{-5} torr, For heating the evaporator material, a source is used which is either in a basket or helix-shaped and made up of refractory metals (W, Mo). Generally, for evaporating the material muffled sources with large capacity and high evaporation rate have been used. The method is useful in the deposition of either single/multi-layered films of metals, alloys and compounds in the range of nano thickness. If the evaporation rate of the source (or sources) can be controlled then the stoichiometric compounds/alloy films may be deposited with the help of this method. CdS, SiO and GaAs or $CuInSe_2$ films of desired thickness have been successfully deposited with required stoichiometry. If an electron beam is used for material evaporation; the technique is named electron beam evaporation (EBE). The electron beam is accelerated by applying 5 to 10 KV potential and the high energy electron beam is focused on the evaporator material; the electrons lose their kinetic energy as heat continuously and increase the temperature around 3000°C at the focused spot [9, 10]. The temperature is suitable to evaporate the refractory metals and compounds. The tungsten filament is heated to generate high energy electrons that are accelerated to form a high energy electron gun. For focusing the electrons beam on the evaporator material both electro-static and electro-magnetic focusing is used. The temperature at the focused spot is very high and the rest of the material remains cool. The interaction between the material and the support is not possible because of the temperature difference; hence the samples are contamination-free and of pure materials. Since the extremely high rate of evaporation (a few microns/second) can be achieved, hence the technique is suitable for high melting point materials. If the evaporated material is transported through reactive gas called plasma, the technique is named activated reactive evaporation (ARE) [9 - 13]. This technique is more suitable for deposit oxides and carbides such as SnO_2, In_2O_3, TiC films [10]. The schematic representation of evaporation techniques is shown in Fig. (**3**). The evaporation techniques on the basis of the different evaporation sources (which are used to convert the source material into vapour phase) named as thermal evaporation, electron beam evaporation, RF heating evaporation, activated reactive evaporation (ARE) *etc.* a lit bit modification has been done in schematic diagram according to the nature of the material to be deposited and the required evaporant source.

Sputtering Techniques

Energetic particles bombarded on a solid surface, the surface atoms are removed due to collision between surface atoms and energetic particles; called erosion solid surface. The phenomenon is named sputtering [9, 10]. The factors which can be influenced the sputtering phenomena are the nature of target material and crystal structure, the energy of incident particles and the angle by which particles incident on a solid surface. Sputtering can be used to produce films of carbides, nitrides, oxides and hydrides. Sputtering techniques are of two types; DC cathode, RF and magnetron sputtering.

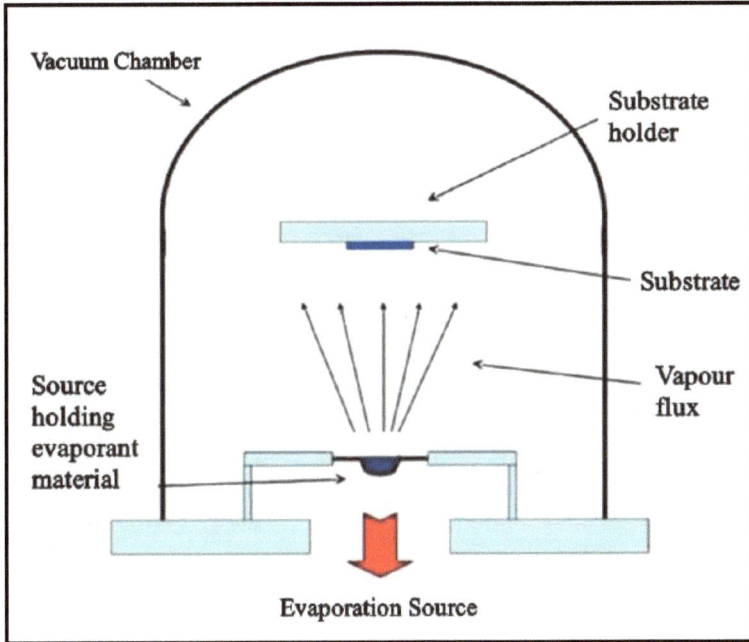

Fig. (5). The most common schematic diagram of evaporation techniques the figure is redrawn from reference [51].

Generally, DC sputtering is used to deposit metals and semiconductors whereas AC or RF sputtering is for insulators. DC Cathode sputtering is also called glow-discharge sputtering or impact evaporation. In the sputtering process source material (metal) is placed at cathode and bombardment of high-energy particles released the source material atoms. The released atoms are accumulated on the substrate and form a thin layer. In this process, the pressure is about 10^{-2} to 10^{-1} Torr [9 - 13]. Table **1** gives the comparative details of the sputtering technique and Fig. (**6**) represents the simplest schematic setup for the sputtering technique. During the deposition process, a small amount of reactive gas such as oxygen, nitrogen or hydrogen is added to the argon gas; causing modification in the chemical composition of the deposited films. This process is known as reactive sputtering. For example, tantalum nitrides sputtered using a tantalum cathode in a nitrogen/argon atmosphere [12]. RF sputtering can be used for the deposition of the films of the dielectric and insulator materials.

Table 1. The comparative details of sputtering techniques.

Name of the Technique	Sputter Source	Deposited Material	Vacuum Conditions
DC sputtering	DC power supply is applied between the chamber and sputtering target	Metals and semiconductors	10^{-1} Torr
RF sputtering	RF power supply is applied between the chamber and sputtering target	Dielectric and Insulators	10^{-3} to 10^{-5} Torr

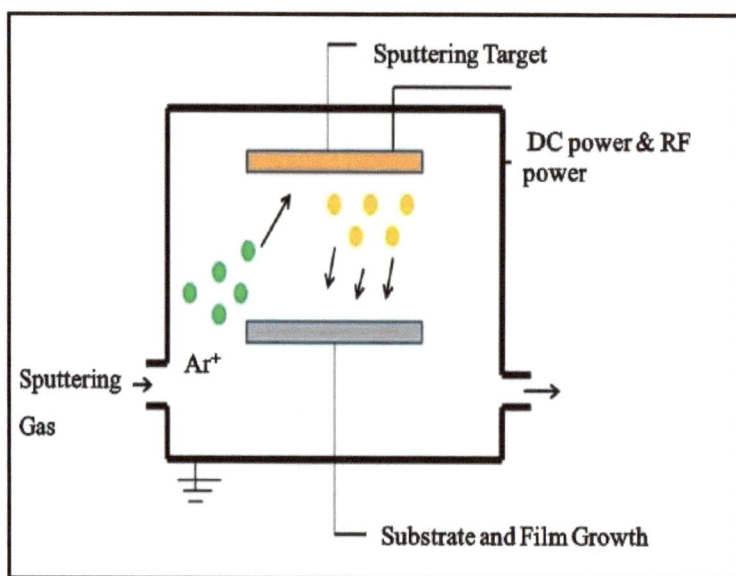

Fig. (6). The simplest schematic setup for the sputtering technique the figure is redrawn from reference [52].

Pulsed Laser Ablation Technique

Pulsed laser ablation deposition (PLAD) is a specific PVD technique in which a high-power density laser light with fine frequency bandwidth is used as a vaporizing source for the formation of vapours of desired material [10 - 14]. The technique is useful in synthesizing the nanotubes and quantum dots. The technique is useful in such situations; when PVD techniques create a problem or have failed. The importance of PLD in the consequence of other techniques is that there is no restriction of the type of target material. The PLD is used for Matrix-assisted hybrid metal-organic materials, biomaterials, polymer complexes, coordinative and other complex compounds, the thickness of the deposited films on various substrates is in the range of 10–500 nm. The PLAD technique enables researchers to modify the synthetic parameters of multicomponent oxide thin films with controlled composition and epitaxial orientation. Fig. (7) shows a schematic PLD setup for nanomaterials synthesis. The ablated flux distribution uniform controls the growth process of films on large substrates [14]. Some parameters like the during deposition process the background gas pressure, target to substrate distance and power and wavelength of laser source affects the film composition.

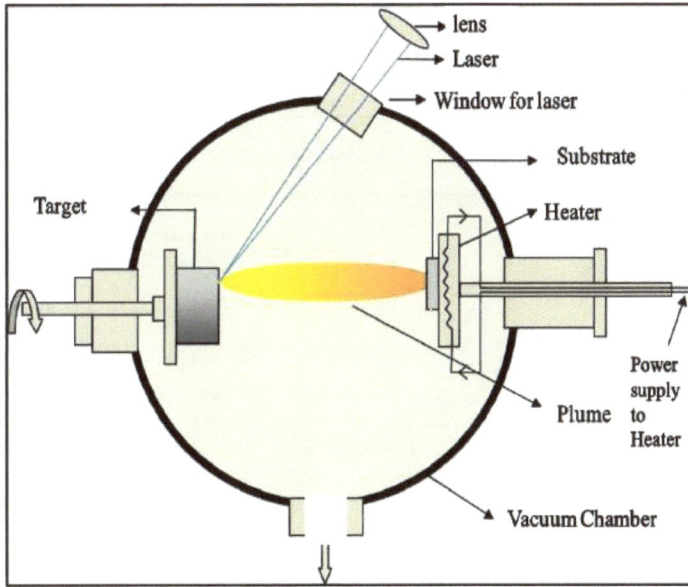

Fig. (7). The simplest Schematic representation of PLAD technique technique the figure is redrawn from reference [53].

Molecular Beam Epitaxy

The epitaxial films can be deposited by the condensation of one or more beams of atoms or molecules from effusion sources under UHV conditions are molecular beam epitaxy. A metallic chamber with a small orifice containing the evaporator material called Knudsen effusion source [9]. The orifice dimension of the Knudsen effusion source is smaller than the mean free path of the vapour species of evaporant material in the chamber. The mean free path of the effusing molecular beam is large enough in comparison to the source-substrate distance. The partial pressure of the vapour species within the chamber, source temperature orifice dimension *etc.* are some important parameters that are helpful in precisely determining the flux of the beam. Fig. (8) shows the simplest Schematic representation of molecular beam epitaxy. The beam is directed towards the substrate by the Knudsen cell orifice, slits and shutters. The growth temperature in MBE is low nearly equal to 600°C, which minimizes the unwanted thermal effects like diffusion. The growth rate is small (1-10 A/sec) which make it possible to deposit film with precise thickness control [9 - 15]. UHV system is required for any MBE to provide a clean ambient. MBE makes it possible to deposit epitaxial films of metals, alloys, compound semiconductor with precisely controlled properties.

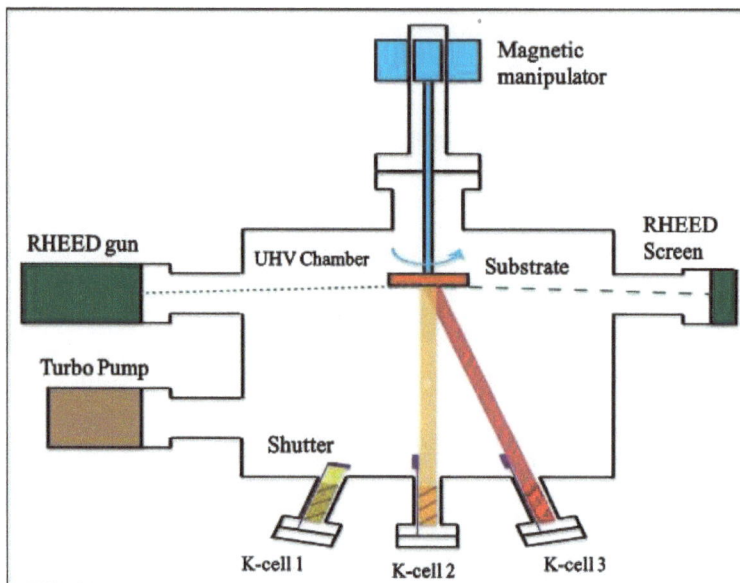

Fig. (8). The simplest Schematic representation of molecular beam epitaxy technique the figure is redrawn from reference [54].

SOLID-PHASE TECHNIQUES

The solid-phase synthetic techniques belong to the top-down method which involves the reduction of the size of materials by either mechanical process or etching method. It is a kind of top-down approach. Generally, two solid-phase methods are commonly used for nanomaterials synthesis are;

- Lithographic technique
- Ball milling technique

Lithographic Technique

The process of transferring patterns of geometric shapes to a thin layer of radiation-sensitive material (resist) in a mask covering the surface of a semiconductor wafer is known as Lithography [9, 10]. Fig. (**9**) depicts the lithographic technique. The masked portions of the underlying are employed to be selectively removed by an etching process. The three parameters: resolution, registration, and throughput determine the performance of a lithographic exposure. The minimum feature dimensions that can be transferred to a resist film on a semiconductor wafer with high fidelity are called Resolution. The measure of accurate patterns on successive masks can be aligned with respect to previously defined patterns on the same wafer called Registration

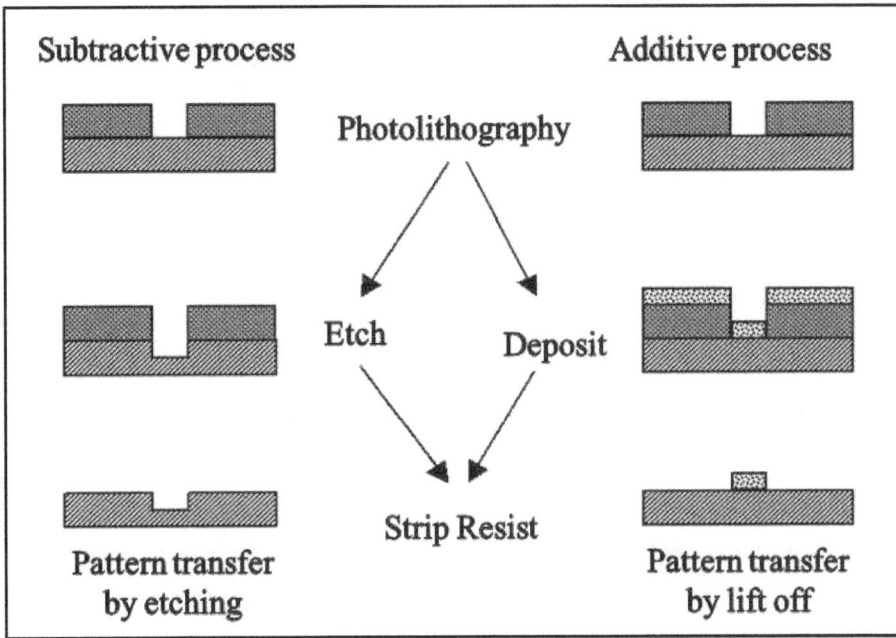

Fig. (9). Illustrates schematically the lithographic process the figure is redrawn from reference [55].

The number of wafers that can be exposed per hour for a given mask level is Throughput. It measures the efficiency of the lithographic process [12, 13]. The lithographic process is subtractive means removal of the unmasked area and additive type means to add a layer of material on the unmasked area.

Ball Milling Technique

A mill uses balls to grind or blend bulk material into nanosized particles by the use of various sized balls called Ball Mill and the process of grinding is ball milling technique [1, 3]. The principle of working is the reduction in size by the continued dropping of balls near the top of a mechanically rotating hollow cylindrical shell. The material is crushing and grinding into an extremely fine form using Ball mills. The material used for the balls, size of the balls and number of balls, rotating speed of cylindrical shell can vary the size of the synthesized nanostructures Fig. (10) shows a schematic representation of the ball milling technique. The cylinder is filled with balls as shown in the figure. The balls are made up of stainless steel, rubber, quartz *etc*. Horizontal, planetary, high energy or shaker ball mills are commonly used for nanomaterials synthesis [8].

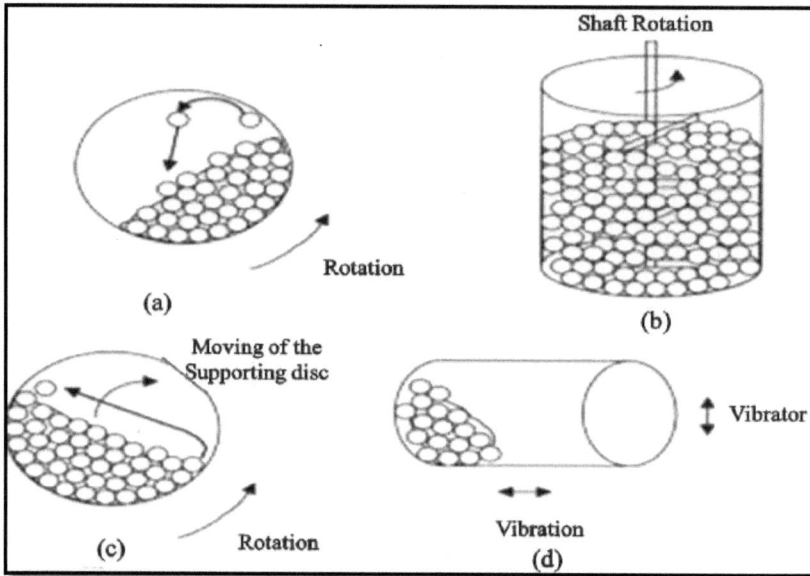

Fig. (10). The Schematic representation of ball milling technique the figure is redrawn from reference [3].

LIQUID PHASE SYNTHESIS METHODS

Liquid phase techniques are the most important growth methods for nanomaterials and nanofilms because of their versatility for synthesizing a variety of compounds at relatively low temperatures. The various chemical methods are as follows-

- Sol-gel
- Hydrothermal and Solvothermal Method
- Reverse micelle or microemulsion methods
- Co-precipitation
- Chemical bath deposition
- Polyol synthesis
- Electrodeposition
- Green or biomimetic synthesis

Sol-Gel Method

A variety of nanomaterials can be synthesized using the sol-gel method. It is a versatile and wet-chemical method that uses either a chemical solution or a colloidal suspension [1, 3]. The sol-gel method includes the hydrolysis and polymerization reactions of the precursors and reactants. The sol is formed by initiating nucleation in a colloidal solution. Condensation of sol contains small nuclei of precipitates, forming porous and viscous solution called Gel. When the wet gel is cast into a mould and the wet gel is converted into a dense ceramic. The heat treatment is given to the casted gel for forming ceramic powder of nanograins [16]. The spin-coating technique is used to convert the wet gel into thin films on a piece of substrate. Fig. (**11**) shows involved steps in sol–gel process to synthesize nanoparticles.

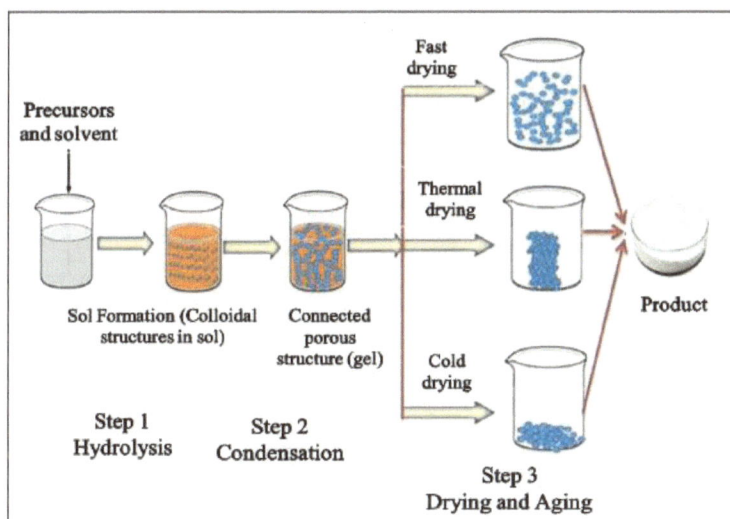

Fig. (11). Steps involved in sol–gel process to synthesize nanoparticles.

For obtaining an aero-gel the solvent of wet gel was removed under supercritical drying conditions. The obtained material is highly porous and extremely low-density called Aero-gel. The liquid phase is removed from the gel and the calculations process can be done for enhancing mechanical properties [16 - 19]. The calcinated powders were ground to obtain ultra-fine and uniform ceramic materials.

Hydrothermal and Solvothermal Method

The hydrothermal method is a liquid phase chemical reactions method in which solution is contained in a closed Teflon-lined stainless steel vessel called autoclave [1, 3, 20]. The reaction takes place under controlled temperature or pressure in an aqueous solution. The liquid source in which the reaction takes place is water and hence the process is named hydrothermal. The autoclave was sealed and the temperature can be elevated above the boiling point of the solvent for a definite time duration. The reaction takes place at constant temperature for certain hours; the liquid solvent is converted into vapours and generated high pressure inside the vessel which is responsible for the reaction and quality of the product.

The process is called high-pressure synthesis at constant elevated temperatures. The centrifugation process is used for separating the liquid phase from products. The product was washed to remove unwanted byproducts, many times with distilled water and ethanol. The product was dried in a vacuum. Fig. (**12**) shows the steps involved in the hydro/solvothermal process of synthesizing nanomaterials. The difference in solvothermal and hydrothermal methods is only of the solvent used. In solvothermal, the organic liquid is used whereas in hydrothermal the aqueous solution is used as a solvent. The elevated is much higher than the hydrothermal method because the boiling points of organic solvents are high enough to the water [20-21]. The solvothermal method has better control of the size and shape distributions of NPs than the hydrothermal method. The solvothermal method is more suitable for the synthesis of a variety of nanomaterials with narrow size distribution. The quality of products depends on the reaction temperature and corresponding pressure and organic solvent used.

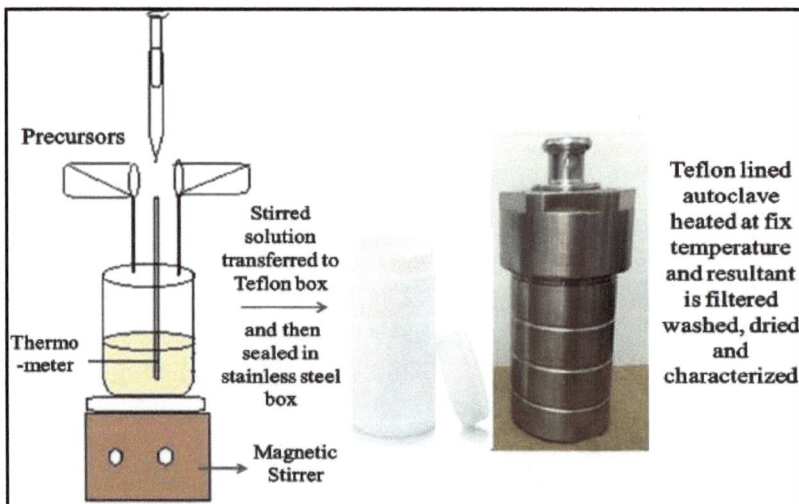

Fig. (12). Steps involved in hydrothermal/solvothermal process to synthesize nanomaterials.

Chemical Bath Deposition (CBD)

The chemical bath deposition (CBD) method is useful in depositing thin films and nanomaterials on any substrate [9, 10]. CBD process includes two steps; one is nucleation and the other one is the growth of particles. The CBD technique is mainly based on the formation of a solid phase from a solution. In CBD method the substrate was immersed in the precursor solution. The parameters like pH of the solution, precursor concentrations, deposition time, nucleation agent and bath temperature affect the quality of the synthesized material. CBD does not create any physical damage to the substrate. The chemical bath is the simplest process in which the substrate is dipped in reaction bath solution and the final product after reaction material is self deposited on it. Fig. (13) shows the simplest chemical bath deposition setup for the synthesis of nanomaterials. During the growth mechanical movements of substrate modify the growth rate and the particle size; the method is named mechanically agitated CBD [21, 22]. The incorporation of ultrasonic waves during the growth; the method named ultrasonically agitated CBD. The minor modification in the CBD method not only modify the growth rate of films but also affects the particle shape and size and in some cases modify the structural phase of the material.

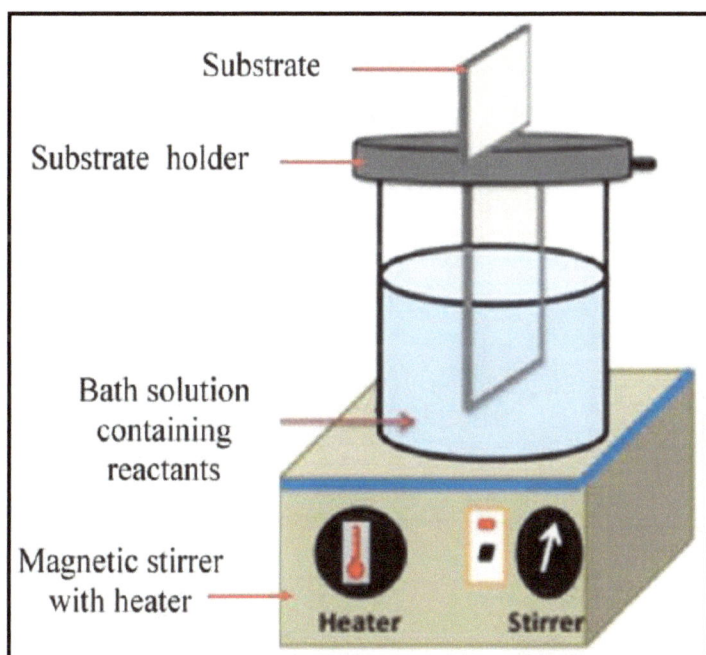

Fig. (13). The simplest chemical bath deposition setup for nanomaterials.

Chemical Vapor Deposition

The process of condensation of compound vapours from a carrier gas onto a substrate where the reaction takes place to produce a solid deposit is known as chemical vapour deposition (CVD) [9]. The gaseous compounds bearing the deposited material is either in liquid form or in solid form. The gaseous compounds can flow either by a pressure differential or the use of carrier gas to the substrate. The chemical reaction takes place at/or near the substrate surface and produces the desired material which is deposited on the surface of the substrate. CVD is similar to the physical vapour deposition (PVD) processes. In fact, the boundaries between chemical and physical vapour deposition processes become hard to define. In physical vapour deposition, the evaporation takes place by heating or sputtering in a reactive atmosphere and the conversion of the deposited material by chemical reaction activated by electron bombardment in a vacuum. But in CVD the decomposition of the vapour phases species in a neutral atmosphere and the deposition of product on the substrate after the reaction among the vapour species. This method has been used for the formation of a variety of layers, compound semi-conductors Ex. GaAs, GaP, SiC, SnO_2 (which is formed during this dissertation) and other various conductors, insulators and resister materials [10, 21]. Fig. (**14**) shows the Simplified CVD setup for deposition of nanomaterials on substrates.

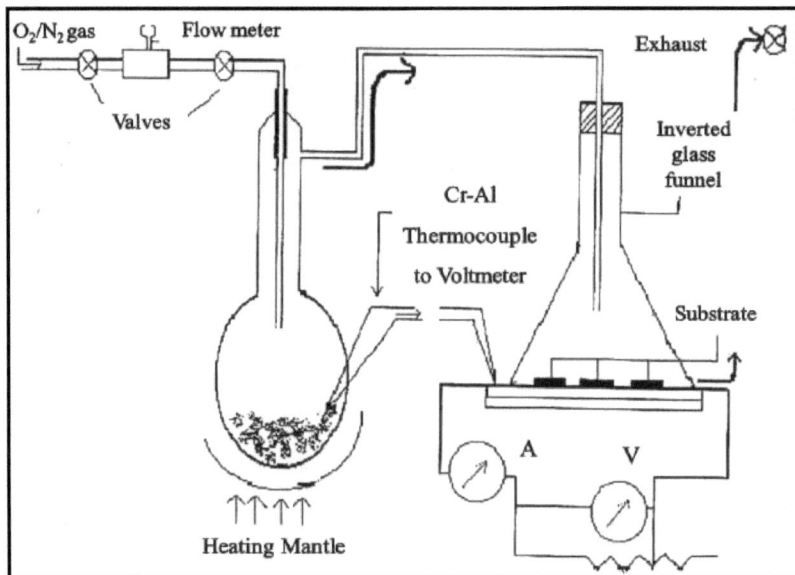

Fig. (14). Simplified CVD setup for deposition of nanomaterials on substrates.

Electrodeposition Method

Electrodeposition is a very common process in which at least two electrodes are involved to produce a coating of metallic material by the reduction at the surface of the cathode [9, 10]. The cathode is generally the substrate on which the material is to be deposited. The reaction solution contains a salt of the metal that is to be deposited on the cathode surface. The transport of ions takes place either by diffusion because of concentration gradient or migration due to the applied electric field and corresponding currents in the electrolyte. The positive ions are attracted towards the cathode and reduced by receiving an electron from the cathode and getting deposited on it. The Electro-deposition processes may be influenced by the current density passes to the electrolyte, pH of the electrolyte, electrolytic bath temperature, the shape of electrodes and the bath composition [21-23]. Metallic films can be grown only by dipping in a suitable electrolytic solution even in the absence of an electric field. Fig. (15) shows the three-electrode experimental arrangement for electrodeposition under potentiostat control. If the deposition is on the catalytic surface; named electroless or autocatalytic deposition. Silvering and galvanization are the most widely used. The other metals which can be deposited include Cu, Au, Ni, Co, Pd *etc* or compounds of Pb, Hg, Zn, Cd has been deposited by this method. Deposition starts spontaneously on Ni, Co Pb *etc*. Other metal surfaces require activation, which is usually by dipping in 0.1 $PdCl_2$ and then rinsing under running water. Enough $PdCl_2$ is left to activate the surface.

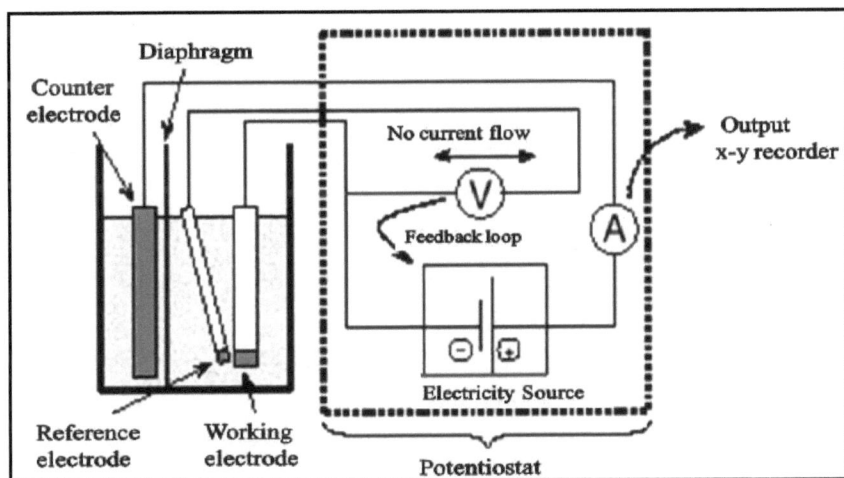

Fig. (15). Three electrodes experimental arrangement for Potentiostatic controlled electrodeposition of nanomaterials.

Spray Pyrolysis

Spray Pyrolysis is the conversion of a liquid solution into a small droplet by the carrier gas and the reaction takes place at/or near the thermally heated substrate surface and gets deposited on it. The spraying of a solution containing soluble salts of the constituent atoms of the desired compound on a substrate maintained at elevated temperatures is called spray pyrolysis. The pyrolytic decomposition of sprayed droplets on the hot substrate takes place and form either a single crystal or cluster of crystalline products [21, 24]. The spray nozzle and flow of carrier gas affect the atomization of the chemical solution into fine droplets. The method must be very efficient for the growth of desired materials when the desired thin-film material. The process involves a thermally activated reaction between various species dissolved in the spray solution. The solution of constituents of chemicals should be volatile at the pyrolysis temperature. Fig. (16) shows the schematic setup of the spray pyrolysis technique. The films obtained by the process are stable with time and temperatures, adherent and free from defects. The film texture depends on the reactivity of chemicals, droplet mobility and the flow rate of carrier gas.

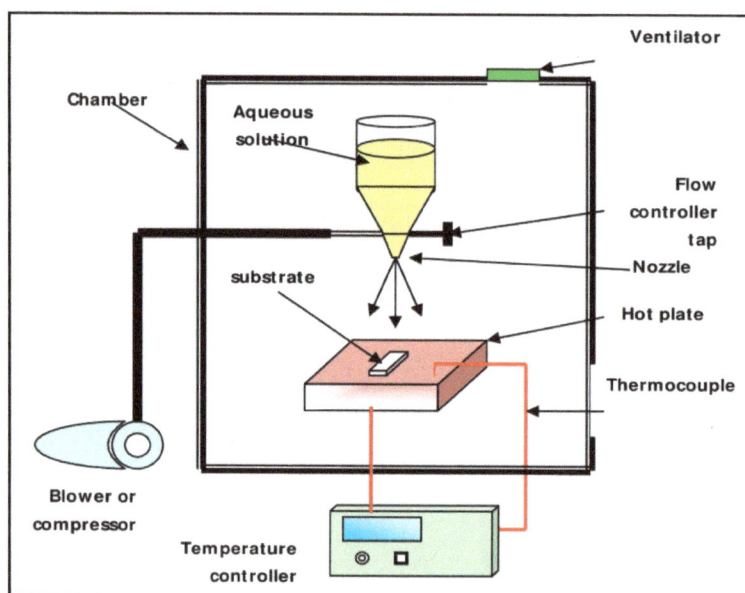

Fig. (16). Setup of spray pyrolysis technique the figure is redrawn from reference [26].

Co-precipitation

Co-precipitation techniques are generally used to separate the trace elements from very dilute solutions, such as natural water [21, 25]. The solubility of the materials in the solvent is governed by the pH value. The change in the pH value is important for tracing elements from the solution. During co-precipitation, a number of factors, affect the growth process and particle size of nanostructures. These factors are; bath temperature and pressure, concentration, stirring rate and time, nature of the precursors,, surfactants *etc*. Generally, pH 9–10 is more suitable for the removal of precipitate ions from a solution. The amounts of precipitants of the coexisting salts in the solution and the ageing time of precipitation affect the co-precipitation yield. Fig. (**17**) depicts the co-precipitation processes for nanomaterials. Chemical co-precipitation is a well-studied, popularly used method for nanomaterials synthesis because of its easiness, economic and sometimes eco-friendly.

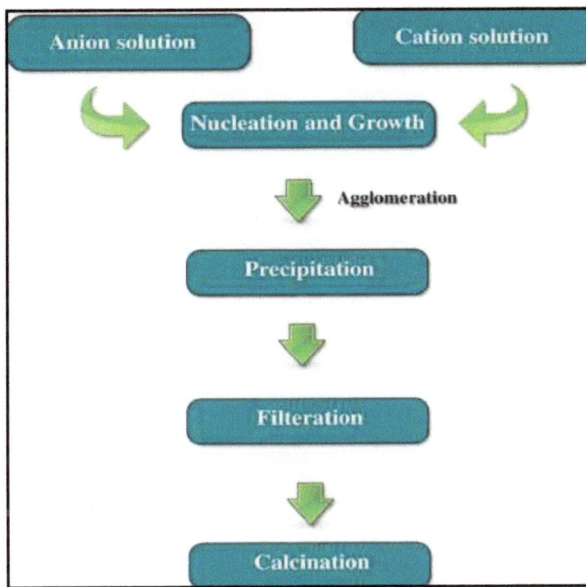

Fig. (17). Co-precipitation processes for nanomaterials the figure is redrawn from reference [56].

Reverse Micelle or Microemulsion Method

Nanomaterials with desired dimensions are very important for applications in various fields like medication, bio-imaging, optoelectronic devices. The shape and morphology of nanomaterials depend on the chosen synthesis methods. The reverse micelle method is very prominent for the synthesis of desired nanoparticles [3]. There are three important components two are immiscible and the third one is a surfactant. The surfactant is amphiphilic in nature. Surfactant molecules are having two parts; one is hydrophobic (water-hating) and the other one is hydrophilic (water-loving). The hydrophilic part of the surfactant molecule is named as a head group and is generally depicted as a circle. The hydrophobic part of the molecule is named as tail and it is attached with hydrocarbon chains, Water, benzene, hexanol, and k-oleate form a simple homogenous microemulsion system [30 - 39]. The hydrocarbon chain may be linear or branched. The tail may be depicted either as a straight line or a wavy tail. The microemulsion process is also known as inverse micelle method. The micelle and inverse micelle solutions may contain three or four components; water, oil, surfactant and co-surfactant.

The surfactant creates a separating layer between the polar water and non-polar oil phase. The interfacial layer is responsible for the formation of microstructures or nanostructures between oil and water and vice versa. Fig. (**18**) shows the normal micelle and reverse micelle surface-active molecules (b) schematic process of the microemulsion method. Reverse micelle is formed by adding water to oil and is very suitable for the synthesis of nanoparticles. In general, the reverse micelles are self-organized molecular aggregates of surfactants in polar media.

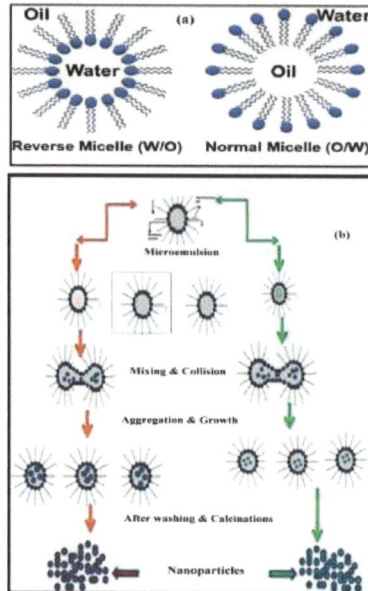

Fig. (**18**). (**a**) Surface-active molecules (**b**) the schematic diagram of nanomaterials synthesis using microemulsion method the figures are redrawn from reference [30, 57].

Polyol Synthesis

The method in which metal precursors are suspended in a glycol solvent and heated to a refluxing temperature to obtain oxide nanomaterials is named as polyol method. In the polyol method the glycolic solvent, reducing agent, and ligand are used [21 - 29]. The ligand plays an important role as it prevents NP agglomeration The polyol type and its reflux temperature and heating time are important parameters that affect the NPs properties. The technique is found suitable for all types NPs such as metallic, oxides, semiconductors *etc*. Mono-metallic and metallic alloy NPs have been also synthesized with the technique. The method is suitable for the preparation of well-defined metal NCs of desired qualities like crystallinity, size, shape, and composition. The basic principle of the polyol technique is the reduction of metal precursors by polyolic solvent at elevated temperature in presence of a suitable capping agent.

Some important factors such as polyol solvents, reducing agents, metal precursor concentrations, capping agents, *etc*. are responsible for the suitable and required size and shape of NPs. Fig. (**19**) represents the schematic process of the polyol technique for nanoparticles. The highly uniform and well-defined nanomaterials are synthesized by the polyol method. The homogeneous nucleation of metal particles was capped with a suitable capping agent to stop the growth process. The stabilization of the formed nuclei at the initial stage of reaction and the required morphology and size is obtained by adding a capping agent. For the polyol method, the most typical capping agent is PVP which enables producing a wide range of metal NCs such as Ag, Pt, Au, Pd, Cu, and Rh [37, 38]. Because of simplicity and low cost; the polyol route is attracted great attention for the controlled synthesis of metal NCs with desired sizes and shapes.

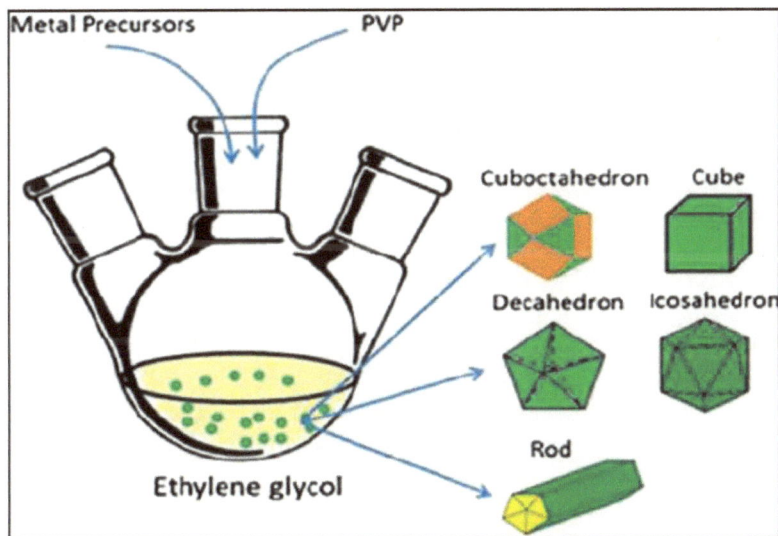

Fig. (19). Schematic representation of polyol synthesis technique for nanoparticles the figure is redrawn from reference [48].

Green or Biomimetic Synthesis

In green or Biomimetics synthesis a biological entity or plant extracts are used for materials formation. Recent scenario prefers nature favourable synthesis techniques for material design [3]. Bioreduction is the fundamental process in biomimetic synthesis. Microorganisms or plant extracts are the main components of all kinds of Biological methods [38 - 40].

The green or biomimetic synthesis techniques have been found an ecofriendly alternative to chemical and physical methods. The synthesis of nanoparticles using plants or parts of plants is giving biocompatible and non-toxic nanoparticles. Fig. (**20**) shows the schematic diagram for the green synthesis of nanoparticles. The future requirement is the development of new, easy, environment favourable processes for nanoparticles synthesis.

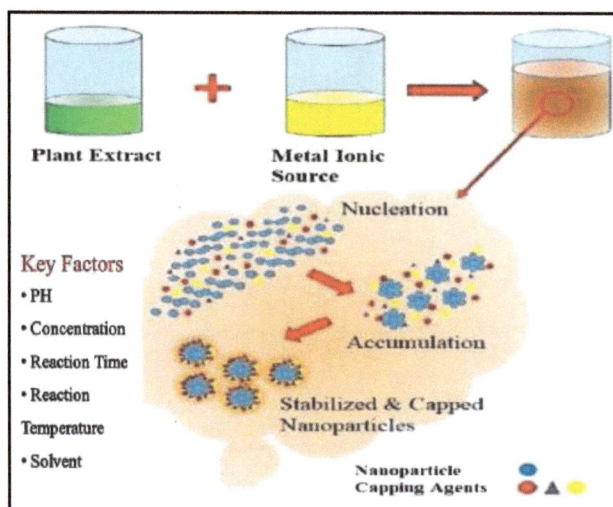

Fig. (20). Schematic diagrams for green synthesis of nanoparticles the figure is redrawn from reference [39].

CHARACTERIZATION TECHNIQUES FOR NANOMATERIALS

X-ray diffraction (XRD) technique is used to identify the microstructural properties of nanoparticles. XRD is a primary characterization technique that can be used to investigate the crystal structure of the synthesized materials. It is helpful in both the ways of analysis as qualitative and quantitative of nanomaterials. It is not only helpful in identifying the crystal structure, but also crystallite size and micro-strain of the samples [1, 41 - 43]. It is the signature technique of material identification. It is working on the principle of Bragg's diffraction ;

$$2dsin\theta = n\,\lambda$$

Where, d is the inter planner separation of atomic arrangement planes, θ is the incidence angle of x-rays on the sample, λ is the wavelength of x-rays and n is the order of diffraction. The XRD is the preliminary technique that is used to get the basic idea of the material quality, whether the material is synthesized or not. XRD technique is the most commonly used characterization tool for the detection of unknown crystalline materials. It is also used in the investigation of unknown solids of material science, environmental science, geology, engineering, chemistry and biology [9, 43]. The XRD technique is used to characterize single crystals, crystalline materials, in the identification of grain size and shape of minerals such as clays and mix layered clays which can be difficult to identify by optical manner. The technique is also suitable for the determination of unit cell parameters and the purity of the phase of the material.

Raman spectroscopy is another important technique that confirms material synthesis. Raman technique is observing vibrational, rotational, and other low-frequency modes of a material. It commonly provides chemical and structural information about samples. Raman scattering is an inelastic scattering process in which energy transfer takes place between photons and molecules of the material. If the photon loses energy during the scattering to the molecule, then the photon is scattered with a large wavelength called Stokes Raman scattering. Inversely, if the photon gains energy from the molecule during scattering, then the photon scattered with higher frequency and lower wavelength called Anti-Stokes Raman scattering. Quantum mechanically Stokes and Anti-Stokes both are equal processes. In general, the majority of molecules are lying at the ground vibrational level and the Stokes scattering process is probably the preferred process. Since, the Stokes Raman scattering lines are always more intense than the anti-Stoke lines [44, 45]. Hence the Stokes Raman scatter is measured in Raman spectroscopy. It is also a basic technique and is used as a signature to material analysis.

UV–Visible spectroscopic technique is used for the analysis of optical properties of nanoparticles. The technique helps determine the formation mechanism and stability of nanoparticles in an aqueous solution. A beam of light including the visible and UV parts of the electromagnetic spectrum passes through a prism or diffraction grating then is separated into its component wavelengths. A half-mirrored device is used to split each monochromatic beam into two equal intensity beams. One beam passes through a reference cuvette and the other beam passes through the cuvette containing sample material. The light beam intensities are measured by electronic detectors and compared. The intensities of the reference beam and sample beam are denoted as I_0 and I respectively. The spectrometer automatically scans all the components of wavelengths in a short period of time. It works on the principle of Beer-Lambert law; according to that, the absorbance is directly proportional to the concentration of the solution and t length of the light path, which is equal to the width of the cuvette [46]. If A is absorbance, c is the concentration of sample material and l is the cuvette width then;

$$A = log\frac{I_o}{I} = \in lc$$

Atomic force microscopy (AFM) is used to study the surface morphology of synthesized nanoparticles. A sharp tip is used to scan the surface with the help of a feedback loop and give the surface image. In AFM the tip and sample surface interaction is mapped. The van der Waals, electrical, magnetic, and thermal techniques are used to measure the force interaction between the tip of AFM and the surface molecules. AFM tips and cantilevers are typically micro-fabricated from Si or Si_3N_4. The typical tip radius is from a few nm. A laser beam deflection system is used in an AFM microscope and the laser is reflected from the back of the reflective AFM lever and onto a position-sensitive detector [47 - 50]. A magnified and fine image of the surface of the sample is obtained with AFM.

Scanning Electron Microscopy (SEM) image is formed by electrons used instead of light. The SEM images figure out the shape, size, morphology and distribution of synthesized nanoparticles. A high-energy electron beam is to scan the sample and produces a magnified image by SEM microscope [47 - 50]. When the electron beam hits the sample surface, it penetrates a few microns deep; that depends on the accelerating voltage and the sample density. The interaction of high energy electrons to the sample produces various phenomena such as secondary electrons emission, backscattered electrons and produce characteristic X-rays. These signals are collected by one or more detectors to form images, which are then displayed on the computer screen.

The transmission Electron Microscopy (TEM) technique is used to get an idea of morphological, compositional, and crystallographic information about the sample. TEM can give the inter-planar lattice spacing, morphology and crystal diffraction pattern. It has an additional tendency to work in diffraction mode and hence, give crystal structure details. The multiple electromagnetic lenses are used to focus the filament emitted electrons to the sample. The sample is thin enough and hence, the electron beam penetrates the sample, during penetrating it interacts with the sample material. The energy of electrons is directly related to the electron wavelength and determines the image resolution. TEM can be operated in various modes such as; imaging mode and diffraction mode. The working of TEM in diffraction mode is two types; one is selected area electron diffraction (SAED) and the other one is nanobeam diffraction (NBD) mode [48]. The diffraction patterns give information about the related crystallographic axis of the crystal structure of the specimen which gives the characteristic properties of the material.

Energy-dispersive X-ray spectroscopy (EDX) is a technique useful for the identification of constituent elements of the sample. It also gives quantitative information about the sample composition. The atoms within the sample are ionized by the high-energy electrons and eject inner-shell electrons. The ejected electrons are of high energy and move to the higher energy state, reside there for 10^{-8} sec and then lose energy with the characteristic energy of specific electronic transitions within the target atoms and coming back to a stable energy state. During the process of excitation of electrons and relaxation process characteristic X-rays are emitted from the specimen. The x-rays collected by the solid-state detectors generate X-ray energy dispersive spectrum (EDX) [47 - 49]. The collected spectrum contains a wealth of chemical information about the specimen. This technique can be used for obtaining the composition of the sample. It is the fingerprinting technique for detecting the elements present in the sample. In general, the EDX instrument is attached with SEM or TEM for giving composition with surface morphology.

All the above-described techniques are the keys to predicting the nanomaterial properties and also show whether the material is properly synthesized, good quality and suitable for desired applications.

CONCLUSION AND FUTURE REQUIREMENTS

The synthesis of advanced nanomaterials is a promising field and could bring various new discoveries and novelties in the next decades. In this chapter, we have included the most popular physical methods and the most recent chemical methods for the synthesis of advanced materials and nanosystems. Each preparative method predicts both advantages and disadvantages. A suitable preparative method can be chosen and controlling the synthesis parameters such as required instrumentation, suitable chemicals, desired composition, operating temperature, reducing agents, nucleation agents, surfactants, *etc.*, produce desired size nanoparticles. The current requirement for the synthesis of nanoparticles is the simplest, eco-friendly, economic and non-toxic methods with multifunctional properties. The liquid phase synthetic approaches are more suitable in this scene. There are promising applications of nanomaterials in various fields, like electrical & electronics, energy storage & conversion, biomedicine, antibacterial and anti-fungal, catalysis, bio-imaging for disease detection as well as in treatment. The future synthetic methods of nanostructures will focus on producing application-oriented properties. Some of the important aspects of future synthetic techniques are that they should be eco-friendly, non-toxic and have low energy consumption. The day by improvements and invention of new liquid phase synthetic approaches make the field very of synthesis of nanomaterials.

CONSENT FOR PUBLICATION

Not applicable.

CONFLICT OF INTEREST

The authors declare no conflict of interest, financial or otherwise.

ACKNOWLEDGEMENTS

Declared none.

REFERENCES

[1]   Pradeep, T. *A text book of Nanoscience and Nanotechnology*; Tata McGraw Hill Education Pvt. Ltd: New Delhi, **2017**.

[2]   Nikolais, L. *Metal-Polymer nanocomposites*; Wiley-Inter-science: New Jersey, **2005**.

[3]   Sulabha, K. *Kulkarni, Nanotechnology: Principles and Practices*; Springer: New York, **2015**.

[4]   Charles, P. *Poole & Jr. Frank J. owens, Introduction to Nanotechnology*; John Wiley & Sons: New Jersey, **2003**.

[5]   Kuno, M. Introduction to Nanoscience and Nanotechnology: A Workbook. **2005**.

[6]   Delerue, C.; Lannoo, M. Nanostructures: theory and Modeling. **2004**,

[7]   John, R. *Solid state Physics*; Tata McGraw Hill Education: New Delhi, **2014**.

[8]   Yadav, T.P.; Yadav, R.M.; Singh, D.P.; Milling, M. a Top Down Approach for the Synthesis of Nanomaterials and Nanocomposites. *Nanoscience and Nanotechnology,* **2012**, *2*(3), 22-48. [http://dx.doi.org/10.5923/j.nn.20120203.01]

[9]   Chopra, K.L.; Malhotra, L.K. *Thin film technology and applications*; Tata McGraw Hill Education Pvt. Ltd: New Delhi, **1985**.

[10] Chopra, K.L. *Thin film phenomena*; McGraw Hill: New York, **1969**.

[11] Rointan, F. *Handbook of Deposition Technologies for Films and Coatings*; Noyes publication: New Jersy, **1994**.

[12] Maissel, L.I.; Glang, R. *Handbook of Thin Film Technology*; McGraw Hill Book Company: New York, **1970**.

[13] Seshan, K. *Handbook of Thin-Film Deposition Processes and Techniques: Principles, Methods, Equipment and Applications,* 2nd ed; Noyes Publications: New York, **2002**.

[14] *Austin Chambers*; Chapman & Hall/CRC Press Company: Modern Vacuum Physics, New York, **2005**.

[15] Cho, A.Y. Advances in molecular beam epitaxy (MBE). *J. Cryst. Growth,* **1991**, *111*(1-4), 1-13. [http://dx.doi.org/10.1016/0022-0248(91)90938-2]

[16] Reiss, P. ZnSe based colloidal nanocrystals: synthesis, shape control, core/shell, alloy and doped systems. *New J. Chem.,* **2007**, *31*(11), 1843-1852. [http://dx.doi.org/10.1039/b712086a]

[17] Siva Prasanna, S. R.V.; Balaji, K. Shyam Pandey, Sravendra Rana, metal oxide based nanomaterials and their polymer nanocomposites: nanomaterials and polymer nanocomposites Elsevier Inc. **2019**.

[18] Sharma, S.; Kurashige, W.; Niihori, Y.; Negishi, Y. Nanocluster Science: Supra-materials Nanoarchitectonics. In: *Ed. William Endrew is an imprint of Elsevier*; , **2017**; pp. 2-32.

[19]   Parashar, M.; Shukla, V.K.; Singh, R. Metal oxides nanoparticles *via* sol–gel method: a review on synthesis, characterization and applications. *J. Mater. Sci. Mater. Electron.,* **2020**, *31*(5), 3729-3749.
[http://dx.doi.org/10.1007/s10854-020-02994-8]

[20]   Gan, Y.X.; Jayatissa, A.H.; Yu, Z.; Chen, X.; Li, M. Hydrothermal Synthesis of Nanomaterials. *J. Nanomater.,* **2020**, *2020*, 1-3.
[http://dx.doi.org/10.1155/2020/8917013]

[23]   Solanki, R.G.; Rajaram, P.; Bajpai, P.K. Growth, characterization and estimation of lattice strain and size in CdS nanoparticles: X-ray peak profile analysis. *Indian J. Phys. Proc. Indian Assoc. Cultiv. Sci.,* **2017**, *92*(5), 595-603.
[http://dx.doi.org/10.1007/s12648-017-1134-8]

[24]   Phuong Nguyen Tri, Claudiane Ouellet-Plamondon, Sami Rtimi, Aymen Amine Assadi and Tuan Anh Nguyen, Methods for Synthesis of Hybrid Nanoparticles. In: *Noble Metal-Metal Oxide Hybrid Nanoparticles*; Elsevier, **2019**; pp. 51-63.

[25]   Solanki, R.G. Electrochemical growth and studies of CdTe thin Films. *Indian J. Pure Appl. Phy.,* **2010**, *48*, 133-135.

[26]   Wafaa, K. Khalef, Eklas K. Hamza. Amenah A. Salman, "Morphology, Optical and Electrical Properties of Tin Oxide Thin Films Prepared by Spray Pyrolysis Method. *Eng. & Tech. Journal,* **2015**, *33*, 539-546.

[27]   Solanki, R.G.; Rajaram, P. Structural, optical and morphological properties of CdS nanoparticles synthesized using hydrazine hydrate as a complexing agent. *Nanostructures and Nano-objects,* **2017**, *12*, 157-165.
[http://dx.doi.org/10.1016/j.nanoso.2017.10.003]

[28]   Sumanth Kumar, D.; Jai Kumar, B.; Mahesh, H.M. Quantum Nanostructures (QDs): An Overview Synthesis of Inorganic Nanomaterials. woodhead publishing, **2018**; pp. 59-88.

[29]   Tran Thi Ngoc Dung. Synthesis of nanosilver particles by reverse micelle method and study of their bactericidal properties. **2009**J. Phys. *Conf. Ser,* , pp. 1-9.

[30]   K. Eid, H. Wang; L., Wang Nanoarchitectonic Metals: Supra-materials Nanoarchitectonics. *William Endrew is an imprint of Elsevier,* **2019**, 135-171.

[31]   https://en.wikipedia.org/wiki/Coprecipitation

[32]   Nagi R. E. Radwana, Jeenat Aslama and Arifa Akhter, "concept of reverse micelle method for the synthesis of nano-structured materials. *Curr. Nanosci.,* **2018**, *14*, 1-8.

[33]   Nagarajan, R. Micellization, mixed micellization and solubilization: The role of interfacial interactions. *Adv. Colloid Interface Sci.,* **1986**, *26*, 205-264.
[http://dx.doi.org/10.1016/0001-8686(86)80022-7]

[34]   Prince, M.L. *Microemulsion Theory and Practice*; Academic Press: New York, **1977**.

[35]   Holmberg, K.; Jönsson, B.; Kronberg, B.; Lindman, B. *Surfactants and polymers in aqueous solution,* 2nd ed; John Wiley & Sons: England, **2003**.

[36]   Tran, T.N.D.; Buu, N.Q.; Quang, D.V.; Ha, H.T.; Bang, L.A.; Chau, N.H.; Ly, N.T.; Trung, N.V. Synthesis of nanosilver particles by reverse micelle method and study of their bactericidal properties. *J. Phys. Conf. Ser.,* **2009**, *187*, 1-9.

[35]   Pileni, P.M. Reverse micelles as microreactors'. *J. Phys. Chem.,* **1993**, *97*(27), 6961-6973.
[http://dx.doi.org/10.1021/j100129a008]

[36]   Wiley, B.; Herricks, T.; Sun, Y.; Xia, Y. Polyol Synthesis of Silver Nanoparticles: Use of Chloride and Oxygen to Promote the Formation of Single- Crystal, Truncated Cubes and Tetrahedrons. *Nano Lett.,* **2004**, *4*(9), 1733-1739.
[http://dx.doi.org/10.1021/nl048912c]

[37] Baig, N.; Kammakakam, I.; Falath, W. Nanomaterials: a review of synthesis methods, properties, recent progress, and challenges. *Mater. Adv.,* **2021**, *2*(6), 1821-1871. [http://dx.doi.org/10.1039/D0MA00807A]

[38] Forough, M.; Farhadi, K. Biological and green synthesis of silver nanoparticles. *J. Eng. Env. Sci.,* **2010**, *34*, 281-287.

[39] Zikalala, N.; Matshetshe, K.; Parani, S.; Oluwafemi, O.S. Biosynthesis protocols for colloidal metal oxide nanoparticles. *Nano-Structures & Nano-Objects,* **2018**, *16*, 288-299. [http://dx.doi.org/10.1016/j.nanoso.2018.07.010]

[40] Brozek-Pluska, B.; Beton, K. 'Oxidative stress induced by *t*BHP in human normal colon cells by label free Raman spectroscopy and imaging. The protective role of natural antioxidants in the form of β-carotene. *RSC Advances,* **2021**, *11*(27), 16419-16434. [http://dx.doi.org/10.1039/D1RA01950C]

[41] Malhotra, S.; Nguyen, T.A.; Nguyen-tri, P. Noble metal oxide Hybrid Nanoparticles: Fundamentals and applications. Elsevier, **2019**.

[42] Cullity, B.D.; Stock, S.R. *Elements of X-ray diffraction,* 3rd edition; Printice hall Publication: New Jersey, **1967**.

[43] Raman, C.V.; Krishnan, K.S. a New Type of Secondary Radiation. *Nature,* **1928**, *121*(3048), 501-502. [http://dx.doi.org/10.1038/121501c0]

[44] Smith, Ewen; Dent, Geoffrey; Spectroscopy, Modern Raman A Practical Approach John Wiley & sons. **2005**.

[45] Banwell, C.N. *Fundamentals of molecular spectroscopy,* 3rd ed; McGraw-Hill Book Company: Tata, England, **1983**.

[46] Amelinckx, S.; Dyck, D.; Landuyt, J.; Tendeloo, G. *Hand book of microscopy methods II*; New York, Basel, Weinheim, Cambridge, Tokyo, **1997**.

[47] Deepak Advanced Transmission Electron Microscopy, F.L. *Applications to Nanomaterials*; Springer International Publishing: Switzerland, **2015**.

[48] Bhattarai, Nabraj Jesus Velazquez-Salazar, and Miguel Jose-Yacaman, Advanced Electron Microscopy in the Study of Multimetallic Nanoparticles, Advanced Transmission Electron Microscopy: Applications to Nanomaterials. Springer International Publishing: Switzerland, **2015**.

[49] Goodhew, P.J. *Electron microscopy and Analysis*; Wykeham publications: London, **1975**.

[50] Brkić, S. *Applicability of Quantum Dots in Biomedical Science: Ionizing Radiation Effects and Application*; Intech Open, **2018**. [http://dx.doi.org/10.5772/intechopen.71428]

[51] Raúl, J. *Martín-Palmaa and Akhlesh Lakhtakiab, Vapor-Deposition Techniques: engineered biomimicy*; Elsevier, **2013**.

[52] https://upload.wikimedia.org/wikipedia/commons/5/5c/Sputtering2.gif

[53] Mittra, J.; Abraham, G.J.; Kesaria, M.; Bahl, S.; Gupta, A.; Shivaprasad, S.M.; Viswanadham, C.S.; Kulkarni, U.D.; Dey, G.K. S. M. Shivaprasad, Chebolu Subrahmanya Viswanadham,Ulhas Digambar Kulkarni and Gautam Kumar Dey, "Role of substrate temperature in the pulsed laser deposition of zirconium oxide thin film. *Mater. Sci. Forum,* **2012**, *710*, 757-761. [http://dx.doi.org/10.4028/www.scientific.net/MSF.710.757]

[54] https://capricorn.bc.edu/wp/zeljkoviclab/research/molecular-beam-epitaxy-mbe/

[55] https://www.memsnet.org/about/processes/lithography.html

[56] https://en.wikipedia.org/wiki/Coprecipitation

[57] https://www.mdpi.com/1996-1944/12/12/1896/html

CHAPTER 7

Effect of Nanostructure-Materials on Optical Properties of Some Rare Earth ions (Eu³⁺,Sm³⁺&Tb³⁺) Doped in Silica Matrix

S. Rai[1,*]

[1] *Physics Department, Mizoram University, Mizoram, Aizawl – 796004, India*

Abstract: Nanoparticles of CdS incorporated in Rare Earth doped silica xerogel (RE^{3+}:SiO_2) matrix have been prepared by sol-gel method to study its various aspect. The prepared materials have been characterized by physical and optical technique, such as XRD, SEM, TEM and Photoluminescence (PL). We can conclude from TEM that the particle size of the materials 8 nm and an average particle dimension of 5 nm. It is also found consistent with the theoretical calculation performed based on the Scherrer equation and effective mass approximation (EMA) model. The optical properties of these materials depend on various parameters such as dimension and surface characteristics, doping and interaction with the surrounding environment. Enhancements of Rare Earth (RE) ions luminescence have been observed with the presence of CdS NPs in RE^{3+}:SiO_2 matrix. A twenty time more intense dominating orange peaks (616 nm) from the characteristic peak of Eu^{3+} ions are observed for CdS/Eu^{3+}:SiO_2 matrix compared to the sample without CdS NPs. The efficient energy transfer (ET) from CdS NPs to RE ions is primarily responsible for this boost in the luminescence intensity. The emission intensity in PL spectra decreases with raise in the concentration of CdS NPs. With an increase in CdS NPs concentration in RE^{3+}:SiO_2 matrix, the emission intensity decreases possibly due to the increase in the concentration of "oxygen vacancy "and "Si hanging" in the matrix of the silica xerogel. Thus, photoluminescence properties of the material are greatly influenced by site symmetry and hence the concentration of dopant ions.

Keywords: Luminescence, Nanoparticles, Quenching, Radiative parameters, Site symmetry, Sol-gel.

INTRODUCTION

In this paper, we review our recent work in nanostructure material along with optically active Rare Earth (RE) ions for photonic devices application [1]. The fabrication of Photonic devices is performed from various materials. The semi-

---
* **Corresponding author Rai S.:** Physics Department, Mizoram University, Mizoram, Aizawl – 796004, India; Tel:+918732853277; E-mail: srai.rai677@gmail.com

Dibya Prakash Rai (Ed.)

conductors and glasses are the two major constituents for core components such as oscillators, amplifiers, frequency converters, modulators switches, and routers and so on. A rapid advance in nano technologies, particularly material synthesis development, grows a solid interest in the luminescence dynamics of Lanthanide ions doped nonmaterial's [2]. We can reduce the size of a material to transform it into Quantum wells, quantum wires and quantum dots depending on its dimension. The dimensional manipulation imports lots of interesting intermediate properties to the material that differ from the properties bulk and atom can possess. The optical properties are one of the most lucrative and valuable aspects of nanomaterial. Applications based on optical properties of nanomaterial comprise lasers, optical detectors, quantum memory, sensor, phosphor display, solar cell and biomedicine. Researchers are actively focusing on the optical properties of different RE ions doped in low dimensional semiconductors to employ in optoelectronic devices [3 - 5]. The optical properties of these materials can be tuned with varying different parameters such as dimension, shape, surface characteristics, doping and interaction with the surrounding environment. A photo-generated carrier confined in semiconductor crystal recombine and transfer energy to RE ion in or near the nanocrystals. The strong cum broad peak at lower energy has slow relaxation rates, which must be associated with some trapping centres in deep energy levels of the microcrystal's [5 - 8]. This work focuses on the role of the particle dimensions, composition, processing and morphology of a host on optical properties of $Eu^{3+,}$ Tb^{3+} and Sm^{3+}. From a technological point of view, understanding the optical properties of these materials is very crucial for optimization.

EXPERIMENTAL

The sol-gel process allows the synthesis of glass samples at relatively low temperatures as compared with melt quench glasses. In addition to the benefits presented by low temperatures, dopant concentrations can be very accurately controlled in the solution stage in which one can insert metallic, organic and inorganic additives [9]. This method enables the possibility of modifying different parameters such as refractive index, phonon energy and transparency of the material. It can be achieved by choosing a suitable matrix either individually or in combination, such as ZnO, CdS, TiO_2, Al_2O_3, ZrO_2 *etc.* [6, 7]. Sol-gel has the potential to hold much higher rare earth concentrations, beyond the threshold where melt quench glasses lose their structure. The sol-gel process empowers the synthesis of materials of various shapes and sizes along with thin films and fibers [10 - 12].

Synthesis of Rare-earth Ions and CdS NPs

RE–doped glass preparation was well documented in our earlier work [2, 13, 14]. We prepared Silica gels containing $Eu^{3+}Tb^{3+}$ and Sm^{3+} ions with different concentrations of CdS by the Sol-Gel Method. Here, tetraethylorthosilicate (TEOS) was used as a glass precursor in the solution along with methanol, deionized water and HNO_3 as a catalyst. The molar ratio of TEOS to water was kept 1:16. Cadmium nitrate [Cd $(NO_3)_2$] and thiourea [SC $(NH_2)_2$] were employed as cadmium and sulfur sources, respectively. Final solution was swirled for 30 min before cast into capped 12x75mm disposable polystyrene test tubes for aging at ambient temperature over 75h. The gel was baked for 48h at 90^0c over period of 5 h. Finally, these xerogel were annealed in air at higher temperature range from 100^0C to 250^0C. At this instant, the gels had shrunk to a volume that was one third of its original value [9].

Spectrofluoremeter (Horiba Jobin Yvon, Fluoro Max-4) was used to measure the Absorption and Fluorescence spectra of the prepared sample at room temperature (RT).

RESULT WITH DISCUSSION

Photoluminescence (PL) and Absorption Spectra of Glass with Only CdS NPs by Sol-Gel Method

Photoluminescence and Absorption spectra of pure CdS NPs are presented in Fig. **1** (**a** and **b**). The absorption of a composite is dependent on the dopant and host matrix. In the case of NPs dispersed in a glass matrix, we observed a strong absorption in the IR domain due to atomic vibration. The transition of electrons from valance band to conduction band as well as exciton state at the edge of the conduction band is responsible for visible absorption by NPs [15]. A shift in the blue side of the optical spectrum of a nanomaterial represents the quantum confinement effect. The CdS NPs formation is confirmed by a blue absorption peak around 424nm as the diameter of the particles is associated with the absorption edge. Effective Mass Approximation (EMA) can be employed qualitatively to explain the quantum confinement effect [16]. We can consider a particle spherical in shape with radius R, the effective bandgap $[E_{g,E_{ff}}(R)]$, is specified by

$$E_{g,E_{ff}}(R) = E_g(\infty) + \frac{h^2\pi^2}{2R^2}\left(\frac{1}{m_e^*} + \frac{1}{m_h^*}\right) - \frac{1.786s^2}{sR} \tag{1}$$

Where $E_g(\infty)$ represents bulk bandgap, m_h^* and m_e^* stand for the effective masses of the hole and electron, and is obviously the relative permittivity or bulk optical dielectric constant of the sample. From the second term on the right-hand side of this equation, it is clearly evident that the effective bandgap is inversely proportional to the square of the radius and hence decreases as size increases. On the other hand, it is visible from the third term that the Coulombic interaction increases as R of the particle decrease as a result effective bandgap energy decreases. However, at a small radius, the second term turns out to be dominant, hence the effective bandgap is looking forward to increasing with decreasing R, especially when the value of R is small. The particles size estimated using the effective mass approximation (EMA) model was around 4.8nm for the prepared sample.

A broad luminescence emission spectrum for bulk CdS is reported by various research groups with an emission maximum of around 500-700nm range. The observed emission in CdS is occurred due to the hole-electron recombination after relaxation (band edge emission). The feature like broad emission peak of CdS NPs is expected due to increased surface area of the NPs and the probability of emission from defect. The PL spectra in Fig. (**1a**) shows emission close to the band edge having a maximum at 445nm.

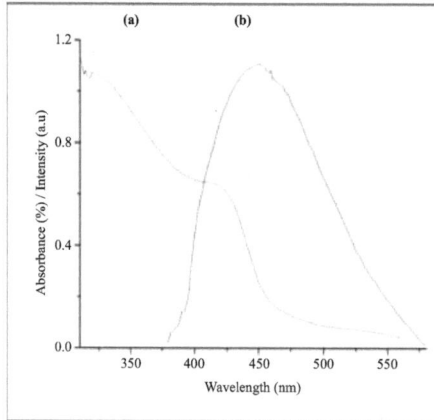

Fig. (1). (a) Optical absorption **(b)** Photoluminescence spectra of CdS alone doped glass.

To determine NPs' properties one should consider its size as an important parameter. A researcher's understanding of NPs' size and their distribution helps him to illuminate the enormous properties of NPs. It arises from the significant size-dependent character of physical and chemical properties of NPs hold. A variety of methods are employed to determine the NPs size. One of the best methods at present, as well as well-known among researchers, is TEM. TEM provides a very high-resolution picture of the sample. Another popular method for determining NPs structure is based on XRD. Although XRD provides an approximate idea of NPs structure, in absence of TEM it becomes a valuable alternative. This method empowered to determine the size of the crystals, but fail to estimate the dimension of the particle. The crystalline dimension is typically smaller than the dimension of the particle. Fortunately, if an NP is crystalline, it typically appears as a single crystal so the crystalline dimension is the same as the particle dimension. Scherrer and Warren's formulas are combined to determine the size of the nanocrystal with the help of XRD data [15]. The XRD diffraction lines have a width, which is strongly connected to the size, size distribution, defects and strain in the nanocrystal. As the size of the nanocrystal reduces, it loses its long-range order related to its bulk state. It reciprocated with a broadening in the line width. Thus, XRD line width can be utilized to approximate the dimension of the particle.

Fig. (2) Illustrates the XRD pattern of CdS NPs embedded in silica glass which was entirely amorphous at RT (inset of Fig. 2) and crystallization initiated at 800°C. The different values of $2\theta(hkl)$ are 28.21° (101), 43.73° (110), 47.8° (103), 50.94° (200) and 52.86° (201), which are recognized for hexagonal CdS phase (JCPDS card-772306). Using Scherrer's relation, the average sizes (D; in Å) are evaluated from the line broadening.

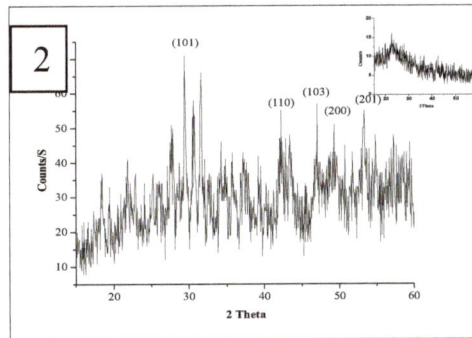

Fig. (2). XRD pattern of Sol-Gel xerogel for CdS/Sm^{3+}:SiO$_2$ sample (inset) and after annealing at 800°C.

$$D = \frac{0.9\lambda}{\beta Cos\theta},$$ (2)

Where D is the nanocrystal diameter, λ represents the "wavelength of CuKα radiation", stands for the "half-width of the diffraction peak" θ and is the angle. Finally, consequent D for nanocrystal obtained in this calculation is 8.1nm.

The TEM image (transmission electron microscopy) of the prepared sample is presented in Fig. (3). An average nanoparticle dimension of 5 nm was obtained from TEM. It was found consistent with the value calculated by the Scherrer equation (8.1 nm) and EMA (particles size 4.8nm). Singh *et al.*, mentioned in their report about the use of SAED (Selective area electron diffract) pattern to study are specific crystal properties of the prepared sample. In simple words we can differentiate a material into single amorphous, crystalline, polycrystalline, or polycrystalline textures. Single crystalline materials in the SAED pattern have only spot patterns while polycrystalline materials have ring pattern form as shown in Fig. (4). In polycrystalline materials, the observed ring in the SAED pattern may be either broad or sharp and sometimes even diffuse on the basis of the particle dimension. It can roughly be anticipated from the line width of the rings with the relation mentioned elsewhere [17, 18]. It reveals almost spherical shapes of NPs. Particles are well distributed homogeneously. Due to the occurrence of rings with discrete spots in the SAED pattern, one can authenticate the larger grain dimension along with polycrystalline nature.

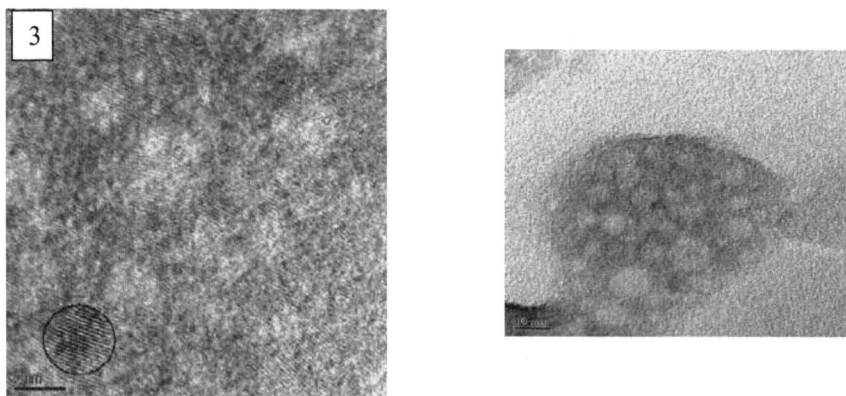

Fig. (3). TEM images of CdS NPs.

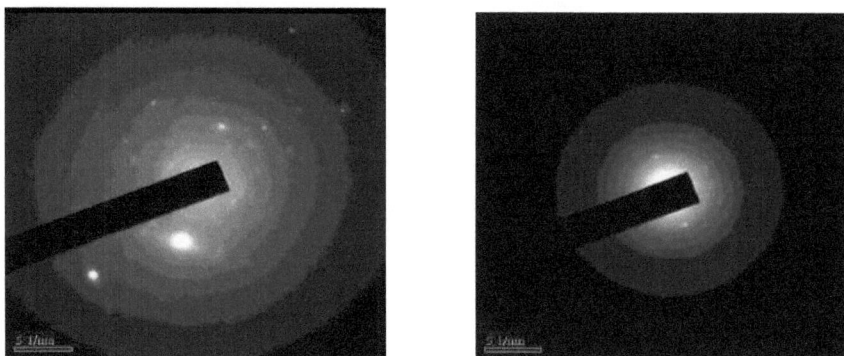

Fig. (4). SAED pattern, due to presence of ring with discrete spot.

Fluorescence Studies of Eu³⁺: SiO₂sol-gel Glass Doped with CdS NPs

The fluorescence spectra of Eu^{3+}:SiO_2 sol-gel glass doped with CdS NPs under the irradiation of 399nm excitation is represented in Fig. (5). Four bands corresponding to transition $^5D_0 \rightarrow {}^7F_J$ (J=1, 2, 3,4)are observed in between 540-700nm. A transition for J=1, is allowed by a magnetic dipole, whereas J=2, 3, 4transitions are electric- dipole allowed. A twenty time more intense dominating orange peaks (616 nm) from a characteristic peak of Eu^{3+} ions are observed for CdS/Eu^{3+}:SiO_2 matrix with compare to the sample without CdS NPs. We can possibly justify intense domination as modification put forward by CdS NPs in the network of Eu^{3+}:SiO_2xerogel which helps to boost the "oxygen vacancy "and "Si hanging" in the silica xerogel matrix. It creates an opportunity for a number of hole and electron to effortlessly excite and hence radiant recombination are amplified. Thus the fluorescence emission intensity for the doped sample is strikingly amplified.

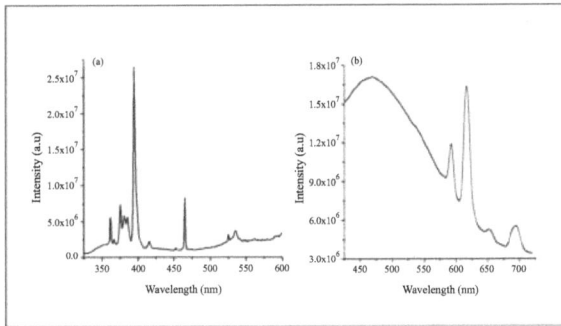

Fig. (5). Fluorescence spectra of CdS/Eu^{3+}:SiO_2 matrix.

The Blue peak at 464nm is due to CdS NPs and Orange-I, Orange-II and Red-I & II peaks (592, 616, 652 and 694nm) are due to Eu^{3+} ions. The origin of blue emission is verified by oxidizing the prepared sample by heating at 150°C in air. An unstable Eu^{2+}ion in an oxidizing atmosphere oxidized easily to Eu^{3+} ion ($Eu^{2+} \rightarrow Eu^{3+} + e$). The spectral overlap is the required explanation for the ET (energy transfer) from CdS NPs to Eu^{3+} [13, 14]. As most of the excitation lines of Eu^{3+} overlap the broadband emission spectrum of CdS (range 350 to 550 nm), the energy emission from CdS NPs is positively absorbed by the Eu^{3+} ions. The4f-4f transitions of rare-earth elements and ET from CdS to Eu^{3+} make it possible. In europium, the Orange-I(592nm) transition is observed due to magnetic dipole. The Orange-II (614nm) is forced hypersensitive electric dipole transition which is permissible only at host matrix with low symmetries and "no inversion centre". The Orange-II to Orange-I intensity ratio (I_{OII}/I_{OI}) is used as a spectroscopic probe to determine site symmetry. The probe indicates lower site symmetries for the host along with no inversion centre for a higher ratio. In our CdS/Eu^{3+}:SiO_2

sample, Eu^{+3} location at low symmetry sites is evident from Fig. (**5**). The 4f–4f parity forbidden transitions that occur from the forced electric dipole is become partially allowed with the presence of the ions in a low symmetry site. Thus the 4f-states inter-mixed with higher electronic configuration to initiate parity forbidden transition [19, 20] and consequently there is an increase in the probability of optical transition, in short, "the rate of radiative emission increases". Thus changes in the luminescence intensity relative to the various doping concentration of dopants are well explained. These results disclose how site symmetry of ions crucially depends on the dopant ions' concentration and the important role it plays to decide the photoluminescence properties [14].

Fluorescence Studies of CdS/Tb^{3+}: SiO$_2$ Matrix

CdS/Tb^{3+}:SiO$_2$ glass sample by Sol-Gel method. Total five-band correspond to the electronic transition of Tb^{3+} ions from its ground state $^7F_6$ to excited states $^5G_4$(342 nm), $^5D_2$,(351 nm),$^5L_{10}$(359 nm),$^5D_3$(369 nm) and $^5D_4$(379 nm), respectively. The excitation was monitored at 544nm.

The emission spectra of CdS/Tb^{3+}:SiO$_2$ sample were recorded under the excitation wavelength (λ_{ex}= 370nm), as presented in Fig. (**6**).

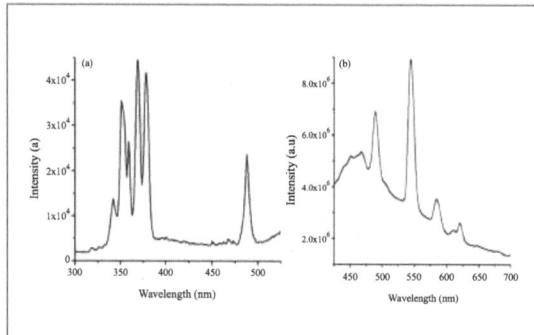

Fig. (6). (a) Excitation spectra CdS/Tb^{3+}: SiO$_2$ **(b)** Fluorescence spectra of CdS/Tb^{3+}: SiO$_2$ [21].

Emission spectra of CdS/Tb^{3+}:SiO$_2$ glass comprise of two groups. The transition from the $^5D_4$ and $^5D_3$ excited level to the ^7F_J manifold of the ground state. The highly probable multiphonon relaxation from higher excited level $^5D_3$ to $^5D_4$ contribute to a much higher emission intensity from $^5D_4 \rightarrow ^7F_J$ manifolds compared to $^5D_3 \rightarrow ^7F_J$ manifolds. The CdS/Tb^{3+}:SiO$_2$xerogel is observed with various luminescence bands belonging to the $^5D_3 \rightarrow ^7F_4$ (450nm), $^5D_3 \rightarrow ^7F_3$ (466nm), and $^5D_4 \rightarrow ^7F_6$ (489nm), $^5D_4 \rightarrow ^7F_5$ (544nm), $^5D_4 \rightarrow ^7F_4$ (585nm) and $^5D_4 \rightarrow ^7F_3$ (621nm). A drastic change in the luminescence intensity with varying CdS concentrations is observed. Increase in the CdS concentration from 0.1M% to 0.3M%, we observe a boost in the luminescence intensity while intensity drops down with the further increase in the CdS concentration. Again we can possibly explain that CdS NPs doped in the matrix of Tb^{3+}:SiO_2xerogel capable of increasing the "oxygen vacancy" and "Si dangle" in silica xerogel network. It creates an opportunity for more electrons and holes to simply excite and hence radiant recombinations are amplified. Thus the emission intensity for sample embedded with CdS strikingly amplified [9, 21, 22]. The band-edge emission of CdS NPs among themselves ought to be one additional contributing factor. The cause of a dramatic decrease in intensity of emission from $^5D_3$ lines comparative to $^5D_4$ lines is the increasing occurrence of cross-relaxation induced quenching. We observed a declining trend in emission intensity in CdS (0.5M)/Tb^{3+}: SiO_2, which might possibly be due to the formation of the cluster by CdS NPs. The faint emission observed in CdS/Tb^{3+}: SiO_2sample with above 0.3M CdS concentration perhaps owing to defect concentration in the silica xerogel network which blocks the ET among the Tb^{3+} ion and defect state.

Photoluminescence Spectrum of CdS/Sm³⁺: SiO₂ Matrix

The fluorescence spectra of CdS/Sm^{3+}:$SiO_2$is are recorded for the sample with excitation (λ_{ex}=399nm) radiation at room temperature as in Fig. (7). Green, Orange, Red colour emission peaks centred at 561nm ($^4G_{5/2}\rightarrow^6H_{5/2}$), 595nm ($^4G_{5/2}\rightarrow^6H_{7/2}$) and 641nm ($^4G_{5/2}\rightarrow^6H_{9/2}$) respectively. The Yellow emission peak has maximum intensity. From the emission spectra, variation of intensities with the different CdS ions concentrations is evident with the same peak positions. The fluorescence intensity of Sm^{3+}indicates the same trend we notice for the other two rare-earth doping. The fluorescence intensity increases as the CdS concentration changes from 0.1M to 0.3M. We monitored similar emission spectra as reported Sm^{3+}system elsewhere [21]. The same fall of intensity in emission spectra was observed for CdS concentration above 0.3M. With the same concluding remark, we suggest that CdS NPs embedded in the matrix of Sm^{3+}:SiO_2xerogel are capable of increasing the "oxygen vacancy" and "Si hanging" in the network of the silica xerogel. It generates an opportunity for more number of hole or electron to effortlessly excite and hence radiant recombination are amplified. Thus emission intensity for sample doped with CdS NPs strikingly improved. Addition factor such as band edge emission among CdS NPs is also contributing. An obvious fluorescence quenching occurred due to an excessive increase in CdS concentration as evident in the report [2].

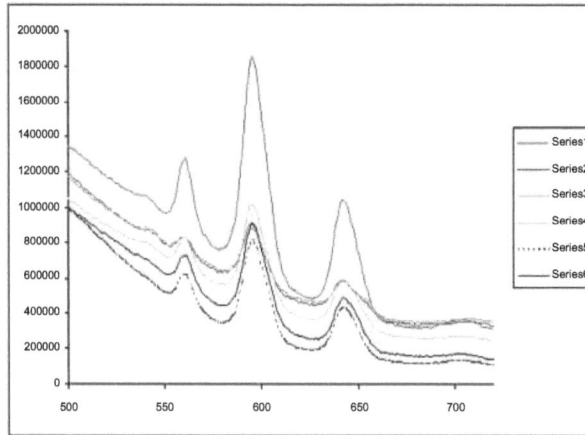

Fig. (7). Fluorescence spectra of CdS/Sm^{3+}:SiO$_2$.

Annealing Effect on CDS NPs Doped RE^{3+}: SiO$_2$ Glass Matrix

The intensity of Fluorescence spectra of CdS/RE^{3+}:SiO$_2$ emission was observed as a fluctuating trend with annealing temperature. The intensity increased with an increase in annealing temperature and reaches utmost at 150^{0}C. It decreases with a further increase in the annealing temperature. This outcome can be interpreted as a consequence of heat treatment where the hydroxyl groups (OH) concentration in the glass will reduce, along with diverse defect centres created by the process of hydrolysis and condensation of alkoxysilane precursor. As the influence of high energy vibration of OH group on the emission of RE ions was declined and the energy released from the hole and electron created by defect recombination was transmitted to the RE^{3+} ions that were implanted in the network of silica xerogel and thus initiating the enhancement of RE^{3+}: CdS environment [2]. Thus, increase in annealing temperature create a more efficient luminescent centre along with enhanced emission intensity [23]. The further rise in the annealing temperature may produce the opposite result by creating a cluster of RE ions or may shrink the concentration of defects in the silica xerogel network, which prevents the ET among the defects and Sm^{3+} ions. This outcome became predominant in high annealing temperatures, and finally, the intensity of emission declined [24].

CONCLUSION

In summary, we synthesized silica glass with and without CdS along with Eu^{3+}/Tb^{3+}/Sm^{3+} ions (anyone) by sol-gel techniques. From the Absorption and Excitation spectra of the prepared sample, it is evident that all these glasses can be successfully excited by a UV source. In the luminescence properties of

$CdS/(RE^{3+}):SiO_2$ matrices, we observed that the fluorescence of this amorphous sample was appreciably dependent on CdS NPs concentration and temperature at which the sample was annealed. Hence, the role of CdS NPs as a network modifier is established. We can conclude that there is an increase in non-bridging oxygen number in the glass sample deteriorating the Si-O-Si bond.

Photoluminescence properties of RE-doped (RED) glasses are the key factor for the development of some components, like integrated optical amplifiers, and biosensors. RE ions doped NPs are gifted materials for phosphor devices.

CONSENT FOR PUBLICATION

Not applicable.

CONFLICT OF INTEREST

The author declares no conflict of interest, financial or otherwise.

ACKNOWLEDGEMENTS

Declared none.

REFERENCES

[1]   Binnemans, K. Lanthanide-based luminescent hybrid materials. *Chem. Rev.,* **2009,** *109*(9), 4283-4374.
        [http://dx.doi.org/10.1021/cr8003983] [PMID: 19650663]

[2]   Rai, S.; Bokatial, L.; Dihingia, P.J. Effect of CdS nanoparticles on fluorescence from Sm3+ doped SiO2 glass. *J. Lumin.,* **2011,** *131*(5), 978-983.
        [http://dx.doi.org/10.1016/j.jlumin.2011.01.006]

[3]   Singh, R.; Rangari, V.K.; Sanagapalli, S.; Jayaraman, V. Nano-structured CdTe, CdS and TiO2 for thin film solar cell applications. *Sol. Energy Mater. Sol. Cells,* **2004,** *82*(1-2), 315-330.
        [http://dx.doi.org/10.1016/j.solmat.2004.02.006]

[4]   Yoon, M. "Surface Modifications and Optoelectronic Characterization of TiO 2 -Nanoparticles : Design of New Photo-Electronic Materials," *J. ofthe Chinese Chem. Soc.,* **2009,** *56*, 449-454.

[5]   Psuja, P.; Hreniak, D.; Strek, W. Rare-Earth Doped Nanocrystalline Phosphors for Field Emission Displays. *J. Nanomater.,* **2007,** *2007*, 1-7.
        [http://dx.doi.org/10.1155/2007/81350]

[6]   Lakshminarayana, G.; Qiu, J. Enhancement of Pr3+ luminescence in TeO2–ZnO–Nb2O5–MoO3 glasses containing silver nanoparticles. *J. Alloys Compd.,* **2009,** *478*(1-2), 630-635.
        [http://dx.doi.org/10.1016/j.jallcom.2008.11.146]

[7]   Shih, H.R.; Chang, Y.S. Structure and Photoluminescence Properties of Sm3+ Ion-Doped YInGe2O7 Phosphor. *Materials (Basel),* **2017,** *10*(7), 779.
        [http://dx.doi.org/10.3390/ma10070779] [PMID: 28773139]

[8]   Lišková, P.; Konopásek, I.; Fišer, R. Simple Way to Detect Trp to Tb³⁺ Resonance Energy Transfer in Calcium-Binding Peptides Using Excitation Spectrum. *J. Fluoresc.,* **2019,** *29*(1), 9-14.
        [http://dx.doi.org/10.1007/s10895-018-2326-0] [PMID: 30471022]

[9]   Silversmith, A.J.; Boye, D.M.; Brewer, K.S.; Gillespie, C.E.; Lu, Y.; Campbell, D.L. 5D3→7FJ

emission in terbium-doped sol–gel glasses. *J. Lumin.,* **2006**, *121*(1), 14-20.
[http://dx.doi.org/10.1016/j.jlumin.2005.09.009]

[10]   Ehrhart, G.; Capoen, B.; Robbe, O.; Beclin, F.; Boy, P.; Turrell, S.; Bouazaoui, M. Energy transfer between semiconductor nanoparticles (ZnS or CdS) and Eu3+ ions in sol–gel derived ZrO2 thin films. *Opt. Mater.,* **2008**, *30*(10), 1595-1602.
[http://dx.doi.org/10.1016/j.optmat.2007.10.004]

[11]   Ruan, Y.; Xiao, Q.; Luo, W.; Li, R.; Chen, X. Optical properties and luminescence dynamics of Eu [3+] -doped terbium orthophosphate nanophosphors. *Nanotechnology,* **2011**, *22*(27), 275701.
[http://dx.doi.org/10.1088/0957-4484/22/27/275701] [PMID: 21597160]

[12]   Rai, S.; Fanai, A.L. Optical properties of Ho3+ in sol-gel silica glass co-doped with Aluminium. *J. Non-Cryst. Solids,* **2016**, *449*, 113-118.
[http://dx.doi.org/10.1016/j.jnoncrysol.2016.07.023]

[13]   Dihingia, P.J.; Rai, S. Synthesis of TiO2 nanoparticles and spectroscopic upconversion luminescence of Nd3+-doped TiO2–SiO2 composite glass. *J. Lumin.,* **2012**, *132*(5), 1243-1251.
[http://dx.doi.org/10.1016/j.jlumin.2011.12.008]

[14]   Bokatial, L.; Rai, S. Photoluminescence and energy transfer study of Eu3+ codoped with CdS nanoparticles in silica glass. *J. Fluoresc.,* **2012**, *22*(1), 505-510.
[http://dx.doi.org/10.1007/s10895-011-0984-2] [PMID: 21953436]

[15]   Abdullah, M.; Iskandar, F.; Okuyama, K. Semiconductor Nanoparticle-Polymer Composites In: *Nanocrystalline Materials*; Elsevier: Great Britain, **2006**; pp. 275-310.
[http://dx.doi.org/10.1016/B978-008044697-4/50022-3]

[16]   Hayakawa, T.; Tamil Selvan, S.; Nogami, M. Influence of adsorbed CdS nanoparticles on 5D0→7FJ emissions in Eu3+-doped silica gel. *J. Lumin.,* **2000**, *87-89*, 532-534.
[http://dx.doi.org/10.1016/S0022-2313(99)00280-X]

[17]   Bokatial, L.; Rai, S. Structural and upconversion studies of Er3+ codoped with CdS nanoparticles in sol-gel glasses. *J. Fluoresc.,* **2012**, *22*(6), 1639-1645.
[http://dx.doi.org/10.1007/s10895-012-1108-3] [PMID: 22872435]

[18]   Singh, V.; Sharma, P.K.; Chauhan, P. Synthesis of CdS nanoparticles with enhanced optical properties. *Mater. Charact.,* **2011**, *62*(1), 43-52.
[http://dx.doi.org/10.1016/j.matchar.2010.10.009]

[19]   Liu, L.; Zhang, Z.; Kang, S.Z.; Mu, J. Effect of SnO 2 Nanocrystals on the Emission of Eu [3+] Ions in Silica Matrix. *J. Dispers. Sci. Technol.,* **2007**, *28*(5), 769-772.
[http://dx.doi.org/10.1080/01932690701341918]

[20]   Tarafder, A.; Annapurna, K.; Chaliha, R.S.; Tiwari, V.S.; Gupta, P.K.; Karmakar, B. Nanostructuring and fluorescence properties of Eu3+:LiTaO3 in Li2O–Ta2O5–SiO2–Al2O3 glass-ceramics. *J. Mater. Sci.,* **2009**, *44*(16), 4495-4498.
[http://dx.doi.org/10.1007/s10853-009-3659-5]

[21]   Rai, S.; Bokatial, L. Effect of CdS nanoparticles on photoluminescence spectra of Tb 3 + in sol – gel-derived silica glasses. In: *bull. mater. sci*; , **2011**; 34, pp. 227-231.

[22]   Jyothy, P.V.; Amrutha, K.A.; Gijo, J.; Unnikrishnan, N.V. Fluorescence enhancement in Tb3+/CdS nanoparticles doped silica xerogels. *J. Fluoresc.,* **2009**, *19*(1), 165-168.
[http://dx.doi.org/10.1007/s10895-008-0398-y] [PMID: 18648752]

[23]   Devi, H.J.; Singh, W.R.; Loitongbam, R.S. *Red, Yellow, Blue and Green Emission from Eu 3 +, Dy 3 + and Bi 3 + Doped Y 2 O 3 nano-Phosphors*; J. Fuorescence, **2016**.
[http://dx.doi.org/10.1007/s10895-016-1776-5]

[24]   Bokatial, L.; Rai, S. Optical properties of Sm3+ ions in sol-gel derived alumino-silicate glasses. *J. Opt.,* **2012**, *41*(2), 94-103.
[http://dx.doi.org/10.1007/s12596-012-0069-x]

CHAPTER 8

Nd$_2$Fe$_{14}$B and SmCo$_5$ a Permanent Magnet for Magnetic Data Storage and Data Transfer Technology

Abeer E. Aly[1,2,*]

[1] *Basic science department, institute of engineering and technology, Cairo, Egypt*

[2] *Physics Department, Higher Institue of Engineering Shorouk Academy, Cairo, Egypt*

Abstract: We present first-standards estimations on Fe, Nd, and SmCo$_5$ utilizing the self-predictable maximum capacity linearized increased plane wave (FPLAPW) strategy. The attractive snapshots of Fe, Nd, and Smco$_5$ were determined utilizing the WIEN2K code. The minutes for BCC Fe and HCP Nd are 2.27µB and 2.65µB separately in great concurrence with test esteems. For Smco$_5$, we efficiently study the impact of considering the twist circle coupling and Coulomb connections in the Sm f shell on the attractive properties, electronic construction, and twist thickness maps. The determined attractive second and magneto-crystalline anisotropy like anisotropy are in acceptable concurrence with test esteems. The twist thickness maps in the (001) plane show that the impact of the twist circle coupling on the twist thickness design of Sm particles is more grounded than that of Coulomb connection. We additionally study the impact of the polarization heading on the energy groups by looking at the highlights of band structure when the charge bearing is along or opposite to the c-axis. The determined outcomes are in acceptable concurrence with the exploratory information.

Keywords: Density functional theory, Linearized augmented plane waves, Spin-density maps, Spin-orbit coupling.

INTRODUCTION

Intermetallic mixtures of uncommon earth molecules and progress metal particles are vital both for mechanical application just as from the perspective of fundamental examination. To begin with, the most remarkable lasting magnets are among this class of materials [1, 2]. Second, they address a major test for the electron hypothesis in light of the fact that their properties are dictated by two very surprising sorts of electronic states (*i.e.*, the profoundly associated and firmly limited 4f conditions of the uncommon earth molecules (R) and the valence cond-

---

* **Corresponding author Abeer E. Aly:** Basic science department, institute of engineering and technology, Cairo, Egypt; Tel:002-01008166968; E-mail: abeerresmat782000@yahoo.com

Dibya Prakash Rai (Ed.)

conditions of the change metal particles which are nearly feebly connected and more delocalized). The revelation of the new elite perpetual magnet dependent on uncommon earth(R) and progress metals has altered the attractive business and opened another skyline for essential examination on attractive materials [3]. Attractive materials of uncommon earth components with change metals have been a difficult class of materials [4 - 8], with equivalent accentuation on the hypothetical portrayal [9 - 16], trial examination of natural genuine design properties [16 - 25], and mechanical applications [16, 23 - 26]. $SmCo_5$ is viewed as an astounding material for high-coercivity magnet applications and is of impressive interest as a result of its huge magneto-glasslike anisotropy energy [27 - 29]. The revelation of the great hard attractive properties of $SmCo_5$ during the 1960s [30] and very nearly 20 years after the fact, the disclosure of less expensive materials for example $Nd_2Fe_{14}B$ [31, 32] have started the improvement of super-magnets with sudden properties, permitting new development standards of electronic motors and different gadgets [23, 26].

The Crystal Structure of $SmCo_5$

$SmCo_5$ solidifies in the $CaCu_5$ structure (space bunch P6/mmm, No. 191) [33]. The unit cell comprises the equation unit (for example 6 iotas) as demonstrated in Fig. (**1**). The test upsides of a and c utilized in our computation are a = 5.00, c= 3.96Å. One generally faces challenges in managing the band construction of f-electron materials utilizing the LDA band computations. This is because of the restricted nature off-electrons just as to the unobtrusive interaction between the twist polarization and the twist circle cooperation. The orbital attractive second is brought about by the twist circle coupling and furthermore by the relationship impacts, the last impacts being more significant.

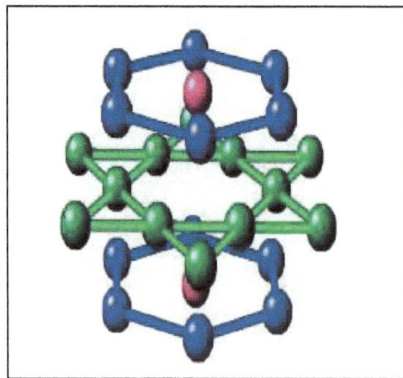

Fig. (1). Crystal Structure of $SmCo_5$.

Electronic Band Structure Calculation

The reason for examining numerous properties of solids is the basic electronic band construction of the material emerging from the arrangement of the many-body Schrödinger condition within the sight of the intermittent capability of the grid, which is talked about in a large group of strong state physical science course readings. The electronic arrangements within the sight of the occasional capability of the cross-section are as Bloch capacities

$$\Psi_{n,k}=U_n\,(k)\,e^{ik.r}$$

where k is the wave-vector, and n marks the band record compared to various answers for a given wave-vector. The cell-intermittent capacity, Un (k), has the periodicity of the grid and regulates the voyaging wave arrangement related to free electrons. Electronic band structure estimation techniques can be gathered into two general classifications [34]. The main class comprises stomach muscle initio techniques, like Hartree-Fock or Density Functional Theory (DFT), which figure the electronic construction from first standards, for example without the requirement for exact fitting boundaries and this technique will be depicted in subtleties in section 2. All in all, these strategies use a variational method to deal with the ground state energy of a many-body framework, where the framework is characterized at the nuclear level. The first estimations were performed on frameworks containing a couple of molecules. Today, estimations are performed utilizing roughly 1000 molecules yet are computationally costly, some of the time requiring hugely equal PCs. As opposed to stomach muscle initio approaches, the subsequent class comprises exact strategies, for example, the Orthogonalized Plane Wave (OPW) [35], tight-restricting [36] (otherwise called the Linear Combination of Atomic Orbital (LCAO) strategy), the k.p technique [37], and the neighbourhood or the non-nearby [38] observational pseudo-expected strategy (EPM). These techniques include observational boundaries to fit trial information, for example, the band-to-band changes at explicit high-balance focuses on optical ingestion tests. The allure of these techniques is that the electronic design can be determined by addressing a one-electron Schrödinger wave condition (SWE). In this way, exact strategies are computationally more affordable than abdominal muscle initio estimations and give a moderately simple method for producing the electronic band structure. Because of their widespread utilization, in this part, we will audit probably the most usually utilized ones, to be specific the exact pseudopotential technique, the tight-restricting, k.p hypothesis and almost free-electron guess.

Spin-Orbit Coupling

Before continuing with the depiction of the different exact band structure techniques, it is helpful to present the twist circle communication Hamiltonian. The impacts of twist circle coupling are most handily considered in regards to the twist circle association energy HSO as an annoyance. In its most broad structure,

HSO working on the wave-capacities Ψk is then given by $H_{SO} = \dfrac{\hbar}{4m^2c^2}[\nabla V \times P].\sigma$

Where V is the potential energy term of the Hamiltonian, P is the momentum and σ is the Pauli spin tensor. It can also be written in the following form as an operator on the cell-periodic function

$$H_{SO} = \frac{\hbar}{4m^2c^2}[\nabla V \times P].\sigma + \frac{\hbar^2}{4m^2c^2}[\nabla V \times k].\sigma$$

The first term is k-independent and is analogous to the atomic spin-orbit splitting term. The second term is proportional to k and is the additional spin-orbit energy coming from the crystal momentum. Rough estimates indicate that the effect of the second term on the energy bands is less than 1% of the effect of the first term. The relatively greater importance of the first term comes from the fact that the velocity of the electron in its atomic orbit is very much greater than the velocity of a wave packet made up of wave-vectors in the neighbourhood of k.

The Empirical Pseudo-Potential Method

The idea of pseudo-possibilities was acquainted by Fermi [39] with concentrated high-lying nuclear states. Thereafter, Hellman recommended that pseudo-possibilities be utilized for ascertaining the energy levels of the soluble base metals [40]. The widespread use of pseudo-possibilities did not happen until the last part of the 1950s, when action in the space of consolidated matter material science started to speed up. The principle benefit of utilizing pseudo-possibilities is that lone valence electrons must be thought of. The centre electrons are treated as though they are frozen in a nuclear-like arrangement. Thus, the valence electrons are thought to move in a powerless one-electron potential. The pseudo-potential strategy depends on the orthogonalized plane wave (OPW) technique because of Herring [34].

The Tight-Binding Method

Tight restricting (TB) is a semi-exact strategy for electronic design computations. While it holds the fundamental quantum mechanics of the electrons, the Hamiltonian is defined and streamlined before the computation, as opposed to developing it from the first standards. The technique is point by point by Slater and Koster [41], who laid the underlying preparation. Theoretically, close restricting works by hypothesizing a premise set which comprises of nuclear-like orbital (for example they share the precise force segments of the nuclear orbital, and are effectively parted into a spiral and rakish parts) for every molecule in the framework, and the Hamiltonian is then defined as far as different high evenness communications between these orbital.

The Nearly Free Electron Approximation

According to Bloch's theorem, electronic wave functions can be expanded as:

$$\psi(x) = \sum_G C_k - Ge^{i(\vec{K}-\vec{G}).\vec{x}}$$

In the almost free-electron estimation, we expect that electronic wave capacities are given by the superposition of a few plane waves. This estimate is legitimate, for example, when the intermittent potential is feeble and contains a set number of complementary cross-section vectors. To sum up, the almost free-electron estimation gives energy groups which are basically free-electron groups from the Brillouin zone limits; close to the Brillouin zone limits, where the electronic gem momenta fulfill the Bragg condition, holes are opened. so instinctively engaging, the almost free-electron guess isn't entirely sensible for genuine solids.

Electrons in Condensed Matter

In atoms, electrons involve discrete quantum states depicted by the quantum numbers n,l,m. These electronic states can be degenerate; states with all over turn have similar energy without an attractive field; now and again, precise force states have similar energy also. In materials, the electronic conditions of the molecules associate with one another to deliver expanded states spread all through the precious stone which is involved by the electrons in the material.

Since there are as large numbers of these states as there were nuclear states from the iotas that make up the gem, there are exceptionally near one another and the energy levels are practically constant over a significant part of the reach.

Nonetheless, certain upsides of the energy are prohibited and we get groups of states with band holes similarly to the cross-section vibrations we considered previously. A sketch appear in (Fig. **2**). It is thus advantageous to utilize a thickness of states portrayal, where now the thickness of states D(E) is the number of electron states in the stretch E and (E+dE). As with iotas, we fill the states arranged by expanding energy, however submitting to the prohibition guideline. At zero temperature, we get (for metals) the situation as demonstrated in the concealed piece of the chart above. The energy of the greatest filled state is known as the Fermi energy. The conditions of higher energy are the unfilled states.

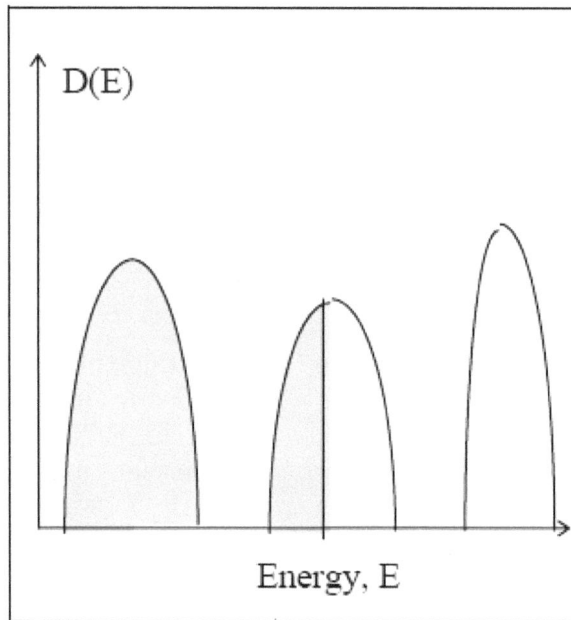

Fig. (2). DOS versus Energy.

Ab Initio Calculations Method

Strong materials and their actual properties are of extraordinary mechanical interest. The actual properties are dictated by the electronic design of the strong. The electronic properties of solids should be depicted by a quantum mechanical treatment. Present-day gadgets in the hardware business give such a model, where the expanded scaling down is one of the key advances. Different applications are found in the space of attractive chronicle or other stockpiling media. One chance to contemplate complex frameworks that contain numerous particles is to perform PC recreations. Estimations for solids by and large (metals, covers, minerals, and so on) can be performed with an assortment of strategies from traditional to quantum mechanical methodologies. The previous are power field or semi-observational plans, in which the powers that decide the collaborations between molecules are defined in such a manner to duplicate a progression of test information. Assuming, in any case, such boundaries are not accessible or if a framework acts in a surprising way, one regularly should depend on stomach muscle initio estimations. They are more requesting as far as PC prerequisites and consequently permit just the treatment of more modest unit cells than what semi-exact computations can do. The upside of first-standards (abdominal muscle initio) strategies lie in the way that they do require just not many test contributions to do such estimations. The way that electrons are indistinct fermions necessitates that their wave capacities should be hostile to symmetric when two electrons are exchanged. Free electrons which submit to the one-electron Schrödinger condition in an intermittent potential are called Bloch electrons and comply with the Bloch hypothesis [42]. The present circumstance prompts the Hartree-Fock and thickness practical hypothesis [43]. The customary plan is the Hartree-Fock (HF) strategy, which depends on a wave work depiction (with one Slater determinant). Trade is dealt with precisely however connection impacts are overlooked by definition. The last can be incorporated by more modern methodologies like setup connection (CI) yet they continuously require more PC time with scaling as extensive as when the framework size (N) develops. As a result, it is simply attainable to concentrate little frameworks which contain a couple of molecules. An elective plan is Density Functional Theory (DFT), which is ordinarily used to figure out the electronic design of complex frameworks containing numerous iotas like huge atoms or solids. It depends on the electron thickness instead of on the wave capacities and treats around both trade and connection. Comparing first standards estimations are mostly done inside Density Functional Theory (DFT), as indicated by which the many-body issue of connecting electrons and cores is planned to a progression of one-electron conditions, the supposed Kohn-Sham conditions [44]. For the arrangement of the KS conditions, a few strategies have been created, with the Linearized-Augmented-Plane-Wave (LAPW) strategy being one among the most precise [45, 46]. The LAPW strategy depends on the purported biscuit tin model of the precious stone potential. The unit cell is separated into two districts by non-covering circles focused on every grid site. Inside the circles, the potential is thought to be roundly symmetric and outside is consistent. In many metals, the district outside the circles is a generally little part of the unit cell volume, and the biscuit tin estimate is excellent. In part 2, we will portray this technique in some detail. In this theory, we do an exhaustive hypothetical examination of electronic and primary properties of $Smco_5$ intensifies dependent on maximum capacity linearized increased plane wave in addition to nearby orbital strategy. Quite possibly the most precise plans of electronic design computations, WIEN2K PC code [47 - 50] has been utilized for this reason. The principle focal point of our hypothetical examinations has been on the estimation of electronic properties of $Smco_5$ and $R_2Fe_{14}B$ utilizing thickness utilitarian hypothesis (DFT) [50 - 54] with summed up slope guess. This exploration work is coordinated as follows: part 2 gives an outline of the fundamental hypothetical information and computational strategies utilized for this reason. Section 3 incorporates consequences of our calculations for electronic band design and thickness of states (DOS) for Nd, Fe and $Smco_5$. Ends have been attracted a different area.

THEORY AND COMPUTATION

What is First-Principles Calculation

In the part of science, the main standards computations – or comparably stomach muscle initio estimations – implies that there are no boundaries to change the outcome aside from the data about the constituent iotas – the nuclear number. Also, the estimations depend on the basic laws of material science instead of on specific model contemplations. Consequently, these computations are not preconditioned by any exploratory discoveries.

Density Functional Theory

Structure of Matter

The advancement of the computation strategies for the electronic construction of materials in the structure of abdominal muscle initio estimation is of incredible premium. This interest comes from the way that it can improve the comprehension of the interaction that decides the naturally visible conduct of issue. The minute conduct of the actual properties of issue is an unpredictable issue. By and large, they manage an assortment of interfacing molecules, which may likewise be influenced by some outside field. This troupe of particles might be in the gas stage (atoms and groups) or in a consolidated stage (solids, surfaces, wires), they could be solids, fluids or nebulous, homogeneous or heterogeneous. Lamentably, the high mathematical precession that can be utilized in little frameworks is frequently difficult to accomplish with regards to compute complex frameworks. To treat such frameworks, a trade off should be found among precision and productivity of the computation. This should be possible by surmised hypotheses, which can be appeared to create dependable outcomes. In this part the thickness utilitarian hypothesis [55 - 57] has become the cutting edge approach for expanded frameworks. The electronic ground condition of any framework, in this hypothesis, can be addressed by its electronic thickness. This technique gives an estimation of actual properties of different frameworks without developing the entire electronic many-body wave work. The beginning stage is the fundamental condition of quantum mechanics, *i.e.*, the Schrödinger condition. It is, thusly, changed to improve on the definitions that can be treated by PC programs. Then, at that point some extra approximations are acquainted with permit the productive estimation of enormous frameworks. At last, the subtleties of the execution that is utilized for this work are clarified. In all cases, the framework is portrayed by various cores and electrons collaborating through coulombic (electrostatic) powers. Officially, the Hamiltonian of such a framework written in the accompanying general structure:

$$\hat{H} = -\sum_{I=1}^{P} \frac{\hbar^2}{2M_I}\nabla_I^2 - \sum_{i=1}^{N} \frac{\hbar^2}{2m}\nabla_i^2 + \frac{e^2}{2}\sum_{I=1}^{P}\sum_{j\neq i}^{P}\frac{Z_I Z_J}{\left|R_I - R_J\right|} + \frac{e^2}{2}\sum_{i=1}^{N}\sum_{j\neq i}^{N}\frac{1}{\left|r_i - r_j\right|} - \frac{e^2}{2}\sum_{I=1}^{P}\sum_{i=1}^{N}\frac{Z_I}{\left|R_I - r_i\right|} \quad (2.1)$$

Where R= $\{R_I\}$, I=1,....., P, is a set of P nuclear coordinates and r = $\{r_i\}$, i=1,..., N, is a set of N electronic coordinates. Z_I and M_I are the P nuclear charges and masses, respectively. All the ingredients are perfectly known and, in principle, all the properties can be derived by solving the many-body Schrödinger equation:

$$\hat{H}\psi_i(r,R) = E_i\psi_i(r,R) \quad (2.2)$$

Where E is the eigenvalue of the many-body Hamiltonian administrator given in nuclear units. Specifically, here the Schrödinger condition is difficult to be decoupled into a bunch of free conditions. This is on the grounds that we are managing 3n+3N coupled level of opportunity. Notwithstanding, a couple and restricted methods have been contrived to take care of this issue. In the accompanying sub-areas, we will depict the most acclaimed approaches. Right off the bat, the adiabatic division of cores and electronic levels of opportunity (adiabatic estimate), which is known as Born-Oppenheimer guess, also; quantum many-body hypothesis (Hartree-Fock guess) and thirdly the Density utilitarian hypothesis (DFT).

Adiabatic (Born- Oppenheimer) Approximation

The Born-Oppenheimer guess [58] depends on the way that the cores are a lot heavier and subsequently much more slowly than the electrons. A further guess is to treat the cores at fixed positions and expect the electrons to be in prompt balance with them. Thusly, the cores follow their elements. The electrons immediately change their wave work as indicated by the atomic wave work.

Subsequent to having applied this estimate on condition (2.1), the cores don't move anymore, their dynamic energy is zero and the initial term vanishes. The third term lessens to steady. So condition (2.1) diminishes to the dynamic energy of the electron gas, the likely energy because of electron-electron collaboration and the possible energy because of electron-cores communications (or outer potential). This officially is composed as:

$$\hat{H} = \hat{T} + \hat{V} + \hat{V}_{ext} \tag{2.3}$$

$$= -\sum_{i=1}^{N} \frac{\hbar^2}{2m} \nabla_i^2 + \frac{e^2}{2} \sum_{i=1}^{N} \sum_{j \neq i}^{N} \frac{1}{|r_i - r_j|} - \frac{e^2}{2} \sum_{I=1}^{P} \sum_{i=1}^{N} \frac{Z_I}{|R_I - r_i|} \tag{2.4}$$

Where

$$\hat{V}_{ext} = -\frac{e^2}{2} \sum_{I=1}^{P} \sum_{i=1}^{N} \frac{Z_I}{|R_I - r_i|} \tag{2.5}$$

The last term of condition (2.4) is the electron-atomic association. It tends to be addressed as a summed up outer potential. The Born Oppenheimer estimation gives a technique to isolate the ionic levels of independence from the electronic ones. This shows that we are left with the issue of tackling the many-body electronic Schrödinger condition for a bunch of fixed atomic positions. In any case, there are as yet important extra improvements and approximations to depict the electrons mathematically.

Quantum Many-Body Theory: Hartree-Fock Approximation

As referenced above, to get data on the construction of the issue it is important to settle the Schrödinger condition for an arrangement of N-interfacing electrons in the outside field that began from an assortment of nuclear cores. Truth be told, the specific arrangement is known uniquely for uncommon cases it is consistently mathematical and it needs a few approximations. The spearheading work that managed this estimate is that proposed by Hartree [59]. In that estimate, the wave work is considered as a direct mix of Slater determinants built from involved and virtual nuclear premise capacities, and the straight coefficients are changed to track down the base of the all-out energy. In this way, the numerous electron wave capacities can be composed as a straightforward result of the one-electron wave work. Every one of them checks a one molecule Schrödinger condition in a compelling potential, which considers the association with different electrons in a mean-field way (we exclude the reliance of the orbitals on R):

$$\phi(R,r) = \Pi_i \phi(r_i) \tag{2.6}$$

$$\left(\frac{-\hbar^2}{2m} \nabla^2 + V_{eff}^{(i)}(R,r) \right) \phi_i(r) = \varepsilon_i \phi_i(r) \tag{2.7}$$

With

$$V_{eff}^{(i)}(R,r) = V(R,r) + \int \frac{\sum\limits_{j \neq i}^{N} \rho_j(r')}{|r - r'|} dr' \tag{2.8}$$

Where

$$\rho_j(r) = |\phi_j(r)|^2 \tag{2.9}$$

Eqn.(2.9) is the electronic thickness related with molecule j. the second term in the right-hand side of condition (2.8) is the old style electrostatic potential produced by the charge dissemination. Notice that this charge thickness does exclude the accuse related to molecule I, so the Hartree estimate is (effectively) self-cooperation free. In this estimate, the energy of the many-body framework isn't only the amount of the eigenvalues of condition (2.7) on the grounds that the detailing as far as a powerful potential tally the electron-electron connection twice. The right articulation for the energy is

$$E_H = \sum_{n=1}^{N} \varepsilon_i - \frac{1}{2} \sum_{i \neq j}^{N} \iint \frac{\rho_i(r)\rho_j(r')}{|r - r'|} dr dr' \tag{2.10}$$

The arrangement of N coupled fractional differential conditions (2.7) can be addressed by limiting the energy regarding a bunch of variational boundaries in a path wave work. Or on the other hand, on the other hand, by recalculating the electronic densities in condition (2.9) utilizing the arrangements of condition (2.7), then, at that point projecting them back into the articulation for the compelling possible condition (2.8), and tackling again the Schrödinger condition. This technique can be rehashed a few times until self-consistency in the info and yield wave capacity or potential is accomplished. This technique is called self-reliable Hartree estimate. The Hartree guess regards the electrons as recognizable particles. A stage forward is to present Pauli Exclusion Principle by proposing an antisymmetrized numerous electron wave work as a Slater determinant:

$$\phi(R,r) = SD\{\phi_j(r_i,\sigma_i)\} \tag{2.11}$$

$$\phi(R,r) = \frac{1}{\sqrt{N_i}} \det[\phi_1(r_1,\sigma_1)\phi_2(r_2,\sigma_2).....\phi_N(r_N,\sigma_N)] \tag{2.12}$$

This wave function introduces particle exchange in an exact manner [60, 61]. The approximation is called Hartree-Fock (HF) or self-consistent field (SCF) approximation. Hartree-Fock (HF) equations look the same as Hartree equations, except for the fact that the exchange integrals introduce additional coupling terms in the differential equations:

$$\left(\frac{-\hbar^2}{2m}\nabla^2 + V(R,r) + \int \frac{\sum\limits_{\sigma',j=1}^{N}\rho_j(r',\sigma')}{|r-r'|}dr\right)\phi_i(r,\sigma) - \sum_{j=1}^{N}\left(\sum_{\sigma'}\int \frac{\phi_j^*(r',\sigma')\phi_i(r',\sigma')}{|r-r'|}dr'\right)\phi_j(r,\sigma) = \sum_{j=1}^{N}\lambda_{ij}\phi_j(r,\sigma) \tag{2.13}$$

The disadvantage of this approximation is that it has a relatively high computational cost, therefore, it is restricted to small systems.

Density Functional Theory

During a similar season of the improvement of the electronic construction hypothesis, Thomas and Fermi proposed a hypothesis that bears their names [62]. They brought up that the full electronic thickness was the major variable of the many-body issue. Further, they determined a differential condition for the thickness without falling back on the one-electron orbital [63, 64]. However, their guess was quite rough since it did exclude trade and connection impacts and couldn't support bound states. Then again, it was the beginning point for the later improvement of thickness utilitarian hypothesis (DFT) which has been the method of the decision in electron structure computations in consolidated matter physical science. In the accompanying passage, we will give a short depiction of this hypothesis. Thickness Functional Theorem (DFT) is thoughtfully not quite the same as the past approaches. In this strategy, the bigger dimensional many-body issue of interfacing electrons are changed into an arrangement of conditions of autonomous electrons. Nonetheless, just the electronic ground state will be thought of. The all-out ground state energy of an inhomogeneous framework made out of n-communicating electrons is given by

$$E = \langle \phi | \hat{T} + \hat{V} + \hat{U}_{ee} | \phi \rangle \tag{2.14}$$

$$E = \langle \phi | \hat{T} | \phi \rangle + \langle \phi | \hat{V} | \phi \rangle + \langle \phi | \hat{U}_{ee} | \phi \rangle \tag{2.15}$$

Where, $|\phi\rangle$ is the n-electron ground state wave function, $\hat{T}$ is the kinetic energy, $\hat{V}$ is the interaction with the external field, and $\hat{U}_{ee}$ is the electron-electron interaction. In fact, the wave function has to include correlations amongst electrons, and its general form is unknown. We are going to concentrate now on the latter, which is the one that introduces many-body effects

$$\hat{U}_{ee} = \langle \phi | \hat{U}_{ee} | \phi \rangle = \langle \phi | \frac{1}{2} \sum_{i=1}^{N} \sum_{j \neq i}^{N} \frac{1}{|r_i - r_j|} | \phi \rangle \tag{2.16}$$

With

$$\rho_2(r,r') = \frac{1}{2} \sum_{\sigma,\sigma'} \langle \phi | \psi_\sigma^+(r) \psi_{\sigma'}^+(r') \psi_{\sigma'}(r') \psi_\sigma(r) | \phi \rangle \tag{2.17}$$

is a two-body density matrix expressed in real space, where ψ and ψ^+ are the creation and annihilation operators for electrons. These operators obey the anti-commutation relations $\{\psi_\sigma(r), \psi_{\sigma'}^+(r)\} = \delta_{\sigma\sigma'}\delta(r - r')$. The two-body direct correlation functions $g(r,r')$ is defined in the following way:

$$\rho_2(r,r') = \frac{1}{2} \rho(r,r)\rho(r',r')g(r,r') \tag{2.18}$$

Where $\rho_2(r,r')$ is the one-body density matrix (in real space) whose diagonal elements $\rho(r) = \rho(r,r)$ correspond to the electron density.

The one-body density matrix is defined as

$$\rho(r,r') = \sum_{\sigma} \rho_{\sigma}(r,r') \tag{2.19}$$

$$\rho_{\sigma}(r,r') = \langle \phi | \psi_{\sigma}^{+}(r)\psi_{\sigma}(r') | \phi \rangle \tag{2.20}$$

With this definition, the electron-electron interaction is written as

$$U_{ee} = \frac{1}{2} \int \frac{\rho(r)\rho(r')}{|r-r'|} dr dr' + \frac{1}{2} \int \frac{\rho(r)\rho(r')}{|r-r'|} [g(r,r')-1] dr dr' \tag{2.21}$$

The underlying term of eq. (2.21) is the old style electrostatic correspondence energy contrasting with a charge spread. The ensuing term joins association effects of both conventional and quantum starting. Basically, considers the way that the presence of an electron at r weaken an ensuing electron to be arranged at a position incredibly close to r because of the coulomb repulsiveness. By the day's end, it says that the probability of finding two electrons at r ~ r' is lessened concerning the probability of finding them at unfathomable detachment from each other. Exchange further decreases this probability because of electrons have a comparative turn projection, inferable from the Pauli dismissal. To appreciate the effect of the exchange, let us imagine that we stay on an electron with wind and we look at the thickness of the other (N-1) electrons. Pauli rule blocks the presence of electrons with a bend toward the start, yet it says nothing in regards to electrons with turn, which can flawlessly be arranged toward the start. Along these lines,

$$g_x(r,r') \to \frac{1}{2} \qquad \text{For} \qquad r \to r' \tag{2.22}$$

Using the HF theory eq. (2.21) for the electron-electron interaction can be written as

$$U_{ee}^{HF} = \frac{1}{2} \int \frac{\rho^{HF}(r)\rho^{HF}(r')}{|r-r'|} dr dr' + \frac{1}{2} \int \frac{\rho^{HF}(r)\rho^{HF}(r')}{|r-r'|} \left[-\frac{\sum_{\sigma} |\rho_{\sigma}^{HF}(r,r')|^2}{\rho^{HF}(r)\rho^{HF}(r')} \right] dr dr' \tag{2.23}$$

Meaning that the exact expression for the exchange depletion (also called exchange hole) is

$$g_x(r,r') = 1 - \frac{\sum_{\sigma} \left| \rho_{\sigma}^{HF}(r,r') \right|^2}{\rho^{HF}(r)\rho^{HF}(r')} \qquad (2.24)$$

The thickness and thickness cross-section are resolved from the HF ground state Slater determinant. The assessment of the relationship opening is a critical issue in many-body theories and, up to the present, it is an open issue in the general example of an inhomogeneous electron gas. The particular responsibility for the homogeneous electron gas is known numerically [55, 56]. There are a couple of evaluations that go past quite far by including bit by bit moving densities through its spatial slants (tendencies cures) and moreover enunciations for the exchange relationship energy that target considering amazingly feeble,non-close by associations of the Vander-Waals type(dispersion interaction) [56]. The energy of the many-body electronic system would then have the option to be written in a going with way:

$$E = T + V + \frac{1}{2} \int \frac{\rho(r)\rho(r')}{|r - r'|} dr\,dr' + E_{XC} \qquad (2.25)$$

Where

$$V = \sum_{I=1}^{P} \left\langle \phi \left| \sum_{i=1}^{N} v(r_i - R_I) \right| \phi \right\rangle = \sum_{I=1}^{P} \int \rho(r)v(r - R_I)dr \qquad (2.26)$$

$$T = \left\langle \phi \left| -\frac{\hbar^2}{2m} \sum_{i=1}^{N} \nabla_i^2 \right| \phi \right\rangle = -\frac{\hbar^2}{2m} \int \left[\nabla_r^2 \rho(r,r') \right]_{r'=r} dr \qquad (2.27)$$

And E_{xc} is the exchange and correlation energy

$$E_{XC} = \frac{1}{2} \int \frac{\rho(r)\rho(r')}{|r - r'|} [g(r,r') - 1] dr\,dr' \qquad (2.28)$$

Hohenberg-Khon Theorem

In 1964, Hohenberg and Kohn [66] figured and demonstrated two hypotheses that put the previous thoughts on strong numerical ground. The principal hypothesis is "The outside potential is univocally controlled by the electronic thickness aside from a paltry added substance constant". According to the primary hypothesis, clearly a given many-electron framework has an extraordinary outer potential, which by the Hamiltonian and the Schrödinger condition yields an exceptional many-molecule ground-state wave work. From this wave work, the comparing electron thickness is effectively found. An outer potential henceforth leads in a clear cut manner to an exceptional ground-state thickness comparing to it. Yet, instinctively it would appear that the thickness contains less data than the wave work. On the off chance that this would be valid, it would not be feasible to track down a novel outer potential if just a ground-state thickness is given. The main hypothesis of Hohenberg and Kohn tells precisely that this is conceivable! The thickness contains as much data as the wave work does (for example all that you might actually think about a particle, atom or strong). All noticeable amounts can be recovered along these lines in an interesting manner from the thickness just, for example, they can be composed as functionals of the thickness. The confirmation of the principal hypothesis, assume the inverse to hold, that the potential isn't univocally dictated by the thickness. Then one would be able to find two potentials v, v' such that their ground state density ρ is the same. Let ψ and $E_0 = \langle \psi | \hat{H} | \psi \rangle$ be the ground state and ground state energy of $\hat{H} = \hat{T} + \hat{U} + \hat{V}$, ψ' and $E_0' = \langle \psi' | \hat{H}' | \psi' \rangle$ the ground state and ground state energy of $\hat{H}' = \hat{T} + \hat{U} + \hat{V}'$. Owing to the variational principle, we have

$$E_0 < \langle \psi' | \hat{H} | \psi' \rangle = \langle \psi' | \hat{H}' | \psi' \rangle + \langle \psi' | \hat{H} - \hat{H}' | \psi' \rangle$$

$$= E_0' + \int \rho(r)(v(r) - v'(r)) dr \tag{2.29}$$

Where we have also used the fact that different Hamiltonians have necessarily different ground states $\psi \neq \psi'$. This is straightforward to show since the potential is a multiplicative operator. Now we can simply reverse the situation ψ and ψ'

(H and H') and readily obtained

$$E_0' < \langle \psi | \hat{H}' | \psi \rangle = \langle \psi | \hat{H} | \psi \rangle + \langle \psi | \hat{H}' - \hat{H} | \psi \rangle$$

$$= E_0 - \int \rho(r) v(r) - v'(r) dr \tag{2.30}$$

Adding these two inequalities, it turns out that $E_0 + E'_0 < E'_0 + E_0$, which is absurd. Therefore, there are no $v(r) \neq v'(r)$ that correspond to the same electronic density for the ground state. The second theorem establishes a variational principle;

" *For any positive definite trial density, ρ, such that $\int \rho(r) dr = N$ then $E[\rho] \geq E_0$* "

Or

"*Let $\tilde{\rho}(r)$ be a non-negative density normalized to N. Then $E_0 < E_v[\tilde{\rho}]$, for*

$$E_v[\tilde{\rho}] = F_{HK}[\tilde{\rho}] + \int \tilde{\rho}(r) v(r) dr \text{ with } F_{HK}[\tilde{\rho}] = \langle \psi[\tilde{\rho}] | \hat{T} + \hat{U} | \psi[\tilde{\rho}] \rangle \text{ where } \psi[\tilde{\rho}] \text{ is}$$

the ground state of a potential that has $\tilde{\rho}$ as its ground state density."

As indicated by the subsequent hypothesis, all-inclusiveness is effortlessly recorded by utilizing the thickness administrator, and assuming the ground-state thickness is known, the commitment to the complete energy from the outside potential can be by and large determined. An unequivocal articulation for the Hohenberg and Kohn practical FHK isn't known. In any case, on the grounds that FHK doesn't contain data on the cores and their positions, it

is a widespread use for any many-electron framework. This implies that on a basic level an articulation exists which can be utilized for each particle, atom or strong which can be envisioned.

The evidence of the subsequent hypothesis, we have

$$\langle \psi[\tilde{\rho}] | \hat{H} | \psi[\tilde{\rho}] \rangle = F_{HK}[\tilde{\rho}] + \int \tilde{\rho}(r) v(r) dr$$

$$= E_v[\tilde{\rho}] \geq E_v[\rho] = E_0 = \langle \psi | \hat{H} | \psi \rangle \tag{2.31}$$

The inequality follows from Rayleigh-Ritz's variational principle for the wave function but is applied to the electronic density. Therefore, the variational principle says

$$\delta \left\{ E_v[\rho] - \mu \left(\int \rho(r) dr - N \right) \right\} = 0 \tag{2.32}$$

And a generalized Thomas-Fermi (TF) equation is obtained

$$\mu = \frac{\delta E_v[\rho]}{\delta \rho} = v(r) + \frac{\delta F[\rho]}{\delta \rho} \tag{2.33}$$

The information infers that one has tackled the full many-body Schrödinger condition. It must be commented that is general use that doesn't rely expressly upon the outside potential. It relies just upon the electronic thickness. In the Hohenberg and Kohn plan, where is the ground state wave work. These two hypotheses structure the premise of DFT.

In Hohenberg and Kohn hypothesis the electronic thickness decides the outer potential, yet it is additionally required that the thickness compares to some ground state antisymmetric wave capacity, and this isn't generally the situation. Notwithstanding, DFT can be reformulated so that this isn't required, by speaking to the compelled search strategy [67]. By defining. For non-negative densities to such an extent that

$\int \rho\,(r)dr = N$ and $\int \left|\nabla \rho^{\frac{1}{2}}(r)\right|^2 dr < \infty$, which arise from an antisymmetric wave function, the search is constrained to the subspace of all the antisymmetric ψ that give rise to the same density ρ. Using DFT, one can determine the electronic

ground state density and energy exactly, provided that $F[\rho]$ is known. A common misleading statement is that DFT is a ground-state theory and that the question of excited states can not be addressed within it. This is actually an incorrect statement because the density determines univocally the potential, and this is, in turn, determines univocally the many-body wave functions, ground and excited states, provided that the full many-body Schrödinger equation is solved. For the ground state, such a scheme was devised by Kohn and Sham and will be discussed in the next subsection.

Khon-Sham Equation

Kohn and Sham [68, 69] proposed the idea of replacing the kinetic energy of the interacting electrons with that of an equivalent non-interacting system. This is because the latter can be calculated. The density matrix $\rho(r,r')$ that derives from the (interacting) ground state is the sum of the spin-up and spin-down density matrices, $\rho(r,r') = \sum_s \rho_s(r,r')$ (s=1,2). The latter can be written as

$$\rho_s(r,r') = \sum_{i=1}^{\infty} n_{i,s}\phi_{i,s}\phi_{i,s}^*(r') \tag{2.34}$$

Where $\{\phi_{i,s}(r)\}$ are the natural spin orbitals and $\{n_{i,s}\}$ are the occupation numbers of these orbitals. The kinetic energy can be written exactly as

$$T = \sum_{s=1}^{2} \sum_{i=1}^{\infty} n_{i,s} \left\langle \phi_{i,s} \left| -\frac{\nabla^2}{2} \right| \phi_{i,s} \right\rangle \tag{2.35}$$

In the following, we shall assume that the equivalent non-interacting system, that is, a system of non-interacting whose ground state density coincides with that of the interacting system, does exist. We shall call this the non-interacting reference system of density $\rho(r)$, which is described by the Hamiltonian

$$\hat{H}_R = \sum_{i=1}^{R} \left(-\frac{\nabla_i^2}{2} + v_R(r_i) \right) \tag{2.36}$$

Where the potential $v_R(r_i)$ is such that the ground state density of $\hat{H}_R$ equals $\rho(r)$ and the ground state energy equals the energy of the interacting system. This Hamiltonian has no electron-electron interactions and, thus, its eigenstates can be expressed in the form of Slater determinants

$$\psi_s(r) = \frac{1}{\sqrt{N_i}} SD\left[\phi_{1,s}(r_1) \phi_{2,s}(r_2) \ldots \phi_{N_s,s}(r_{N_s}) \right] \tag{2.37}$$

Where we have chosen the occupation numbers to be 1 for $i \le N_s$ (s=1,2) and 0 for $i > N_s$. This means that the density is written as

$$\rho(r) = \sum_{s=1}^{2} \sum_{i=1}^{N_s} \left| \phi_{i,s}(r) \right|^2 \tag{2.38}$$

While the kinetic term is

$$T_R[\rho] = \sum_{s=1}^{2} \sum_{i=1}^{N_s} \left\langle \phi_{i,s} \left| -\frac{\nabla^2}{2} \right| \phi_{i,s} \right\rangle \tag{2.39}$$

The single-particle orbitals $\{\phi_{i,s}(r)\}$ are the N_s lowest eigenfunctions of

$\hat{h}_R = -\dfrac{\nabla^2}{2} + v_R(r)$, that is,

$$\left\{ \dfrac{-\nabla^2}{2} + v_R(r) \right\} \phi_{i,s}(r) = \varepsilon_{i,s}\phi_{i,s}(r) \tag{2.40}$$

Using $T_R[\rho]$, the universal density functional can be rewritten in the following form:

$$F[\rho] = T_R[\rho] + \dfrac{1}{2} \iint \dfrac{\rho(r)\rho(r')}{|r-r'|} dr\, dr' + E_{XC}[\rho] \tag{2.41}$$

Where this equation defines the exchange and correlation energy as a functional of the density. The fact that $T_R[\rho]$ is the kinetic energy of the non-interacting reference system implies that the correlation piece of the true kinetic energy has been ignored and has to be taken into account somewhere else. Upon substitution of this expression for F in the total energy functional $E_v[\rho] = F[\rho] + \int \rho(r)v(r)dr$, the latter is usually renamed the KS functional:

$$E_{KS}[\rho] = T_R[\rho] + \int \rho(r)v(r)dr + \dfrac{1}{2} \iint \dfrac{\rho(r)\rho(r')}{|r-r'|} dr\, dr' + E_{XC}[\rho] \tag{2.42}$$

In this way, we have expressed the density functional in terms of the $N = N_\uparrow + N_\downarrow$ orbitals (KS orbitals), which minimize the kinetic energy under the fixed density constraint. In principle, these orbitals are a mathematical objects constructed in order to render the problem more tractable and do not have a sense by themselves, but only in terms of density. The KS orbitals satisfy equation (2.40) while the problem is to determine the effective potential v_R, or v_{eff} as it is also known. This can be done by minimizing the KS functional over all densities that integrate to N particles. For the minimize (*i.e.*, correct) density ρ, we have

$$\dfrac{\delta T_R[\rho]}{\delta\rho(r)} + v(r) + \int \dfrac{\rho(r')}{|r-r'|} dr' + \dfrac{\delta E_{XC}}{\delta\rho(r)} = \mu_R \tag{2.43}$$

The functional derivative $\delta T_R[\rho]/\delta\rho(r)$ can be quickly found by considering the non-interacting Hamiltonian $\hat{H}_R$ (Eqn. 2.36). Its ground state energy is E_0 . We can construct the functional

$$E_{VR}[\tilde{\rho}] = T_R[\tilde{\rho}] + \int \tilde{\rho}(r)v_R(r)dr \qquad (2.44)$$

Then, clearly $E_{VR}[\tilde{\rho}] \geq E_0$, and only for the correct density ρ we will have $E_{VR}[\tilde{\rho}] = E_0$. Hence, the functional derivative of $E_{VR}[\tilde{\rho}]$ must be vanish for the correct density leading to

$$\frac{\delta T_R[\rho]}{\delta\rho(r)} + v_R(r) = \mu_R \qquad (2.45)$$

Where μ_R is the chemical potential for the non-interacting system. To summarize, the KS orbitals satisfay the well-known self-consistent KS equations

$$\left\{\frac{-\nabla^2}{2} + v_{eff}(r)\right\}\phi_{i,s}(r) = \varepsilon_{i,s}\phi_{i,s}(r) \qquad (2.46)$$

Where, by comparison of expressions (2.43) and (2.45), the effective potential v_R or v_{eff} is given by

$$v_{eff}(r) = v(r) + \int \frac{\rho(r')}{|r - r'|}dr' + \mu_{XC}[\rho](r) \qquad (2.47)$$

And the electronic density is constructed with KS orbitals

$$\rho(r) = \sum_{i=1}^{N_s}\sum_{s=1}^{2}\left|\phi_{i,s}(r)\right|^2 \qquad (2.48)$$

The exchange-correlation potential $\mu_{XC}[\rho](r)$ defined above is simply the functional derivative of the exchange-correlation energy $\frac{\delta E_{XC}[\rho]}{\delta\rho(r)}$.

Exchange and Correlation Potentials

If the exact expression for the kinetic energy including correlation effects, $T[\rho]=\langle\psi[\rho]|\hat{T}|\psi[\rho]\rangle$ (with $\psi[\rho]$ being the interacting ground state of the external potential that has ρ as the ground state density), were known, then we could use the original definition of the exchange-correlation energy that does not contain kinetic contributions as follows:

$$E_{XC}^0 = \frac{1}{2}\iint\frac{\rho(r)\rho(r')}{|r-r'|}[g(r,r')-1]drdr' \tag{2.49}$$

Since we are using the non-interacting expression for the kinetic energy $T_R[\rho]$, we have to redefine it in the following way:

$$E_{XC}[\rho]= E_{XC}^0[\rho]+T[\rho]-T_R[\rho] \tag{2.50}$$

It can be shown that the kinetic contribution to the correction energy can be taken into account by averaging the pair correlation function $g(r,r')$ over the strength of the electron-electron interaction that is

$$E_{XC}[\rho]= \frac{1}{2}\iint\frac{\rho(r)\rho(r')}{|r-r'|}[\tilde{g}(r,r')-1]drdr' \tag{2.51}$$

Where

$$\tilde{g}(r,r')=\int_0^1 g_\lambda(r,r')d\lambda \tag{2.52}$$

And $g_\lambda(r,r')$ is the pair correlation function corresponding to the Hamiltonian $\hat{H} = \hat{T} + \hat{V} + \lambda\hat{U}_{ee}$ [15].

2.2.4.3.1. The Local Density Approximation

The LDA has been for a long time the most widely used approximation to the exchange-correlation energy. The main idea is to consider general inhomogeneous electronic systems as locally homogeneous, and then to use the exchange-correlation hole corresponding to the homogeneous electron gas for which there are very good approximations and also exact numerical (quantum Monte Carlo) results [70]. This means that

$$\widetilde{\rho}_{XC}^{LDA}(r,r') = \rho(r)\Big(\widetilde{g}^h\big[\big|r-r'\big|,\rho(r)\big]-1\Big) \qquad (2.53)$$

With $\widetilde{g}^h\big[\big|r-r'\big|,\rho(r)\big]$ the pair correlation function of the homogeneous gas, which depends only on the distance between r and r', evaluated at the density ρ^h, which locally equals $\rho(r)$. Within this approximation, the exchange-correlation energy density is defined as

$$\varepsilon_{XC}^{LDA}[\rho] = \frac{1}{2}\int \frac{\widetilde{\rho}_{XC}^{LDA}(r,r')}{|r-r'|}dr' \qquad (2.54)$$

And the exchange-correlation energy becomes

$$E_{XC}^{LDA}[\rho] = \int \rho(r)\varepsilon_{XC}^{LDA}[\rho]dr \qquad (2.55)$$

2.2.4.3.2. The Local Spin Density Approximation

In magnetic systems or, in general, in systems where open electronic shells are involved, better approximations to the exchange-correlation functional can be obtained by introducing the two spin densities, $\rho_\uparrow(r)$ and $\rho_\downarrow(r)$ such that $\rho(r) = \rho_\uparrow(r) + \rho_\downarrow(r)$ and $\varepsilon(r) = (\rho_\uparrow(r) - \rho_\downarrow(r))/\rho(r)$ is the magnetization density. The non-interacting kinetic energy Eq. (2.39) splits trivially into spin-up and spin-down contributions, and the external and Hartree potential depends on the full density $\rho(r)$, but the approximate XC functional- even if the exact functional should depend only on $\rho(r)$ -will depend on both spin densities independently, $E_{XC} = E_{XC}[\rho_\uparrow(r), \rho_\downarrow(r)]$. KS equations then read exactly as in eq. (2.46), but the effective potential $V_{eff}(r)$ now acquires a spin index:

$$V_{eff}^\uparrow = v(r) + \int \frac{\rho(r')}{|r-r'|}dr' + \frac{\delta E_{XC}[\rho_\uparrow(r),\rho_\downarrow(r)]}{\delta\rho_\uparrow(r)} \qquad (2.56)$$

$$V_{eff}^{\downarrow} = v(r) + \int \frac{\rho(r')}{|r-r'|}dr' + \frac{\delta E_{XC}\left[\rho_{\uparrow}(r),\rho_{\downarrow}(r)\right]}{\delta\rho_{\downarrow}(r)} \tag{2.57}$$

The density given by expression eq.(2.48) contains a double summation over the spin states and over the number of electrons in each spin state, N_s. The latter have to be determined according to the single-particle eigenvalues by asking for the lowest $N = N_{\uparrow} + N_{\downarrow}$ to be occupied. This defines a Fermi energy E_f such that the occupied eigenstates have $\varepsilon_{i,s} < \varepsilon_f$. In the case of non-magnetic systems, $\rho_{\uparrow}(r) = \rho_{\downarrow}(r)$ everything reduces to the simple case of double occupancy of the single-particle orbitals. The equivalent of the LDA in spin-polarized systems is the local spin density approximation(LSDA).Which basically consists of replacing the XC energy density with a spin –polarized expression:

$$E_{XC}^{LSDA}\left[\rho_{\uparrow}(r),\rho_{\downarrow}(r)\right] = \int \left[\rho_{\uparrow}(r) + \rho_{\downarrow}(r)\right]\varepsilon_{xc}^{h}\left[\rho_{\uparrow}(r),\rho_{\downarrow}(r)\right]dr \tag{2.58}$$

Obtained, for instance, by interpolating between the fully polarized and fully unpolarized *XC* energy densities using an appropriate expression that depends on $\varepsilon(r)$. The LDA is very successful an approximation for many systems of interest, especially those where the electronic density is quite uniform such as bulk metals, but also for less uniform systems as semiconductors and ionic crystals. There are, however,a number of known features that the LDA fails to reproduce considering the way this approximation has been obtained, it is obvious that for a uniform system, it is exact. Further, it is expected to be still valid for slowly varying electronic density. However, in other cases, its behaviour is not well controlled.

The Generalized Gradient Approximation

Undoubtedly, and probably because of its computation efficiency and its similarity to the LDA, the most popular approach has been introducing semi-locally the in homogeneities of the density, by expanding $E_{XC}[\rho]$ as series in terms of the density and its gradients. This approach is known as generalized gradient approximation (GGA). Whether gradient corrections are an improvement over LDA or not is still under debate. The exchange-correlation energy has a gradient expansion of the type

$$E_{XC}[\rho] = \int A_{XC}[\rho]\rho(r)^{4/3}\,dr + \int C_{XC}[\rho]|\nabla\rho(r)|^{2}/\rho(r)^{4/3}\,dr + \ldots\ldots \tag{2.59}$$

Which is asymptotically valid for densities that vary slowly in space. The basic idea of GGA is to express the exchange-correlation energy in the following form:

$$E_{xc}[\rho] = \int \rho(r)\varepsilon_{XC}[\rho(r)]dr + \int F_{XC}[\rho(r), \nabla\rho(r)]dr \qquad (2.60)$$

Where the function F_{xc} is asked to satisfy a number of formal conditions for the exchange-correlation hole, such as sum rules, long rang decay and so on. This can not be done by considering directly the bare gradient expansion (2.59). what is needed from the functional is a form that mimics a re-summation to infinite order, and this is the main idea of the GGA, for which there is not a unique recipe. Naturally, not all the formal properties can be enforced at the same time, and this differentiates one functional from another. A thorough comparison of different GGA can be found in reference [71].

LDA+U Total Energy Functional

In the context of strongly correlated systems, for example, those exhibiting narrow d and f bands, where the limitation of the LDA is at describing strong on-site correlations of the Hubbard type, these features have been introduced within the so-called LDA+U approach [72].This theory considers the mean-field solution of the Hubbard model on top of the LDA solution, where the Hubbard on-site interaction U are computed for the d or f orbitals by differentiating the LDA eigenvalues with respect to the occupation numbers [73].

Different Method of Calculating Energy Bands

Solving the Basis Set

$$\left\{\frac{-\nabla^2}{2} + v_{ext}(r) + \int \frac{\rho(r')}{|r-r'|}dr' + v_{XC}[\rho](r)\right\}\phi_{i,s}(r) = \varepsilon_{i,s}\phi_{i,s}(r) \qquad (2.61)$$

For DFT, $v_{XC}[\rho](r)$ is the exchange-correlation operator in the LSDA, GGA or another approximation. Exchange and correlation are both treated, but both approximately. The ϕ_m are mathematical single-particle orbitals. 'Solving' in most methods means that we want to find the coefficient C_p^m needed expressed ϕ_m in a given basis set ϕ_p^b.

$$\phi_m = \sum_{p=1}^{P} C_p^m \phi_p^b \qquad (2.62)$$

The wave functions ϕ_m belong to a function space that has an infinite dimension; P is therefore in principle infinite. In practice, one works with a limited set of basis functions. Such a limited basis will never be able to describe ϕ_m exactly, but one could try to find a basis that can generate a function that is 'close' to ϕ_m. What is a good basis set? If the functions of the basis set are very similar to ϕ_m, one needs only a few of them to accurately describe the wave function, and hence P and the matrix size are small. Such a basis set is called efficient. However, this assumes that you know the solution to your problem almost before you start solving it. Such a basis set can therefore never be very general: For some specific problems it will very quickly yield the solution, but for the majority of cases it will poorly describe the eigenfunctions. In the latter cases the required P is much higher than what is affordable, and limiting P would lead to approximate eigenfunctions that are not acceptable. These approximations carry too many properties from the basis function, and such a basis set is therefore called biased. The art of theoretical condensed matter physics is to find a basis set that is simultaneously efficient and unbiased. Two families of basis sets will be described-plane waves and augmented plane waves.

Solving the Basis Set by (Plane Wave)

2.4.2.1. The Pseudo Potential Method

We formulated two principle requirements for a basis set in which we want to expand the eigenstates of the solid-state Hamiltonian: the basis set should be unbiased and efficient. Furthermore, it would be nice if the basis functions are mathematically simple. This makes both theory development and programming work easier. A basis set that is certainly unbiased and simple is the plane-wave basis set. From Bloch theorem any eigenfunction $\psi_{\vec{k}}^{n}$ of a periodic Hamiltonian can be expressed exactly in this basis set by means of an infinite set of coefficients $C_{\vec{K}}^{n,\vec{k}}$:

$$\psi_{\vec{k}}^{n}(\vec{r}) = \sum_{\vec{K}} C_{\vec{K}}^{n,\vec{k}} e^{i(\vec{k}+\vec{K})\vec{r}} \qquad (2.63)$$

This has to be compared with the general formulation in eq. (2.62), where m stands for $(n, \vec{k})$ and p for $\vec{k} + \vec{K}$. One basis function for $\psi_{\vec{k}}^n(\vec{r})$ or $\vec{K}$ is, therefore:

$$\psi_{\vec{K}}^{\vec{k}}(\vec{r}) = \left| \vec{K} \right\rangle = e^{i(\vec{k} + \vec{K}) \cdot \vec{r}} \tag{2.64}$$

Note that this basis set is $\vec{K}$-dependent: all eigenstates $\psi_{\vec{k}}^n$ that have the same $\vec{K}$ but a different n will be expressed in the basis set with this particular value of $\vec{K}$. For eigenstates with another $\vec{K}$, a new basis set using that other $\vec{K}$ has to be used. Should we conclude now that a plane wave basis set can not be used? The most oscillating part of the wave functions is the tails that reach out into the region close to the nucleus. But this region of the solid is quite shielded from the more outer regions of the atoms where chemistry happens, and the electrons will not behave very differently from free atom electrons here. One can therefore replace the potential in these inner regions by a pseudo-potential that is designed to yield very smooth tails of wave functions inside the atom. More to the outer regions of the atoms, the pseudo-potential continuously evolves into the true potential, such that this region of the crystal behaves as if nothing happened. Although the pseudo potential method is extremely useful, there are reasons why alternatives could be attractive. Is the introduction of the pseudo potential completely innocent? What do you do if you are interested in information that is inherently contained in the region near the nucleus (hyperfine fields for instance, or core level excitations)? Can the basis set be made more efficient. Therefore, we will search for a basis set that uses other functions than plane waves and that does not require the introduction of a pseudo potential. Such a basis set will have to be more efficient, but of course we do not want it to be biased. This basis will be the Augmented Plane Wave (APW) basis set.

The Family of Augmented Plane Wave Method

2.3.3.1. Augmented Plane Wave Method (APW)

In 1937 Slater [74], constructed a set of basis functions, called augmented plane waves using the so-called muffin-tin approximation as a starting point. They consist of plane waves in regions of slowly varying potential but transform into atomic orbital like functions, as soon as the potential demands a description of faster varying wave functions.

Basis Function in a Muffin-tin Potential

The potential in a strong gem shifts consistently all through space, yet two significant districts can be seen. In the biscuit tin guess the gem is separated into the biscuit tin (MT) locale, comprising of non-covering circles focused around each nuclear site, and the encompassing space called the interstitial (I). While the potential is circularly symmetric in the biscuit tin area, it will be genuinely level in the interstitial.

The muffin-tin approximation to the true crystal potential is therefore defined as

$$V_{MT}(r) \equiv \begin{cases} cons\tan t & r \in I \\ V(r_\alpha) & r \in S_\alpha \end{cases} \qquad (2.65)$$

Where α enumerates the muffin-tin spheres, and r_α is the length of the local position vector $r_\alpha = r - \tau_\alpha$, pointing from the sphere centre τ_α as shown in Fig. (1). To start with, let us look at the extreme case of a non-interacting electron gas in an otherwise empty Bravais lattice. The solutions describing this system of non-interacting particles in a constant external potential are the mathematically simple plane waves, $e^{ik.r}$. As we start to insert nuclei into the crystal structure, the constant potential will be repeatedly broken by muffin-tin spheres with a spherical potential. However, very conveniently, the plane waves satisfy the Bloch condition and can therefore still serve as basis functions in this new periodic potential. Consequently, Slater [74] chose plane waves as basis functions in the interstitial region. Inside the muffin-tin spheres, an eigenstate is better described by the solutions to a spherical potential,

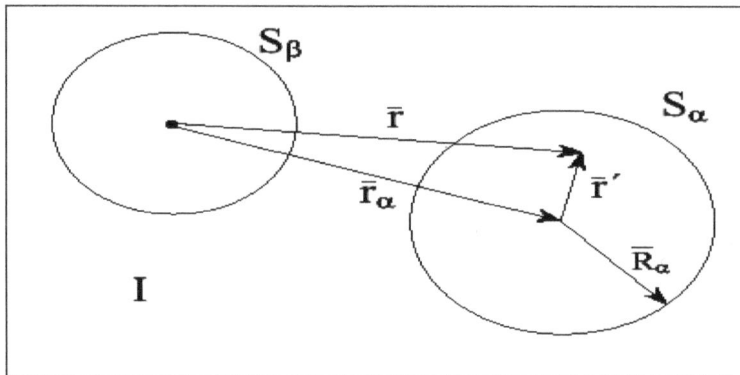

Fig. (1). Division of a unit cell in muffin-tin regions and the interstitial region, for a case with two in-equivalent atoms.

$$u_l\left(r_\alpha, E\right)Y_l^m\left(\hat{r}_\alpha\right) \tag{2.66}$$

Where Y_l^m is the spherical harmonic function of angular momentum quantum numbers l and m, and $\hat{r}_\alpha$ is the angular part of the local vector r_α. The function U_l satisfies the radial Schrödinger equation,

$$-\frac{1}{r^2}\frac{d}{dr}\left(r^2\frac{du_l}{dr}\right) + \left[\frac{l(l+1)}{r^2} + V(r)\right]u_l = Eu_l \tag{2.67}$$

with the spherical potential V from eq.(2.65). The only boundary condition to eq.(2.67), for a fixed energy E, is that U_l *(r)* should be well defined at r=0. Each plane wave is now augmented inside the muffin-tin spheres, by a linear combination of the solutions described by eq .(2.66). Consequently, one augmented plane wave (APW) used in the expansion of $\psi_{\vec{k}}^n$ is defined as

$$\phi_{\vec{K}}^{\vec{k}}\left(\vec{r}, E\right) = \begin{cases} \dfrac{1}{\sqrt{V}}e^{i\left(\vec{k}+\vec{K}\right)\vec{r}_\alpha} & \vec{r} \in I \\[2mm] \sum\limits_{l,m} A_{l,m}^{\alpha,\vec{k}+\vec{K}}u_l^\alpha\left(r', E\right)Y_l^m\left(\hat{r}\right) & \vec{r} \in S_\alpha \end{cases} \tag{2.68}$$

V is volume of the unit cell. Note that, the APW basis set is K-dependent, as was the plane wave basis set. The position inside the spheres is given with respect to the center of each sphere by $\vec{r}' = \vec{r} - \vec{r}_\alpha$. The length of $\vec{r}'$ is r', and angle θ' and ϕ' specifying the direction of $\vec{r}'$ in spherical coordinates are indicated as $\vec{r}'$. The Y_l^m are spherical harmonics. The $A_{lm}^{\alpha,\vec{k}+\vec{K}}$ are yet undetermined parameter as is E. The latter has the dimension of energy. The u_l^α are solutions to the radial part of the Schrödinger equation for a free atom α, and this at the energy E. For a true free atom, the boundary condition that $u_l^\alpha\left(r, E\right)$ should vanish for $r \to \infty$, limits the number of energies E for which solution u_l^α can be found. But as this boundary condition does not apply here, we can find a numerical solution for any E. Hence, the u_l^α themselves are only part of a basis function, not of the searched eigen-function itself. If an eigen function would be discontinuous, its kinetic energy would not be well –defined. Such a situation can therefore never happen, and we have to require that the plane wave outside the sphere matches the function inside the sphere over the complete surface of the sphere (in value, not in slope). A plane wave is oscillating and has a unique direction built in, how can it match another function based on spherical harmonics over the entire surface of a sphere. To see how this is possible, we expand the plane wave in spherical harmonics about the origin of the sphere of atom α.

$$\frac{1}{\sqrt{V}}e^{i(\vec{k}+\vec{K})\vec{r}} = \frac{4\pi}{\sqrt{V}}e^{i(\vec{k}+\vec{K})\vec{r}_\alpha}\sum_{l,m}i^l j_l\left(\left|\vec{k}+\vec{K}\right|\left|\vec{r}'\right|\right)Y_l^{m^*}\left(\vec{k}+\vec{K}\right)Y_l^m\left(\vec{r}'\right) \quad \textbf{(2.69)}$$

j_i (x) is the Bessel function of order ℓ. Requiring this at the sphere boundary to be equal to ℓm –part of eq. (2.68) easily yield:

$$A_{lm}^{\alpha,\vec{k}+\vec{K}} = \frac{4\pi i^l e^{i(\vec{k}+\vec{K})\vec{r}_\alpha}}{\sqrt{V}u_l^\alpha\left(R_\alpha,E\right)}j_l\left(\left|\vec{k}+\vec{K}\right|R_\alpha\right)Y_l^{m^*}\left(\vec{k}+\vec{K}\right) \quad \textbf{(2.70)}$$

This uniquely defines the $A_{lm}^{\alpha,\vec{k}+\vec{K}}$, a part from the still undetermined E. In principle, there are an infinite number of terms in eq.(2.69) which would force us to use an infinite number of $A_{lm}^{\alpha,\vec{k}+\vec{K}}$ in order to create the matching. The condition $R_\alpha K_{max} = l_{max}$ allows to determine a good l_{max} for a given k_{max}. A finite value for l_{max} means that for each APW the matching at the sphere boundaries will not be exact, but good enough to work with. At first sight, it looks like we can now use the APW as a basis se, and proceed in the same way as for the plane wave basis set in order to determine the coefficient $c_{\vec{k}}^{n,\vec{k}}$ in the expansion of the searched eigen-function. However, this does not work. We did not settle the parameter E yet. It turns out that in order to describe an eigenstate $\psi_{\vec{k}}^n(\vec{r})$ accurately with APW one has to set E equal to the eigenvalue (or band energy) $\varepsilon_{\vec{k}}^n$ of that state. But this is exactly what we are trying to determine .We are hence forced to start with a guessed value for $\varepsilon_{\vec{k}}^n$ and take this as E. Now we can determine the APW, and construct the Hamiltonian matrix elements and overlap matrix. The secular equation is determined, and our guessed $\varepsilon_{\vec{k}}^n$ should be a root of it. Usually it is not, hence we have to try a second guess. Due to this new E, the APW have to be determined again, and similarly for all matrix elements. With the help of root determination algorithms, this guessing continues until a root say $\varepsilon_{\vec{k}}^{(n=1)}$ is found. A and then the whole procedure starts over for $\varepsilon_{\vec{k}}^{(n=2)}$, *etc* (see Fig. **2**).

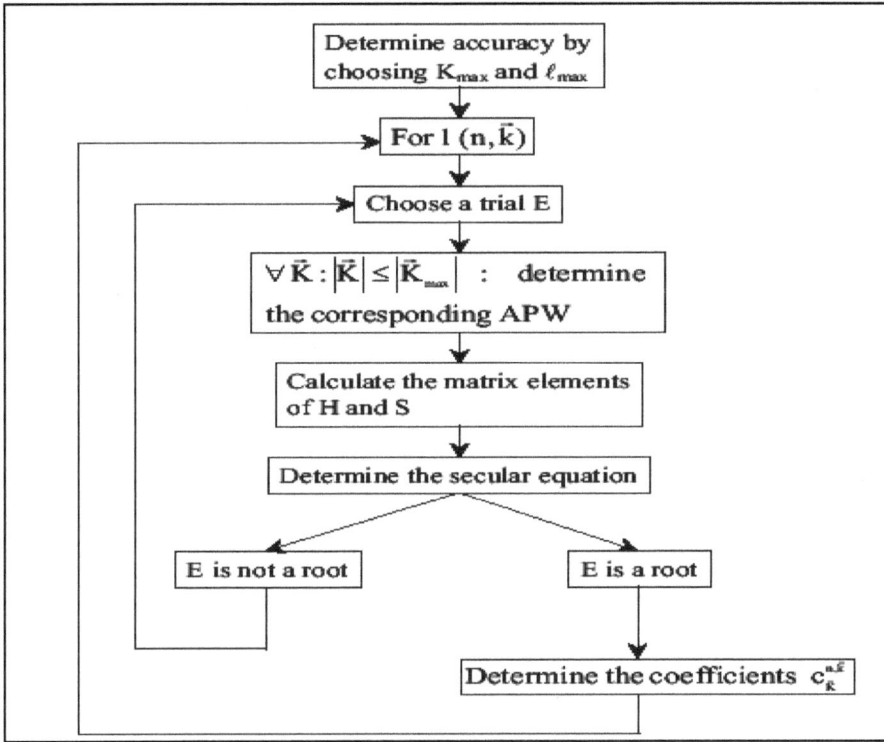

Fig. (2). Flowchart of the APW method.

The LAPW Method

2.3.3.2.1. The Regular LAPW Method

The problem with the APW method was that the $u_l^\alpha(r', E)$ have to be constructed at the yet unknown eigenenergy $E = \varepsilon_{\vec{k}}^n$ of the searched eigenstate. It would be helpful if we were able to recover $u_l^\alpha(r', \varepsilon_{\vec{k}}^n)$ on the fly from known quantities. That is exactly what the Linearized Augmented Plane Wave method enables us to do. If we have calculated u_l^α at some energy E_o, We could make a Taylor expansion to find it at energies not far way from it:

$$u_l^\alpha\left(r', \varepsilon_k^n\right) = u_l^\alpha\left(r', E_0\right) + \underbrace{\left(E_0 - \varepsilon_{\vec{k}}^n\right)\frac{\partial u_l^\alpha\left(r', E\right)}{\partial E}\Bigg|_{E=E_0}}_{\dot{u}_l^\alpha\left(r', E^0\right)} + O\left(E_0 + \varepsilon_k^n\right)^2 \qquad (2.71)$$

Substituting the first two terms of the expansion in the APW for a fixed E_o gives the definition of an LAPW. The energy difference $(E_o + \varepsilon_k^n)$ is unknown, and hence a yet undetermined $B_{lm}^{\alpha,\vec{k}+\vec{K}}$ has to be introduced:

$$\phi_{\vec{K}}^{\vec{k}}(\vec{r}) = \begin{cases} \dfrac{1}{\sqrt{V}} e^{i(\vec{k}+\vec{K})\vec{r}} & \vec{r} \in I \\ \displaystyle\sum_{l,m} (A_{l,m}^{\alpha,\vec{k}+\vec{K}} u_l^{\alpha}(r',E_0) + B_{lm}^{\alpha,\vec{k}+\vec{K}} \dot{u}_l^{\alpha}(r',E_0)) Y_{l\alpha}^m(r') \end{cases} \tag{2.72}$$

In order to determine both $A_{l,m}^{\alpha,\vec{k}+\vec{K}}$ and $B_{lm}^{\alpha,\vec{k}+\vec{K}}$, we will require that the function in the sphere matches the plane wave both in value and in slop at the sphere boundary. This results in a 2x2 system from which both coefficients can be solved. Eq. (2.72) is not the final definition of an LAPW yet. Imagine we want to describe an eigenstate $\psi_{\vec{k}}^n$ that has predominately p-character (l=1) .This means that in its expansion in LAPW, the $A_{l,m}^{\alpha,\vec{k}+\vec{K}}$ are large. It is therefore advantageous to choose E near the center of the p-band. In this way, the $O(E_o - \varepsilon_k^n)^2$ term in eq. (2.71) will remain small, and cutting after the linear term is certainly allowed. We can repeat this argument for every physically important l (s, p, d.and f states) (*i.e.* up to l=3) and for every atom. As a result, we should not choose one universal E_o, but a set of well-chosen E_l^{α} up to l=3, for higher l, a fixed value can be kept.

The final definition of an LAPW [75, 76] is then

$$\phi_{\vec{K}}^{\vec{k}}(\vec{r}) = \begin{cases} \dfrac{1}{\sqrt{V}} e^{i(\vec{k}+\vec{K})\vec{r}} & \vec{r} \in I \\ \displaystyle\sum_{l,m} (A_{l,m}^{\alpha,\vec{k}+\vec{K}} u_l^{\alpha}(r',E_0) + B_{lm}^{\alpha,\vec{k}+\vec{K}} \dot{u}_l^{\alpha}(r',E_0)) Y_l^m(r') & \vec{r} \in S_{\alpha} \end{cases} \tag{2.73}$$

With E_l^{α} the being fixed, the basis functions can be calculated once and for all. The same procedure as used for the plane wave basis set can now be applied. One diagonalization will yield P different band energies for this $\vec{k}$. The accuracy of a plane wave basis set was determined by K_{max}. For the APW or LAPW basis set, it is not incorrect to use the same criterion. However, a better quantity to judge the accuracy here is the product $R_{\alpha}^{\min} K_{max}$ between the smallest muffin-tin radius and K_{max}.

LAPW with Local Orbital

Local orbital [77] were introduced into the LAPW method in order to treat semi-core states. Another type of basis function to the LAPW basis set is called a local orbital (LO). A local orbital is defined as

$$\phi_{\alpha',Lo}^{lm}(\vec{r}) = \begin{cases} 0 & \vec{r} \notin S_\alpha' \\ \left(A_{lm}^{\alpha',Lo} u_l^{\alpha'}\left(r',E_{1,l}^{\alpha'}\right) + B_{lm}^{\alpha',Lo} \dot{u}_l^{\alpha'}\left(r',E_{1,l}^{\alpha'}\right) + C_{lm}^{\alpha',Lo} u_l^{\alpha'}\left(r',E_{2,l}^{\alpha'}\right)\right)Y_l^m(\hat{r}') \end{cases} \quad (2.74)$$

A local orbital is defined for a particular l and m and for a particular atom α'. The "'" indicates that all atoms in the unit cell are considered, not just in equivalent atoms. A local orbital is zero in the interstitial region and in the muffin-tin spheres of other atoms, hence its name local orbital. In the muffin-tin sphere of the atom α', the same $u_l^{\alpha'}\left(r',E_{1,l}^{\alpha'}\right)$ and $\dot{u}_l^{\alpha'}\left(r',E_{1,l}^{\alpha'}\right)$ as in the LAPW basis set are used, with as linearization energy $E_{1,l}^{\alpha'}$ a value suitable for the highest of the two valence states. The lower valence state that is much more free atom-like is sharply peaked at energy $E_{2,l}^{\alpha'}$. A single radial function $u_l^{\alpha'}\left(r',E_{2,l}^{\alpha'}\right)$ at that same energy will be sufficient to describe it. Local orbitals are not connected to plane waves in the interstitial region, they have hence no $\vec{k}$ or $\vec{K}$ dependence. The three coefficients $A_{lm}^{\alpha',Lo}$, $B_{lm}^{\alpha',Lo}$ and $C_{lm}^{\alpha',Lo}$ are determined by requiring that the LO is normalized, and has zero value and zero slopes at the muffin-tin boundary. Adding local orbitals increases the LAPW basis set size.

The Eigen Value Problem

Using the DFT inside the Kohn-Sham plan eq.(2.40) and with the LDA as an estimation for the trade relationship energy, we are left with a solitary molecule eigenvalue issue eq.(2.40). Taking care of eigenvalue issues mathematically is control in science accordingly and there exist various methods, each with various products for various actual issues and mathematical plans.

As a matter of first importance, to make computations plausible, one needs to think about evenness perspectives. In a strong, molecules are not conveyed in space haphazardly however masterminded in unique balances a high balance is frequently vigorously good. To portray a glasslike strong, a boundless occasional exhibit of discrete focuses is determined, a Bravais grid, with game plan and direction that shows up precisely the equivalent, from whichever of the focuses the cluster is seen. The cross-section vector is a vector starting with one grid point then onto the next. A volume that, when interpreted through all conceivable grids, simply occupies the entirety of the space without either covering itself or leaving voids is known as a crude cell of the cross-section. The decision of crude cell isn't remarkable and a typical decision is a Wigner-Seitz cell; that locale around a cross-section point that is nearer to that point than to some other grid point. An actual strong would now be able to be depicted by its basic cross-section, along with a portrayal of the plans of iotas inside the crude cell. Regular high evenness precious stone constructions are the body-focused cubic (bcc) and the face focused cubic (fcc) structures. The strategies examined so far in (Sec 2.2) must be applied to solids with amazing gem evenness. This is on the grounds that the eigenvalue issue eq.(2.40) need just be addressed in the Wigner-Seitz cell and not in the limitlessly enormous space.

The higher the symmetry of a specific structure (keeping the number of atoms fixed) the smaller the numerical problem. Eq. (2.40) is invariant under the translation $r \rightarrow r + R$, where R is a lattice vector, and this periodicity of the lattice is expressed by the so-called Bloch condition,

$$\psi(k, r + R) = e^{ik.R}\psi(k, r)$$

By imposing an appropriate boundary condition (the Born-von Karman boundary condition) the wave vector k is restricted to special allowed values. The Schrödinger equation (2.2) can, using all above-mentioned techniques, be written

$$H_s(r)\psi_\mu(k, r) = \varepsilon_\mu(k)\psi_\mu(k, r) \qquad (2.75)$$

The one-electron Schrödinger equation is thus solved for a specific k-vector. This solution yields a set of energy eigenvalues, ε, indexed by μ. Just as a position in space can be characterized by a vector r, a k-vector characterizes a position in a space called the reciprocal space or simply the k-space. The reciprocal space is more or less the inverse of the real space and can be measured by x-ray spectroscopy. There exists also a lattice in this space, determined by those wave vectors K satisfying the relation $e^{ik.r} = 1$ The correspondence of the Wigner-Seitz cell in the reciprocal space is called the Brillouin zone (BZ). To get a perfect knowledge of the electronic structure of a solid everywhere in space, eq. (2.75) has to be solved for many k points (in principle an infinitely dense mesh of k points in the BZ). The energy below which the one-electron levels are occupied and above which they are unoccupied in the ground state of a metal defines the so-called Fermi level. An important contribution to the total energy is the so-called band-energy,

$$E_{band} = \int_{-\infty}^{E_F} \varepsilon D(\varepsilon) d\varepsilon \qquad (2.76)$$

Where ε are eigenvalues and D (ε) are the DOS. Eigenvalues, ε, are for obvious practical reasons, only calculated for a discrete mesh of k points, ε_μ (μ is introduced in eq.(2.40) as an index for a specific eigenvalue and k-points). Therefore the integral in eq. (2.76) is replaced by a discrete summation. Problems arise at the Fermi level due to the discretization of the k-point mesh. The Fermi level will be somewhere between the highest occupied eigenvalue and the lowest unoccupied eigenvalue. The most straightforward method is to make a linear interpolation between k points (the linear tetrahedron method, LTM) [78]. A more involved method is to let the interpolation mimic some effects of higher-order (the modified tetrahedron method, MTM) [79]. The often-used Gaussian broadening method (GB), has another approach. In this method, the eigenvalues are respectlessly smeared out by a Gaussian function. When solving the differential eq.(2.75) numerically one has to choose a basis set. The choice of basis is, of course, free but since the numerical problems are tremendously large it is of uttermost importance to choose a basis with optimal numerical efficiency.

Magnetism

In Sec. 2.2 we presumed that for an arrangement of electrons in an outer potential (the potential from the cores), the potential and the ground state energy are interestingly dictated by the electron thickness. This hypothesis, the DFT, and the entire method laid out in this part can be applied additionally to attractive frameworks. The definition will rather be that an arrangement of electrons in an outer potential has a potential and a ground state energy that is exceptionally dictated by two electrons densities, in particular one electron thickness for each twist channel, (alluded to as the lion's share or twist up the channel and the minority or twist down channel). Another option – and same – plan is that the potential and the ground state energy are remarkably controlled by the electron thickness and the charge thickness. The element of the framework relating to the Hamiltonian administrator for an attractive framework will be twofold the component of the grid addressing a non-attractive framework. The upper left quarter of the grid addresses one twist channel and the lower right quarter of the lattice addresses the other twist channel. The off corner to corner components relate to hybridization between various twists or (and) pivoted turn organizes. The estimation of the trade relationship potential (see Sec. 2.2.4.3) can likewise be reached out to cover the attractive circumstances. The attractive correspondence to LDA is known as the nearby twist thickness estimation (LSDA)

Self-Consistency

The Kohn-Sham equation eq. (2.40) is highly non-linear and complex. To avoid numerical problems most efficiently, calculations are made iteratively up to self-consistency. The initial potential V_s *(r)* can be derived from a guest electron density, n(r). The potential are thereafter plugged into eq. (2.40) and eigenvalues and eigenfunctions comes out. A new electron density can be calculated from eq. (2.38) which in turn gives a new potential. This is now repeated iteratively until the difference between electron densities from two successive iterations are sufficiently small and we say that self-consistency is achieved. As a starting guess for the electron density, the density of superimposed free single atoms is often used.

The Density of States

The density of states is defined as the number of energy states per unit volume per unit energy range. thus the number of states per unit volume of the crystal having energies in the range E to E+dE is given by

$$N (E) = n (E) dE \tag{2.77}$$

Where n(E) is the density of states.

RESULTS AND DISCUSSION

We start our outcomes and conversation by a part on the calculation strategies utilizing the WIEN2K code followed by our information on components Fe and Nd. The compound $Smco_5$ will be examined in the following area.

Fe and Nd

The magnetic properties of Fe and Nd have been extensively studied by first-principles calculations [80]. We present here the calculation on Nd and Fe using Full-Potential Linearized Augmented Plane Wave (FPLAPW) [81]. The Local Density Approximation (LDA) of Perdew and Wang [82] and the Generalized Gradient Approximation (GGA) of Perdew, Burke and Ernzerhof [83] were used for correlations and exchange potentials.

Results and Discussion

Ferromagnetic BCC Iron

The Ferromagnetic Fe is BCC (space bunch Im-3m #229), the grid steady a= 2.8665Å [84]. We determined the attractive second utilizing RK⁻max =9 and acquired the worth of 2.27 µB which is in great concurrence with the exploratory worth of 2.25 µB [84]. Figs. **(1b to 1d)** show the larger part and minority thickness of states (DOS) for Fe. The tops in the dominant part are discovered to be situated beneath Ef while the tops in minority are discovered to be situated above and underneath Ef. The general DOS structure is steady with crafted by Gu and Ching [85]. Figs. **1 (a and c)** show the larger part and minority band structure for Fe. The larger part d-states for Fe is appeared in Fig. **1(e)** and the orbital-decayed (PDOS) appear in Figs. **1 (f and g)**. The d-state turn-up Dos is overwhelmed by (d-e$_g$) state as demonstrated in Fig. **1(f)** and (d-t$_{2g}$) state as demonstrated in Fig. **1(g)**. While the d-states turn dn for Fe has appeared in Fig. **1(h)** and the orbital-decayed (PDOS) appear in Figs. **1 (i and j)**. The d-state turn dn DOS is overwhelmed by (d-e$_g$) state as demonstrated in Fig. **1(i)** and (d-t2g) state as demonstrated in Fig. **1(j)**.

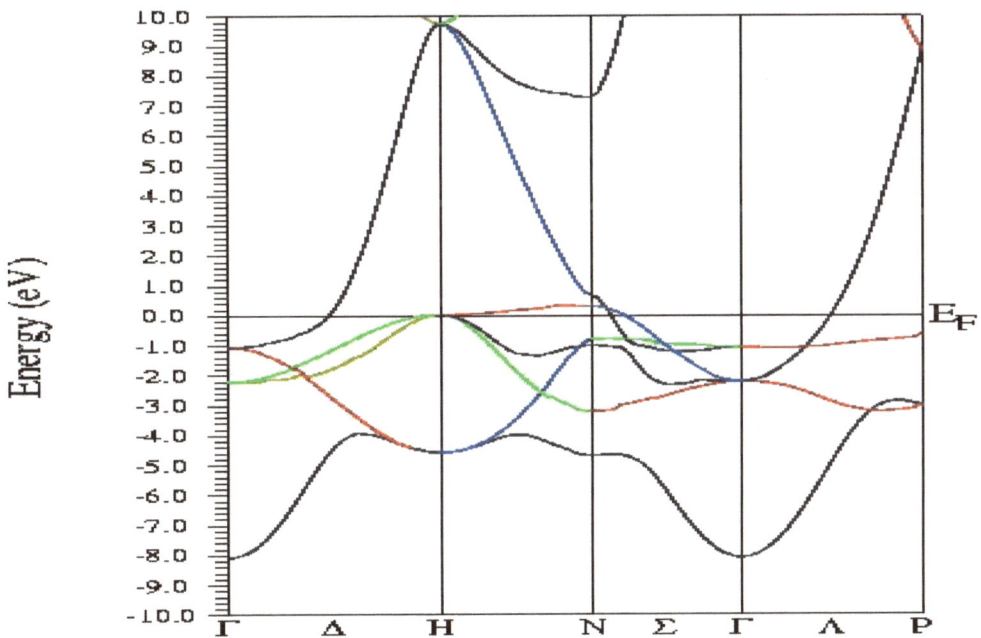

Fig. 1(a). Band structure (total-Fe) (spin-up).

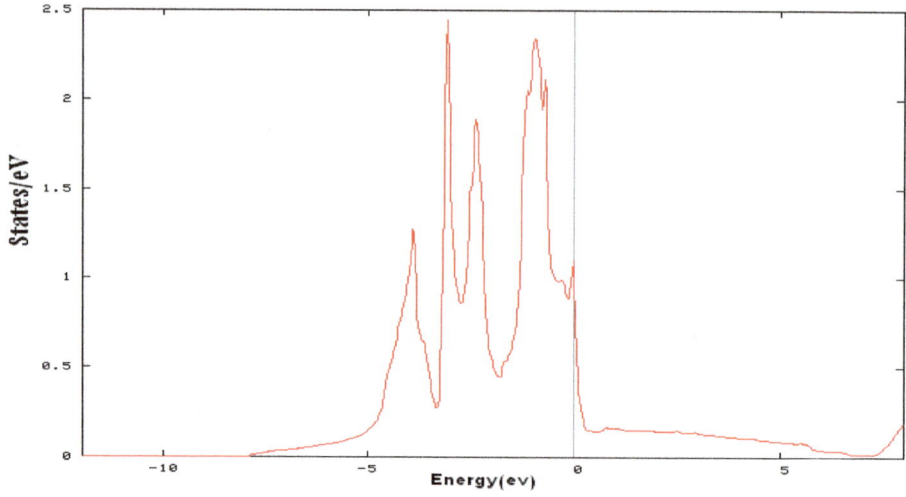

Fig. 1(b). DOS(total-Fe)(spin-up) versus Energy.

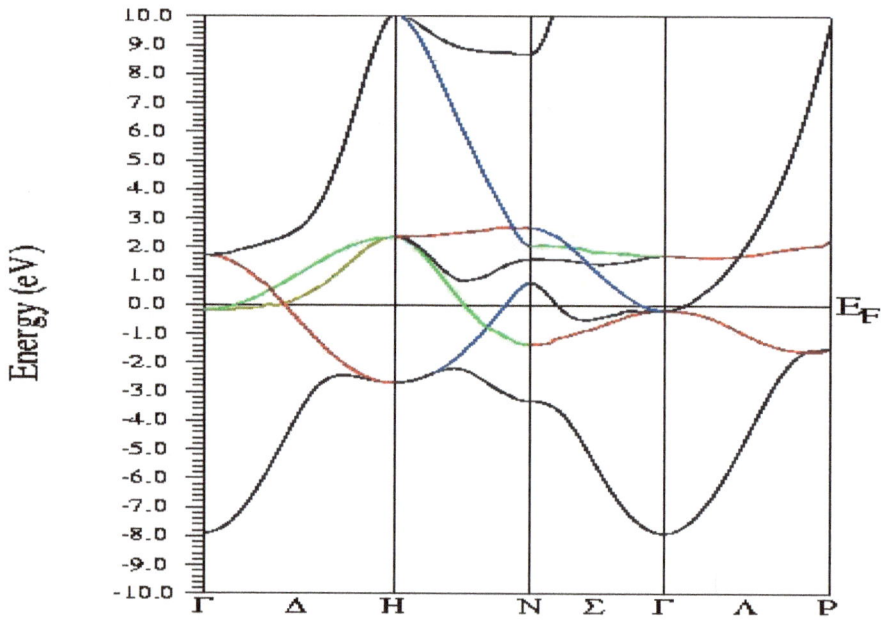

Fig. 1(c). Band structure (total-Fe) (spin-dn).

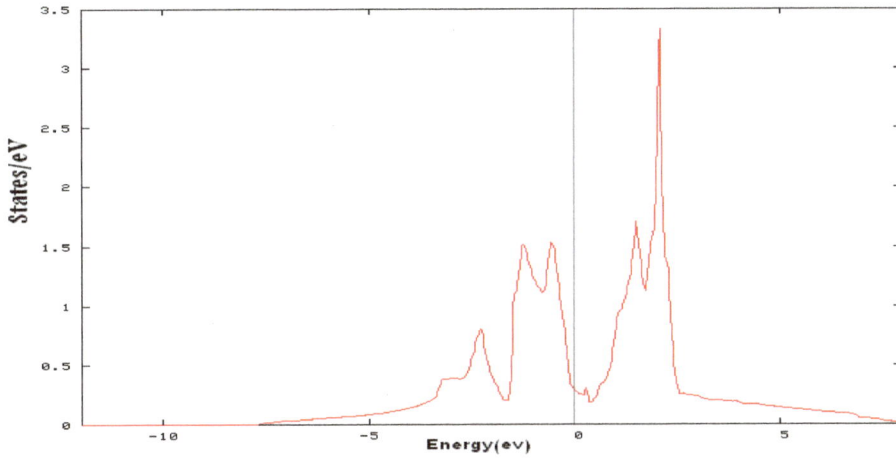

Fig. 1(d). DOS (total-Fe)(spin-dn) versus Energy.

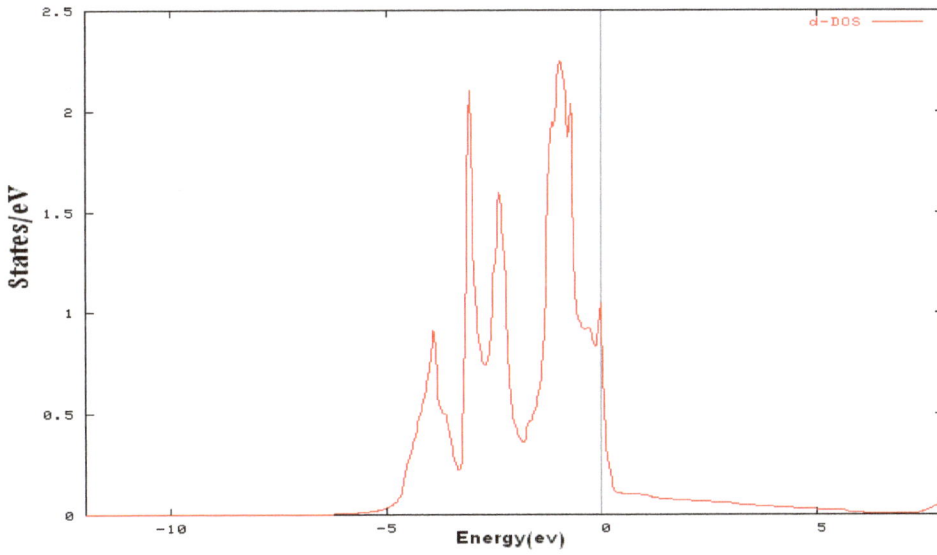

Fig. 1(e). D-state (spin-up) versus Energy.

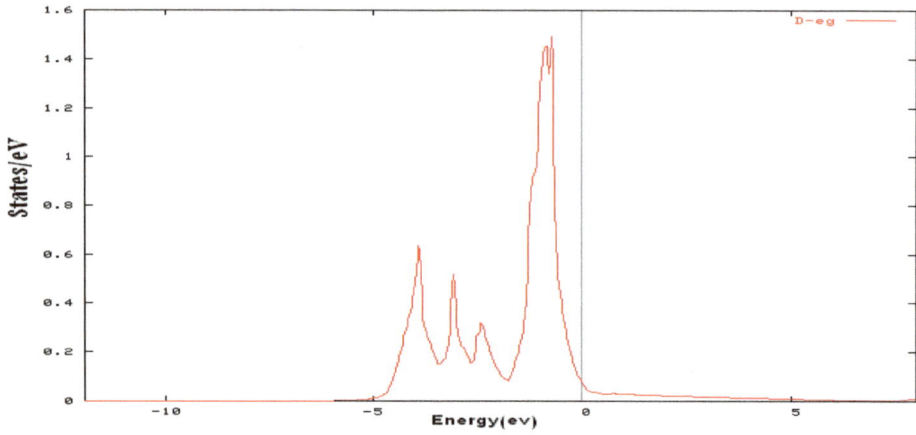

Fig. 1(f). PDOS(D-eg state)(spin-up) versus Energy.

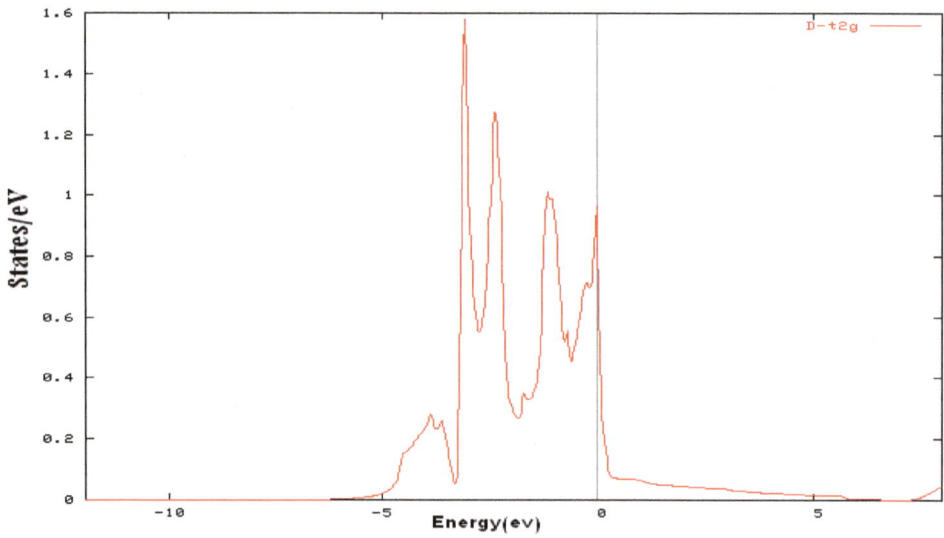

Fig. 1(g). PDOS(D-t2g state)(spin-up) versus Energy.

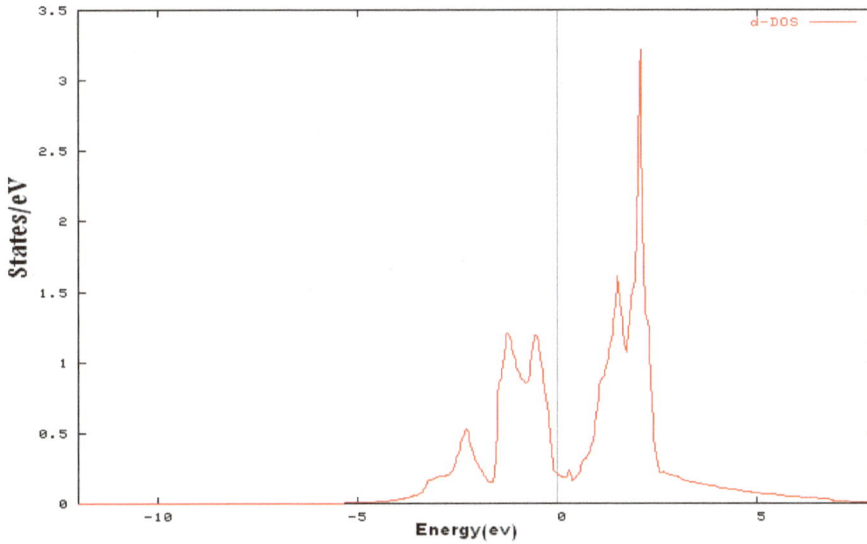

Fig. 1(h). D-state (spin-dn)versus Energy.

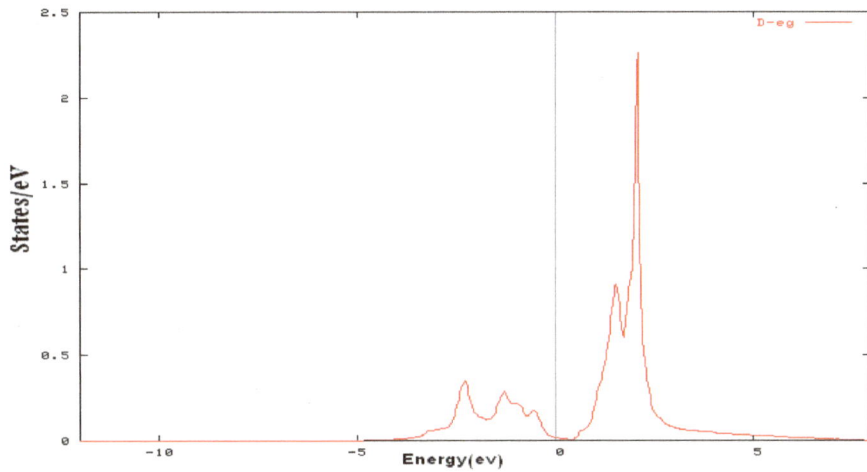

Fig. 1(i). PDOS (D-eg state)(spin-dn)versus Energy.

HCP Neodymium

The Nd component has a hexagonal close pressed (HCP) structure (space bunch P63/mmc #194), the grid constants a, b and c utilized in our estimation are a=b = 3.658 and c=11.799 Å [86]. We determined the attractive second utilizing RK¬max =9 and acquired the worth 2.65 µB which is in concurrence with the exploratory worth 2.7 [86]. Fig. **2(b-d)** show the lion's share and minority thickness of states (DOS) for Nd. In Fig. **2(b)** the pinnacle is discovered to be found bunched around Ef while the tops in Fig. **2(d)** is discovered to be situated above Ef. The general DOS structure is predictable with crafted by Gu and Ching [85]. Fig. **2(a-c)** show the dominant part and minority band structure for Nd . We note that the all-out Nd groups in Fig. **2(a)** for (turn up) case structure a level band around Ef and this level band structure compares to a high thickness of states *i.e* a top in DOS Fig. **2(b)**. Fig. **2(f-h)** and Fig. **2(e-g)** show the dominant part and minority thickness of states (DOS) and band structure for Nd (f-state) individually. Fig. **2(f-e)** are predictable since the top in DOS, found grouped around Ef, implies a high thickness of state as reflected in the band structure while we note that the Nd f groups in Fig. **2(g)** for (turn dn) case structure a thin level band above Ef and this level band structure compares to the thickness of states as demonstrated in Fig. **2(h)**.

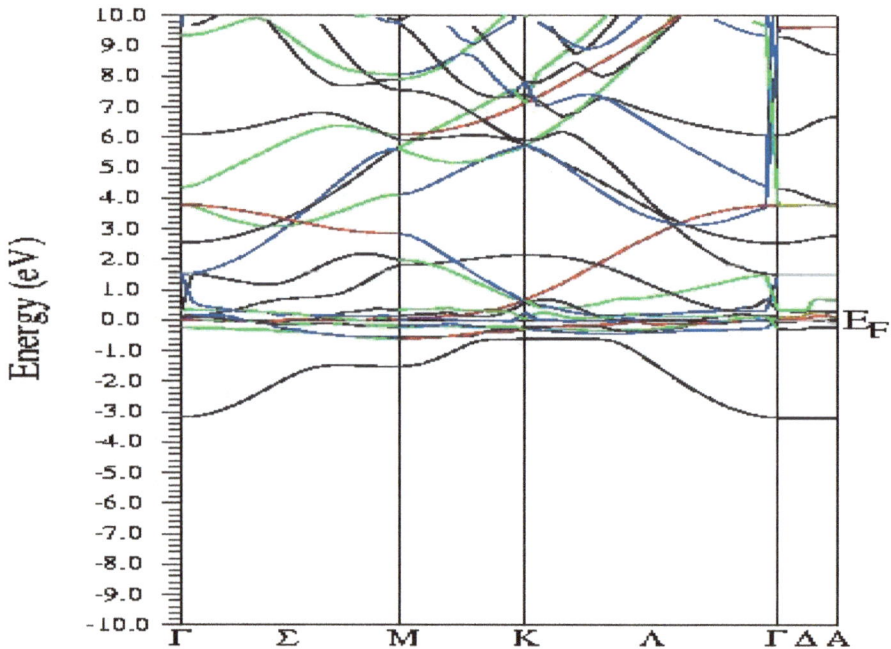

Fig. 2(a). Band structure (total Nd) spin-up.

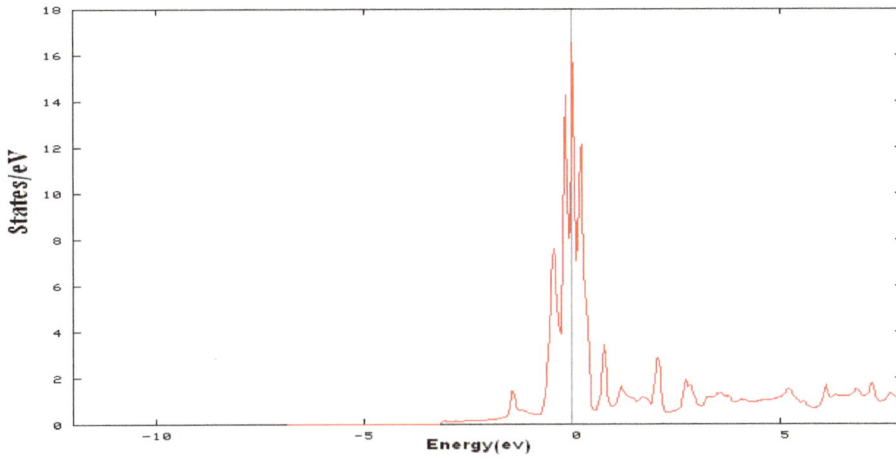

Fig. 2(b). DOS (total Nd)(spin-up) versus Energy.

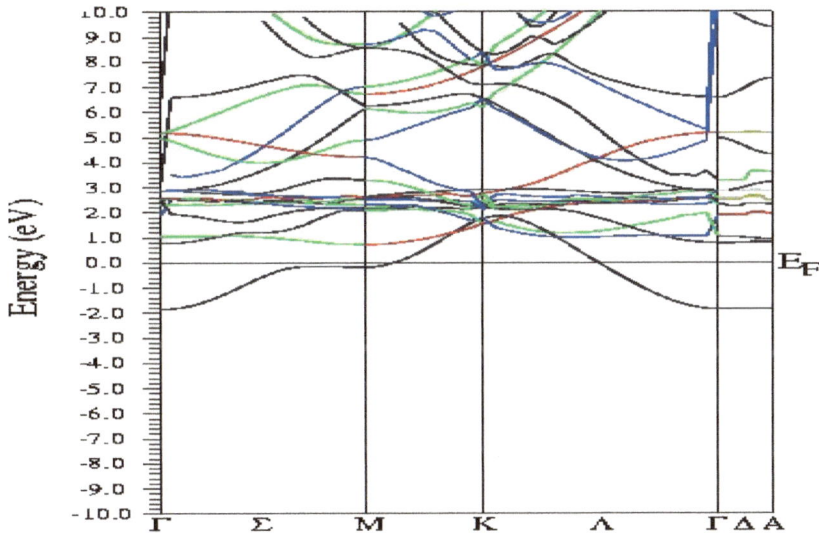

Fig. 2(c). Band structure (Total Nd) spin-dn.

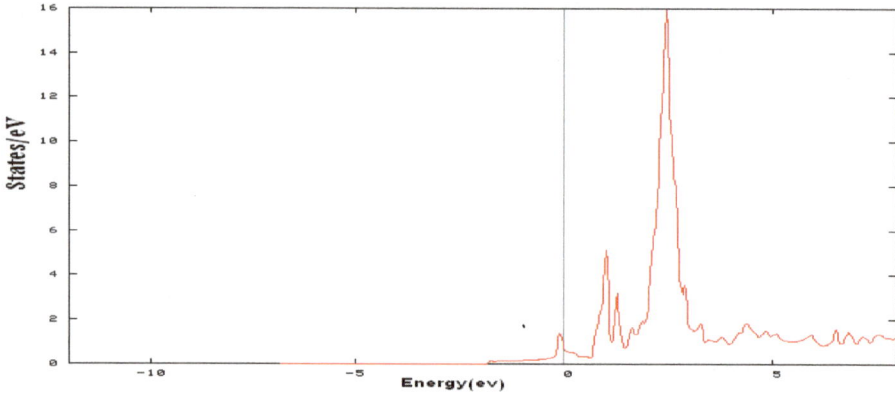

Fig. 2(d). DOS(total Nd)(spin-dn) versus Energy.

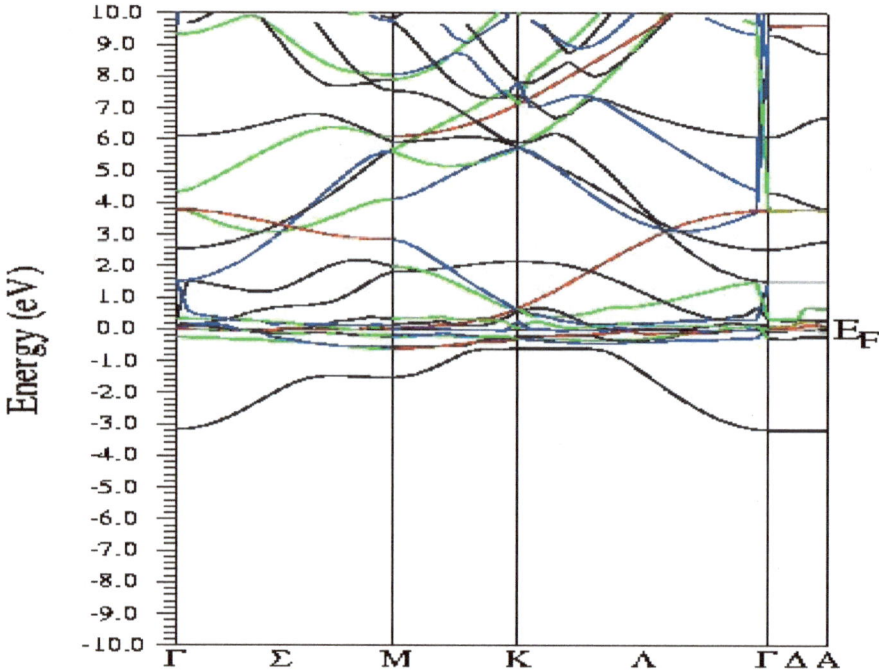

Fig. 2(e). Band structure (f-state) spin-up.

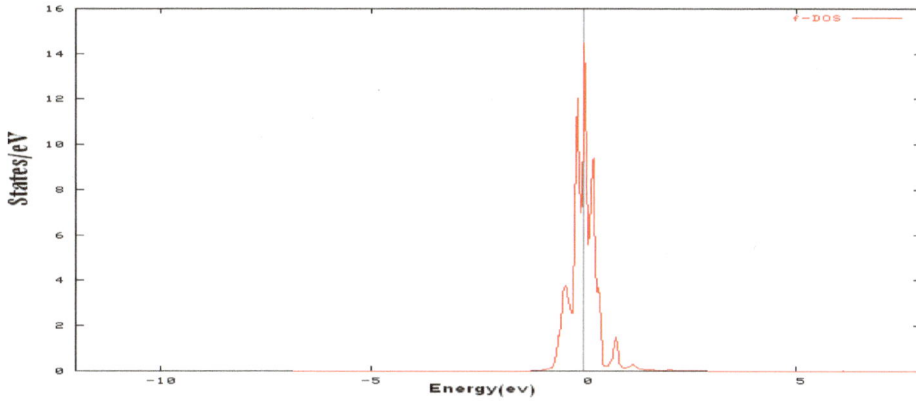

Fig. 2(f). DOS (f-state)(spin-up) versus Energy.

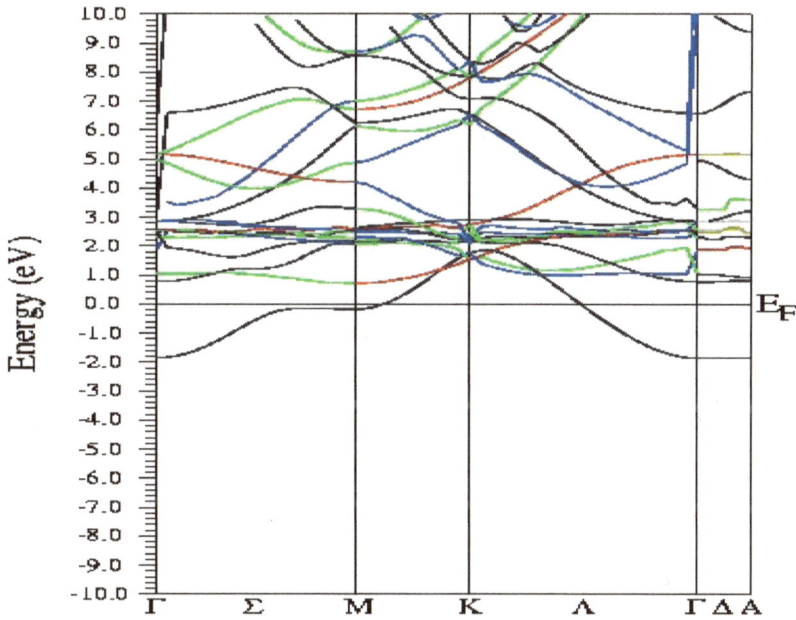

Fig. 2(g). Band structure (f-state) spin-dn.

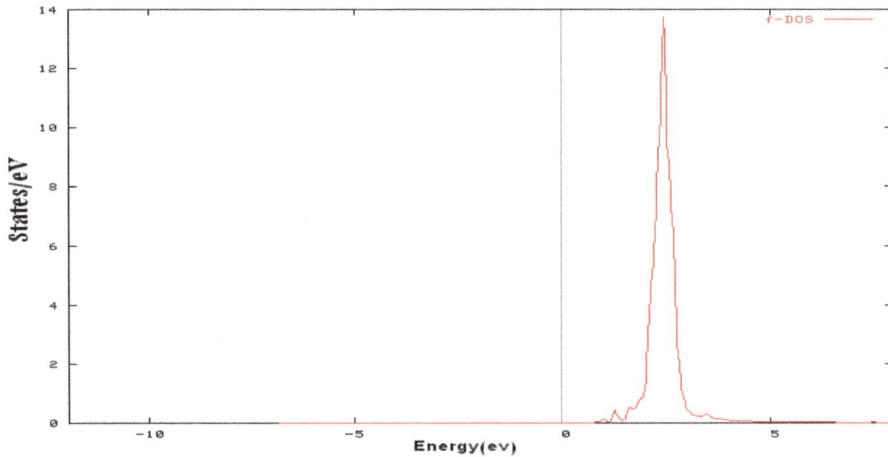

Fig. 2(h). DOS (f-state)(spin-dn) versus Energy.

$SmCo_5$

Uncommon earth change metal mixtures are fascinating from specialized and key perspectives. One generally faces troubles in managing band design of f-electron materials utilizing the neighbourhood thickness estimation (LDA) as a result of the confined idea of f electrons and the unpretentious interchange between the twist polarization and twist circle connection (SO). The orbital attractive second is framed by turn circle coupling and connection impacts. Since the coulomb relationships in the Sm 4f are required to be huge in these frameworks, then, at that point LDA + U strategy [87] will be valuable. The LDA + U, eliminates the lack of LDA by consolidating the Hubbard-like association term for 4f-electrons. There has been a nonstop interest in the electronic construction and attractive properties of $Smco_5$ [87 - 96]. The increased plane-wave (APW) strategy has been utilized in a couple of studies written about $Smco_5$ band structure [91, 92]. No examinations, up as far as anyone is concerned, were committed to contemplating the electronic band design and twist thickness maps in $Smco_5$ utilizing thickness utilitarian hypothesis (DFT) based strategies [82]. Nonetheless, DOS estimation for the two twists bearings for Sm f-orbitals was accounted for by Larson *et al.* [87]. Twist and charge thickness shapes were accounted for, just for some uncommon earth change metal mixtures for example R2T14 B and different frameworks [97 - 100]. We present in this paper a DFT-put together examination with respect to the electronic band structure, turn thickness maps, attractive second and attractive anisotropy in $Smco_5$ utilizing various plans and Brillouin-zone coordination techniques as carried out in the Wien2k bundle [101].

Computation Methods

Smco$_5$ solidifies in the CaCu$_5$ structure (space bunch P6/mmm, No. 191). The unit cell contains one recipe unit. The exploratory cross-section constants utilized in our computations are a = 5.00 and c = 3.96 A0 [87]. The electronic-structure code Wien2k utilizes the Full Potential Linearized Augmented Plane Wave (FPLAPW) in view of DFT [81]. Both centre and valence states are determined self-reliably, the centre states are dealt with completely relativistically for the round piece of the potential, while the maximum capacity is utilized for the valence states. The LDA strategy for Perdew and Wang [82] and the summed up angle estimation (GGA) of Perdew *et al.* [83] are utilized for connection and trade possibilities. Nearby orbital expansions [102] with a united premise of around 900 premise capacities are utilized to lessen the linearization blunders in Sm and Co circles. For the Brillouin-zone combination, we utilize the changed straight tetrahedron (MLT) and Gaussian widening spreading (GB) techniques. The self-steady computations were performed with 80 k-focuses in the final Brillouin-zone. The biscuit tin (MT) circle radii are 2.115 and 2.015 a.u. for Sm and Co, individually. The premise sets are dictated by a plane-wave cut-off of RMT × Kmax = 9.0 which gives a decent assembly. The SO cooperation is incorporated utilizing a second-request variational plot [103] by taking all states underneath the removed energy of 1.5 Ry. The Hubbard and trade boundaries for the f shell are Uf ~5.2 and Jf ~0.75 eV, separately [87]. The twist thickness guides of Smco$_5$ determined from the distinction between turn all overdensities have been plotted in the (001) plane. The plots were finished with and without turn circle coupling in the LDA + U plan. The base and greatest forms utilized in the plots are 0.0 and 2.0, separately, with a time period.

Results and Discussion

Density of States and Band Structure

To methodically contemplate the impact of thinking about various connections, for example SO and Coulomb cooperations on the attractive properties, electronic groups and twist thickness structure we originally performed turn captivated computation and afterwards accordingly consolidated SO and additionally LDA + U plans into the estimation. For instance, Figs. (**3a and 5**) shows the complete thickness of states (DOS), the Sm f-DOS of the twist-up states and the band structure has appeared in Fig. **7(a)** while Fig. **3(b)** show the absolute thickness of states (DOS) of the twist dn states utilizing LDA + U however without including turn circle coupling. Each top in DOS implies that many wave vectors have a similar eigenvalue. For turn up Sm f states Fig. (**5**), we note that

Sm f orbitals structure a tight pinnacle grouped around Ef. A striking contrast is seen between the highlights of Fig. **7(a)** and its partner in the twist energized determined in our past work [104] in particular the undeniable bunching of the groups at Ef in the last case contrasted with the isolated groups appeared in Fig. **7(b)**. The justification for such a distinction is that presenting the Hubbard and trade connections through the LDA + U decays the tops by size of ~U - J = 5 eV [87]. Another element of Fig. **3(a)** is the strength of the f-character of the Sm iota above and beneath Ef. Level groups, connoting high thickness of states, are situated inside ~1 eV (~ 4–6 eV) above (beneath) Fermi energy. Figs. **(4a)** and **6** show the all-out thickness of states (DOS), the Sm f-DOS of the twist-up states and the band structure has appeared in Fig. **7(c)** while Fig. **4(b)** shows the absolute thickness of states (DOS) of the twist dn states along the [001] heading in the LDA + U + SO conspire. The places of the Sm f-levels are simply above Ef and in the 4–6 eV range underneath Ef. We have rehashed the computation in this plan for the twist down the case and tracked down that level groups are situated around 2–4 eV above Ef. Our discoveries are in acceptable concurrence with the DOS structure announced by Larson *et al.* [87]. It could be referenced here that photoemission (PE) spectroscopy utilizing synchrotron radiation has been done on certain 4f frameworks *e.g* $SmAl_2$ and the outcomes were contrasted and the determined Sm f-thickness of states [105]. We are ignorant of comparable PE concentrates on $Smco_5$. We have determined the band structure in the LDA + U + SO conspire for the [100] heading too. The outcomes are shown in Fig. **7(d)**. As far as we could possibly know no DOS estimation has been done along with this precious stone heading and accordingly we can't contrast and past examinations. The band structures in Fig. **7(c and d)** are distinctive in that groups above Ef in Fig. **7(d)** are moved down while those beneath Ef are moved up contrasted with those of (Fig. **7c**). This distinction in the energy length of the f-band for those two opposite precious stone bearings means the commitment of the Sm f-levels to the magneto-glasslike anisotropy energy of $Smco_5$. Another distinction is that centre levels around 6 eV are parted by SO communication when the polarization is corresponding to the c-pivot (Fig. **7c**). We have determined, in the LDA + U + SO plot, a twist attractive snapshot of ~12.18 μB and an orbital snapshot of ~2.78 μB for example a net attractive snapshot of ~9.4 μB/f.u. Utilizing either the MLT or GB technique, in great concurrence with 9.9 μB detailed by Larson *et al.* [87]. Our estimation is likewise in reasonable concurrence with the exploratory second [106, 107] of 8–8.9 μB. Using a turn spellbound plan, we acquired a too-huge worth (~12.8 μB) for the attractive second. We have determined MAE for $Smco_5$, inside the LDA + U plan, utilizing MLT and GB strategies. We acquired upsides of ~16 and ~9.8 meV/f.u. utilizing these two techniques, individually. The previous worth is in acceptable concurrence with the test esteem [108] of 2.62×108 erg/cm3 (around 14 meV/f.u.) at 4.2 K and the qualities detailed by

Sankar *et al*. [90] and Radwanski *et al*. [109]. Rehashing the computations with just SO communication considered, brought about an impressively low worth (~4.6 meV/f.u.) for MAE. This again shows the ampleness of utilizing LDA + U in this framework.

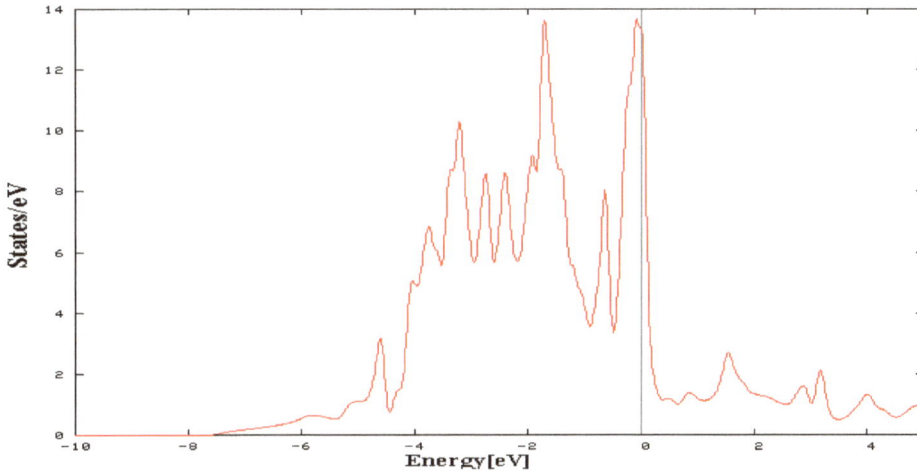

Fig. (3a). Calculated total spin–up density of states (DOS) against energy for SmCo$_5$ using (spin-polarized only).

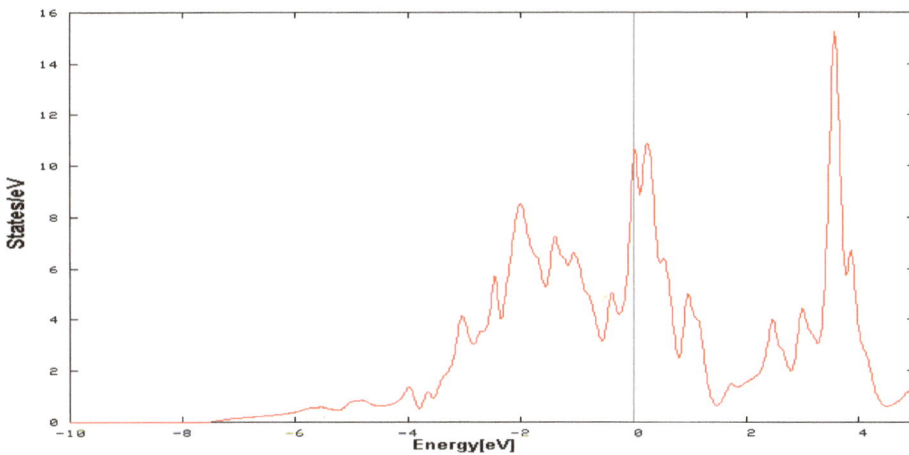

Fig. (3b). Calculated total spin–up density of states (DOS) against energy for SmCo$_5$ using (spin-polarized only).

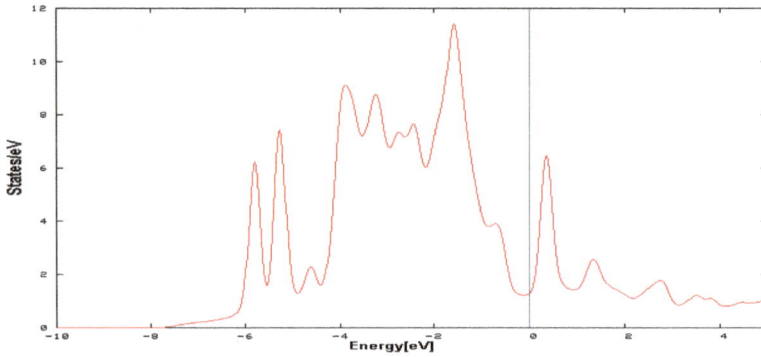

Fig. (4a). Calculated total spin–up density of states (DOS) against energy for SmCo$_5$ using (spin-orbit coupling).

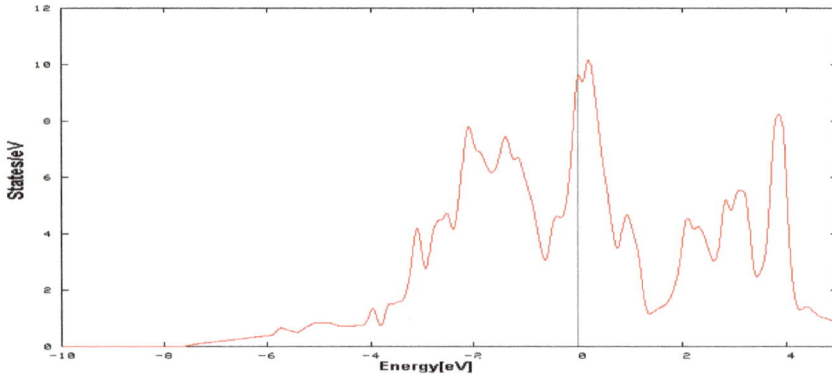

Fig. (4b). Calculated total spin–up density of states (DOS) against energy for SmCo$_5$ using (spin-orbit coupling).

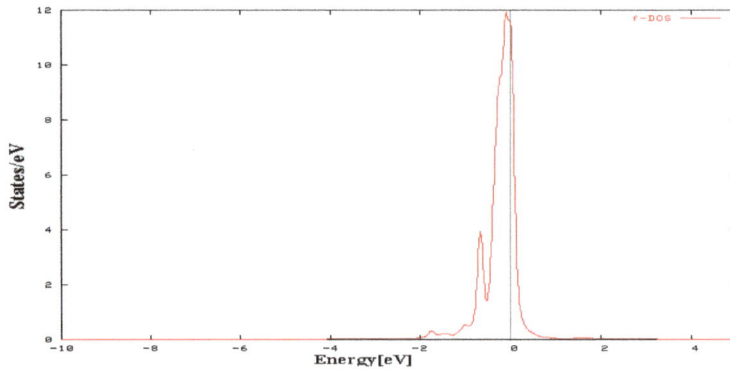

Fig. (5). Calculated spin–up density of states (DOS) against energy for Sm f orbital using spin-polarized only.

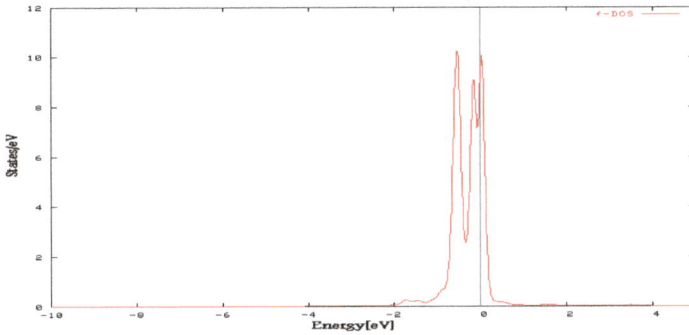

Fig. (6). Calculated spin–up density of states (DOS) against energy for Sm f orbital using spin-orbit coupling.

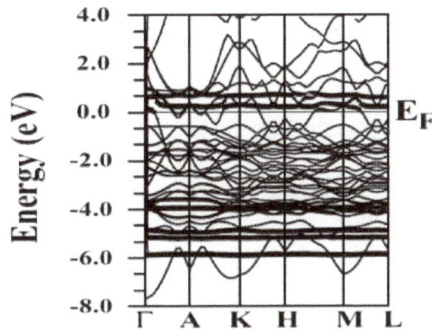

Fig. (7a). GGA and LDA + U schemes (band structure) were used for the Sm f-orbitals (spin-up) without spin–orbit coupling.

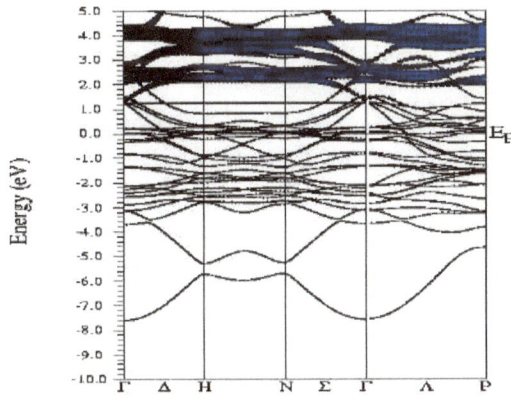

Fig. (7b). GGA and LDA + U schemes (band structure) were used for Sm f-orbitals (spin-dn) without spin–orbit coupling.

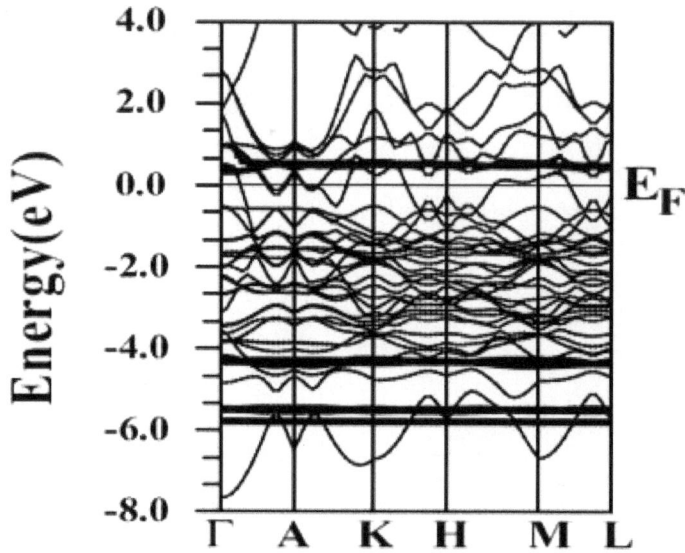

Fig. (7c). GGA and LDA + U schemes (band structure) were used for Sm f-orbitals plus SO is included and the magnetization is along [001] direction.

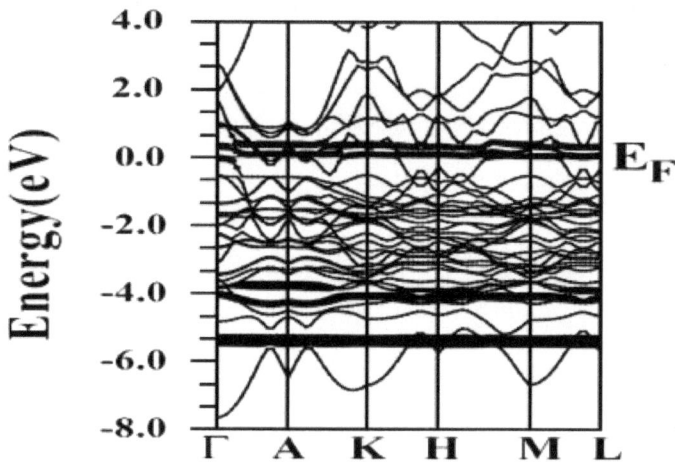

Fig. (7d). GGA and LDA + U schemes (band structure) were used for Sm f-orbitals plus SO is included and magnetization is along [100] direction.

The Spin-Density Maps

We determined the forms of the twist thickness maps in the (001) plane of the Smco$_5$ gem utilizing the accompanying plans: turn energized, turn spellbound with LDA + U + SO, LDA + U or SO as it were. Figs. (**8a** and **b**) are models where turn energized and LDA + U + SO conspires have been utilized, separately. In these two figures, the twist shapes in a part of the (001) plane containing five Sm and six Co(2c) molecules have appeared. A few comments might be drawn from our computations: first, the coulomb relationship energy greatly affects the shape design of Co molecules than on that of the Sm iota. Specifically, we have tracked down that the twirl forms around Co molecules become denser when the coulomb connection is on. Also, the twist circle communication makes greater asphericity the Sm particles Figs. (**8a** and **b**), which is normal in light of the strength of this cooperation on this non-s-state molecule. The general impact of the SO cooperation on the twist maps is more clear than the impact of the relationship collaboration. This exhibits the significance of the previous collaboration in this framework.

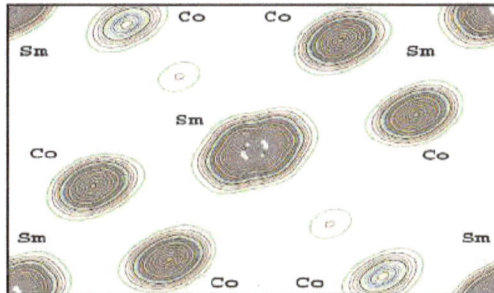

(a): the calculation is spin-polarized

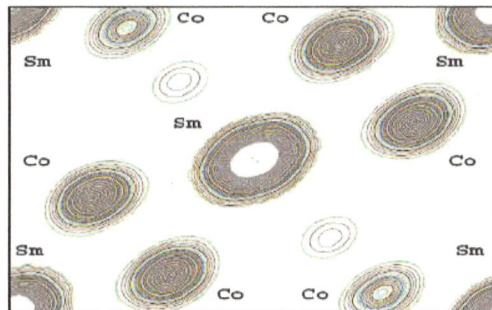

(b): spin-polarized calculation with LDA + U + SO

Fig. (8). Spin-density maps in a portion of the basal-plane of SmCo$_5$. The positions of Sm and Co atoms are indicated in (a and b).

CONCLUSION

- We have performed first-standards estimations on Nd, Fe, and Smco$_5$ utilizing FPLAPW with LDA+U. We have considered the attractive second, thickness of states and band structure for Nd and Fe components and similar properties for Smco$_5$ in addition to band structure.
- The attractive snapshots of the basic Nd and Fe are 2.27 µB and 2.65 µB for BCC Fe and HCP Nd individually in great concurrence with trial esteems. Likewise, we have considered the thickness of states and the relating band design of these 3d and 4f components.
- We have played out a deliberate abdominal muscle initio computation on Smco$_5$ utilizing diverse communication plans accessible in the Wien2k code. The electronic band structure, attractive anisotropy and attractive second are best portrayed in the LDA + U + SO plot as demonstrated by examination with accessible computations and investigations on Smco$_5$.
- We have determined, in the LDA + U + SO plot, a twist attractive snapshot of ~12.18 µB and an orbital snapshot of ~2.78 µB for example a net attractive snapshot of ~9.4 µB/f.u. Utilizing either the MLT or GB strategy, in great concurrence with 9.9 µB . Our computation is likewise in reasonable concurrence with the exploratory snapshot of 8–8.9 µB. Using the turn captivated plan, we acquired a too-enormous worth (~12.8 µB) for the attractive second.
- We have determined MAE for Smco$_5$, inside the LDA + U plan, utilizing MLT and GB techniques. We got upsides of ~16 and ~9.8 meV/f.u. utilizing these two techniques, individually. The previous worth is in acceptable concurrence with the trial worth of 2.62×108 erg/cm3 (around 14 meV/f.u.) at 4.2 K. Rehashing the estimations with just SO communication considered, brought about an extensively low worth (~ 4.6 meV/f.u.) for MAE. This again exhibits the ampleness of utilizing LDA + U in this framework.
- Spin thickness forms for Smco$_5$ show solid reliance on the plan utilized Spin–circle cooperation has the most grounded impact on the construction of the twist thickness guides of Sm particles in this framework. It will bear some significance with studying the twist thickness maps in planes containing the Co particles also.

CONSENT FOR PUBLICATION

Not applicable.

CONFLICT OF INTEREST

The author declares no conflict of interest, financial or otherwise.

ACKNOWLEDGEMENTS

Declared none.

REFERENCES

[1]   Buschow, K.H.J *Rep. Prog. Phys.,* **1991**, *54*, 1123.

[2]   Kronmuller, H.; Durst, K. D.; Hock, S.; Martinek, G. *J. Phys. Colloq.,* **1988**, *49*, C8-623.

[3]   Herbst, J. F.; Lee, R. W.; Pinkerton, F. E. *Ann. Rev. Mater. Sci,* **1986**, *16*, 467.

[4]   Buschow, K. H. J. *Rep. Prog. Phys,* **1977**, *40*, 1179.

[5]   Buschow, K. H. J. *Rep. Prog. Phys,* **1979**, *42*, 1373.

[6]   Buschow, K.H.J.; Wohlfarth, E.P. *Ferromagnetic materials*; Amsterdam: North-Holland, **1980**, 1, pp. 297-414.

[7]   Burzo, E.; Chelkowski, A.; Kirchmayr, H. *Landolt-Bornstein*; Volume, D.Z.; Wijn, H., Eds.; Springer: Berlin, **1990**, 19, .

[8]   Coey, J. *Magn. Magn. Mater,* **1996**, *159*, 80.

[9]   Coehoom, R.; Long, J.; Grandjean, F. (Dordrecht: kluwer) "Supermagnets and Hard Magnetic Materials", Lecture notes NATO-ASI ed. *GP,* **1991**, 133.

[10] Coehoom, R. *Magn. Magn. Mater.,* **1991**, *99*, 55.

[11] Johanssom, B.; Eriksson, O.; Nordstrom, L.; Serverin, L.; Brooks, M.S.S. *Physica B,* **1991**, *172*, 101. [http://dx.doi.org/10.1016/0921-4526(91)90422-B]

[12] Johanssom, B.; Nordstrom, L.; Eriksson, O.; Brooks, M.S.S. *Phys. Scr. T.,* **1991**, *39*, 100.

[13] Brooks, M.; Johansson, B.; Buschow, K.H.J. HandBook of Magnetic Materials. North-Hollaod: Amsterdam, **1993**; 7, pp. 139-230.

[14] Fahnle, M.; Hummler, K.; Liebs, M.; Beuerle, T. *Appl. Phys., A Mater. Sci. Process.,* **1993**, *57*, 67. [http://dx.doi.org/10.1007/BF00331219]

[15] Jaswal, S.S.; Grandjean, F.; Long, G.J.; Buschow, K.H.J. Interstitial Intermatallic Alloys. Dordrecht: Kluwer, **1995**; pp. 411-32.

[16] Fuzii, H.; Sun, H.; Buschow, K.H.J. Handbook of Magnet Materials. Elsevier: Amsterdam, **1995**; 9, pp. 303-404.

[17] Herbst, J. F. *Rev. Mod. Phys,* **1991**, *63*, 819.

[18] Buschow, K.H.J. New developments in hard magnetic materials. *Rep. Prog. Phys.,* **1991**, *54*(9), 1123-1213. [http://dx.doi.org/10.1088/0034-4885/54/9/001]

[19] Li, H.S.; Coey, J.M.D.; Buschow, K.H.J. HandBook of Magnet Materials. Amsterdam: North-Holland, **1991**; 6, pp. 1-83.

[20] Franse, J.; Radwanski, R.; Buschow, K.H.J. HandBook of Magnet Materials. North-Holland: Amsterdam, **1991**; 6, pp. 307-501.

[21] Liu, J. P.; de Boer, F. R.; de chattel, F. R.; Coehoom, R.; Buschow, K.J. *J. Magn. Magn. Mater,* **1994**, *132*, 159.

[22] Buschow, K.H.J.; Grandjean, F.; Long, G.I.; Buschow, K.H.J. Interstitial Intermatellic Alloys. Dordrecht: Kluwer, **1995**.

[23] Kirchmayr, H. *J. Phys D: Appl. Phys,* **1996**, *29*, 2763.

[24]   Coey, J. *Soild State Commun,* **1997**, *102*, 101.

[25]   Buschow, K. H. J. *Soild State Commun,,* **1991**, *54*, 1123.

[26]   Fastenau, R.; Uan, E. *J. Magn. Magn. Mater,* **1996**, *157*, 1581.

[27]   Buschow, K. H. J.; van Diepen, A.M.; de Wijn, H.W. *Solid State Commun,* **1974**, *15*, 903.

[28]   Sankar, S.G.; Rao, V.U.S.; Segal, E.; Wallace, W.E.; Frederick, W.G.D.; Garrett, H.J. Magnetocrystalline anisotropy of SmCo$_5$ and its interpretation on a crystal-field model. *Phys. Rev., B, Solid State,* **1975**, *11*(1), 435-439. [http://dx.doi.org/10.1103/PhysRevB.11.435]

[29]   Radwański, R.J. The rare earth contribution to the magnetocrystalline anisotropy in RCo5 intermetallics. *J. Magn. Magn. Mater.,* **1986**, *62*(1), 120-126. [http://dx.doi.org/10.1016/0304-8853(86)90744-4]

[30]   Strnat, K.; Wohlforth, E.P.; Buschow, K. Amsterdan:North-Holland. *Ferromagnetic Materials,* **1988**, *4*, 131-210. [http://dx.doi.org/10.1016/S1574-9304(05)80077-X]

[31]   Sagawa, M.; Fujimura, S.; Togawa, N.; Yamarnoto, H.; Matsuura, Y. *J. Appl. Phys,* **1984**, *55*

[32]   Croat, J.; Herbs, J. F.; Lee, R. W.; Pinkerton, F. E. *J. Appl. Phys,* **1984**, *55*, 2078.

[33]   Steinbeck, L.; Richter, M.; Eschrig, H. Itinerant-electron magnetocrystalline anisotropy energy of YCo 5 and related compounds. *Phys. Rev. B Condens. Matter,* **2001**, *63*(18)184431 [http://dx.doi.org/10.1103/PhysRevB.63.184431]

[34]   Yu, P.Y.; Cardona, M. *Fundamentals of Semiconductors*; Springer-Verlag: Berlin, **1999**. [http://dx.doi.org/10.1007/978-3-662-03848-2]

[35]   Herring, C. *Phys. Rev,* **1940**, *57*, 1169.

[36]   Chadi, D.J.; Cohen, M.L. Tight-binding calculations of the valence bands of diamond and zincblende crystals. *Phys. Status Solidi, B Basic Res.,* **1975**, *68*(1), 405-419. [http://dx.doi.org/10.1002/pssb.2220680140]

[37]   Luttinger, J.; Kohn, W. *Phys. Rev. 869,* **1955**, *97*

[38]   Chelikowsky, J.R.; Cohen, M.L. Nonlocal pseudopotential calculations for the electronic structure of eleven diamond and zinc-blende semiconductors. *Phys. Rev., B, Solid State,* **1976**, *14*(2), 556-582. [http://dx.doi.org/10.1103/PhysRevB.14.556]

[39]   Fermi, E. *Nuovo Cimento,* **1934**, *11*, 157.

[40]   Hellman, H. J. *Chem J.. Phys,* **1935**, *3*, 61.

[41]   Slater, J. C.; Coster, G. F. *Phys.Rev. 1498,* **1954**, *94*

[42]   Kittle, C. *Introduction to Soild State Physics,* 6th ed; John Wiley & Sons, Inc, **1986**.

[43]   Callaway, J.; March, N. H. Density Functional Method: Theory and Applications. *Soild State Physics,* **1984**, *38*, 135.

[44]   Kohn, W.; Sham, L.J. *Phys. Rev. A,* **1965**, *1133*, 140.

[45]   Singh, O.K. *Plane Wave, PseudoPotentials and the LAPW Method*; Kluwer Academic: Boston, **1994**. [http://dx.doi.org/10.1007/978-1-4757-2312-0]

[46]   Anderson, O.K. *Phys. Rev,B.,* **1975**, *12*, 3060. [http://dx.doi.org/10.1103/PhysRevB.12.3060]

[47]   Schwarz, K.; Blaha, P. B.; Madsen, G. K. H. *Soild State Physics ,* **2002**, *147*, 71.

[48]   Schwarz, K.; Blaha, P.B. Quantum Mechanical Computations at The Atomic Scale Material Sciences

[49]   Blaha, P.B.; Schwarz, K.; Madsen, G.K.H.; Kvasnicka, D.; Luitz, J. *Augmented Plane Wave Plus Local Orbitals Program for Calculating Crystal Properties*; , **2001**.

[50]   Schwarz, K.; Blaha, P. B. *Comp. Mat. Sci.*, **2003**, *140*, 269.

[51]   Hohenberg, P.; Kohn, W. *J., Phys. Rev.*, **1965**, *140A*

[52]   Perdew, J.P.; Wang, Y. Accurate and simple analytic representation of the electron-gas correlation energy. *Phys. Rev. B Condens. Matter*, **1992**, *45*(23), 13244-13249.
[http://dx.doi.org/10.1103/PhysRevB.45.13244] [PMID: 10001404]

[53]   Perdew, J.P.; Burke, K.; Ernzerhof, M. Generalized Gradient Approximation Made Simple. *Phys. Rev. Lett.*, **1996**, *77*(18), 3865-3868.
[http://dx.doi.org/10.1103/PhysRevLett.77.3865] [PMID: 10062328]

[54]   Engel, E.; Vosko, S.H. Exact exchange-only potentials and the virial relation as microscopic criteria for generalized gradient approximations. *Phys. Rev. B Condens. Matter*, **1993**, *47*(20), 13164-13174.
[http://dx.doi.org/10.1103/PhysRevB.47.13164] [PMID: 10005620]

[55]   Ceperley, D. Ground state of the fermion one-component plasma: A Monte Carlo study in two and three dimensions. *Phys. Rev. B Condens. Matter*, **1978**, *18*(7), 3126-3138.
[http://dx.doi.org/10.1103/PhysRevB.18.3126]

[56]   Ceperley, D.M.; Alder, B.J. Ground State of the Electron Gas by a Stochastic Method. *Phys. Rev. Lett.*, **1980**, *45*(7), 566-569.
[http://dx.doi.org/10.1103/PhysRevLett.45.566]

[57]   Zaremba, E.; Kohn, W. Van der Waals interaction between an atom and a solid surface. *Phys. Rev., B, Solid State*, **1976**, *13*(6), 2270-2285.
[http://dx.doi.org/10.1103/PhysRevB.13.2270]

[58]   Born, M.; Oppenheimer, ; Annoder, J.R. *Annoder, Phys,* **1927**, 457.

[59]   Hartree, D. R. *Proc. Cambridge. Philos. Soc,* **1928**, *24*, 89.

[60]   Fock, V. Z. Phys. **1930**, *61*, 126.

[61]   Slater, J. C. *Phys. Rev,* **1930**, *35*

[62]   March, N.H. *In theory of The Inhomogeneous Electron Gas*; Lundquist, S.; March, N.H., Eds.; Plenum publishing: New York, **1983**.

[63]   Thomas, L.H. The calculation of atomic fields. *Math. Proc. Camb. Philos. Soc.*, **1927**, *23*(5), 542-548.
[http://dx.doi.org/10.1017/S0305004100011683]

[64]   Fermi, E. Z . *Phys.*, **1928**, *48*, 73.

[65]   Zaremba, E.; Kohn, W. Van der Waals interaction between an atom and a solid surface. *Phys. Rev., B, Solid State*, **1976**, *13*(6), 2270-2285.
[http://dx.doi.org/10.1103/PhysRevB.13.2270]

[66]   Hohenberg, P. *Phys. Rev, B.*, **1976**, *864*, 136.

[67]   Hevy, M. *Phys. Rev. A,* **1982**, *26*, 1200.
[http://dx.doi.org/10.1103/PhysRevA.26.1200]

[68]   Kohn, W.; Sham, L.J. *Phys. Rev. A,* **1965**, *1133*, 140.

[69]   Langreth, D.C.; Perdew, J.P. Exchange-correlation energy of a metallic surface: Wave-vector analysis. *Phys. Rev., B, Solid State*, **1977**, *15*(6), 2884-2901.
[http://dx.doi.org/10.1103/PhysRevB.15.2884]

[70]   Hammond, B. L.; Lester, W. A.; Reynolds, P. J. Monte Carlo Method in Ab initio Quantum chemistry', (world Scientific, Singappore). **1994**.

[71]   Filippi, C.; Umrigar, C.J.; Taut, M. J., Chem. Phys. **1994**, *100*, 1295.

[72] Anisimov, V.I.; Zaanen, J.; Andersen, O.K. Band theory and Mott insulators: Hubbard *U* instead of Stoner *I*. *Phys. Rev. B Condens. Matter,* **1991**, *44*(3), 943-954. [http://dx.doi.org/10.1103/PhysRevB.44.943] [PMID: 9998274]

[73] Anisimov, V.I.; Gunnarsson, O. Density-functional calculation of effective Coulomb interactions in metals. *Phys. Rev. B Condens. Matter,* **1991**, *43*(10), 7570-7574. [http://dx.doi.org/10.1103/PhysRevB.43.7570] [PMID: 9996375]

[74] Slater, J.C. *Phys. Rev*; , **1937**, 51, pp. 151-156.

[75] Singh, D.J. *Plane Waves, Pseudopotentials, and The LAPW method*; Kluwer Academic: Boston, **1994**. [http://dx.doi.org/10.1007/978-1-4757-2312-0]

[76] Anderson, O. K. *Phys. Rev,* **1975**, *12*, 3060-3083.

[77] Singh, D. *Phys. Rev.,* **1991**, *B.43*, 6388-6392.

[78] Jepsen, O.; Andersen, O. K. *Solid State Com,* **1971**, *9*, 1763.

[79] Blöchl, P.E.; Jepsen, O.; Andersen, O.K. Improved tetrahedron method for Brillouin-zone integrations. *Phys. Rev. B Condens. Matter,* **1994**, *49*(23), 16223-16233. [http://dx.doi.org/10.1103/PhysRevB.49.16223] [PMID: 10010769]

[80] Kohn, W.; Sham, L.J. Self-Consistent Equations Including Exchange and Correlation Effects. *Phys. Rev.,* **1965**, *140*(4A), A1133-A1138. [http://dx.doi.org/10.1103/PhysRev.140.A1133]

[81] Singh, D.J. *Planewaves, Pseudopotentials, and the LAPW Method*; Kluwer Academic: Boston, **1994**. [http://dx.doi.org/10.1007/978-1-4757-2312-0]

[82] Perdew, J.P.; Wang, Y. Accurate and simple analytic representation of the electron-gas correlation energy. *Phys. Rev. B Condens. Matter,* **1992**, *45*(23), 13244-13249. [http://dx.doi.org/10.1103/PhysRevB.45.13244] [PMID: 10001404]

[83] Perdew, J.P.; Burke, K.; Ernzerhof, M. *Phys. Rev. Lett*; , **1996**, 77, .

[84] Danan, H.; Herr, A.; Meyer, A. J. P. *J. Appl. Phys,* **1968**, *39*, 669.

[85] Gu, Z.Q.; Ching, W.Y. Comparative studies of electronic and magnetic structures in Y2Fe. *Phys. Rev. B Condens. Matter,* **1987**, *36*(16), 8530-8546. [http://dx.doi.org/10.1103/PhysRevB.36.8530] [PMID: 9942673]

[86] Johansson, T.; Lebech, B.; Nielson, M. H. BjerrumMoller and A. R. Mackintosh. *Phys. Rev. Lett.,* **1970**, *25*, 524. [http://dx.doi.org/10.1103/PhysRevLett.25.524]

[87] Larson, P.; Mazin, I.I.; Papaconstantopoulos, D.A. Calculation of magnetic anisotropy energy in SmCo 5. *Phys. Rev. B Condens. Matter,* **2003**, *67*(21)214405 [http://dx.doi.org/10.1103/PhysRevB.67.214405]

[88] Novak, P.; Kuriplach, J. Ab initio calculation of crystal field parameters in several RT/sub 5/ (R=rare earth; T=Co, Ni) compounds. *IEEE Trans. Magn.,* **1994**, *30*(2), 1036-1038. [http://dx.doi.org/10.1109/20.312482]

[89] Larson, P.; Mazin, I.I. *J. Appl. Phys.,* **2003**, *93*, 6888.

[90] Shaukat, Composition-dependent band gap variation of mixed chalcopyrites. *J. Phys. Chem. Solids,* **1990**, *51*, 1413-1418. [http://dx.doi.org/10.1103/PhysRevB.11.435]

[91] Malik, S.K.; Arlinghaus, F.J.; Wallace, W.E. Spin-polarized energy-band structure of Y Co_5, $SmCo_5$, and Gd Co 5. *Phys. Rev., B, Solid State,* **1977**, *16*(3), 1242-1248. [http://dx.doi.org/10.1103/PhysRevB.16.1242]

[92] Arlinghaus, F. Spin polarized energy band structure of $SmCo_5$. *IEEE Trans. Magn.,* **1974**,

10(3), 726-728.
[http://dx.doi.org/10.1109/TMAG.1974.1058382]

[93] Daalderop, G.H.O.; Kelly, P.J.; Schuurmans, M.F.H. Magnetocrystalline anisotropy of Y Co_5 and related RE Co 5 compounds. *Phys. Rev. B Condens. Matter,* **1996**, *53*(21), 14415-14433.
[http://dx.doi.org/10.1103/PhysRevB.53.14415] [PMID: 9983240]

[94] Richter, M. *J. Phys.,* **1998**, *D.31*, 1017.

[95] Steinbeck, L.; Richter, M.; Eschrig, H. Itinerant-electron magnetocrystalline anisotropy energy of YCo_5 and related compounds. *Phys. Rev. B Condens. Matter,* **2001**, *63*(18)184431
[http://dx.doi.org/10.1103/PhysRevB.63.184431]

[96] Hummler, K.; Fähnle, M. *Ab initio* calculation of local magnetic moments and the crystal field in scrR 2 Fe 14 B (*scrR* =Gd, Tb, Dy, Ho, and Er). *Phys. Rev. B Condens. Matter,* **1992**, *45*(6), 3161-3163.
[http://dx.doi.org/10.1103/PhysRevB.45.3161] [PMID: 10001880]

[97] Steinbeck, L.; Richter, M.; Nitzsche, U.; Eschrig, H. *Ab initio* calculation of electronic structure, crystal field, and intrinsic magnetic properties of Sm_2Fe_{17}, $Sm_2Fe_{17}C_3$, and Sm_2Co_{17}. *Phys. Rev. B Condens. Matter,* **1996**, *53*(11), 7111-7127.
[http://dx.doi.org/10.1103/PhysRevB.53.7111] [PMID: 9982157]

[98] Gu, Z.Q.; Ching, W.Y. Comparative studies of electronic and magnetic structures in Y2Fe. *Phys. Rev. B Condens. Matter,* **1987**, *36*(16), 8530-8546.
[http://dx.doi.org/10.1103/PhysRevB.36.8530] [PMID: 9942673]

[99] Ching, W.Y.; Xu, Y.N.; Harmon, B.N.; Ye, J.; Leung, T.C. Electronic structures of FeB, Fe_2B, and Fe_3B compounds studied using first-principles spin-polarized calculations. *Phys. Rev. B Condens. Matter,* **1990**, *42*(7), 4460-4470.
[http://dx.doi.org/10.1103/PhysRevB.42.4460] [PMID: 9995976]

[100] Yang, C.Y.; Johnson, K.H.; Salahub, D.R.; Kaspar, J.; Messmer, R.P. Iron clusters: Electronic structure and magnetism. *Phys. Rev. B Condens. Matter,* **1981**, *24*(10), 5673-5692.
[http://dx.doi.org/10.1103/PhysRevB.24.5673]

[101] Blaha, P.; Schwarz, K.; Madsen, G.K.H.; Kvasnicka, D.; Luitz, J. *WIEN2K, An Agumented Plane Wave + Local Orbitals for Calcu- lating Crystal Properties*; K. Schwarz, Techn. Universitat Wien Austria, **2001**.

[102] Singh, D. Ground-state properties of lanthanum: Treatment of extended-core states. *Phys. Rev. B Condens. Matter,* **1991**, *43*(8), 6388-6392.
[http://dx.doi.org/10.1103/PhysRevB.43.6388] [PMID: 9998076]

[103] Koelling, D.D.; Harmon, B. *J. Phys.,* **1977**, *10*, 3107.

[104] Sherif Yehia, S. Aly, Abeer E. Aly, Computational Materials Science, Vol 41, Issue 4, pp.482-485, 2008; Sherif Yehia, S. Aly, A. S. Hamid, Abeer E. Aly, M. Hammam. *International Journal of Pure and Applied Physics,* **2006**, *2*(3), 205-213.

[105] Gotsis, H.J.; Mazin, I.I. Ferromagnetism and spin-orbital compensation in Sm intermetallics. *Phys. Rev. B Condens. Matter,* **2003**, *68*(22)224427
[http://dx.doi.org/10.1103/PhysRevB.68.224427]

[106] Zhao, T.S.; Jin, H.M.; Gua, G.H.; Han, X.F.; Chen, H. *Phys. Rev. B Condens. Matter,* **1991**, *43*, 8593.
[http://dx.doi.org/10.1103/PhysRevB.43.8593] [PMID: 9996492]

[107] Shibata, T.; Katayama, T. *J. Magn. Magn. Mater.,* **1983**, *31*(1029)

[108] Kutterer, R.; Hilzinger, H.R.; Kronmuller, H. *J. Magn. Magn. Mater.,* **1977**, *4*(1)

[109] Radwanski, R.J. *J. Magn. Magn. Mater.,* **1986**, *62*(120)

CHAPTER 9

A Comparative Study on Visible Light Induced Photocatalytic Activity of MWCNTs Decorated Sulfide Based (ZnS & CdS) Nano Photocatalysts

Rajesh Sahu[1], **S.K. Jain**[1,*] and **Balram Tripathi**[2]

[1] *Department of Physics, School of Basic Sciences, Manipal University Jaipur, Jaipur-303007, India*

[2] *Department of Physics, S.S. Jain Subodh P.G. College, Jaipur-302004, India*

Abstract: Sulfide-based semiconductor nano photocatalysts like ZnS and CdS of different particle sizes were prepared by chemical method. These photo catalysts show an excellent photo catalytic activity in the visible region due to their appropriate energy bandgap (Eg). Multiwalled carbon nanotubes (MWCNTs) intercalated sulfide-based photocatalysts like ZnS/MWCNTs and CdS/MWCNTs composites enhance photocatalytic response in comparison to ZnS and CdS NCs. The photocatalytic activity of MWCNTs intercalated ZnS and CdS composites were studied *via* decomposition of organic pollutant. The obtained particle size of the CdS, ZnS, MWNT/CdS, and MWCNT/ZnS crystals were found to be 32.0 nm, 8.48 nm, 38.5 nm, and 13.18 nm, respectively. The FTIR characteristics of MWCNT/ZnS and MWCNT/CdS composites represent bands at 1637 and 3313 cm^{-1} in presence of methylene blue. The intense band at 1637 cm^{-1} could be the stretching vibrations of the C=O group and the other intense band at 3313cm^{-1} was assigned to the stretching vibration of O-H group. The reduction in optical band gap for MWCNTs/CdS (2.39eV) over CdS (2.44eV) and MWCNT/ZnS (3.77 eV) over ZnS (3.88 eV) was observed. Enhancement in photocatalytic activity was verified along with pseudo-first-order chemical kinetics.

Keywords: MWCNTs, Visible light, UV- visible Spectroscopy.

INTRODUCTION

In past years, many scientific efforts have been made on the controlled crystalline size of various nanomaterials to improve their photocatalytic properties. We investigate a new technique to improve environmentally friendly photocatalytic

* **Corresponding author S.K. Jain:** Department of Physics, School of Basic Sciences, Manipal University Jaipur, Jaipur -303007, India; Tel: 9828034055; E-mail: sushilkumar.jain@jaipur.manipal.edu

Dibya Prakash Rai (Ed.)

activity by chemical method. Most semiconductors are used in scientific research for their versatile properties. Similarly, ZnS and CdS NCs are group II-IV semiconductors both are used frequently due to their size dependence, bandgap, chemical and physical properties [1 - 5]. In the past few years, ZnS and CdS NCs have been the most extensively investigated material because of their excellent optical and electronic properties like solar cells, light-emitting diodes, display devices, flat-panel devices, *etc*. ZnS and CdS NCs provide a good result as a photocatalyst because of their suitable potential for organic pollutant elimination and energy bandgap for solar hydrogen production 3.7 eV and 2.42eV, respectively [6 - 13].

On the other hand, MWCNTs show good electronic conductivity, high mechanical strength, and optical properties. High electronic conductivity represents MWCNTs that are capable of accepting, storing and transporting electrons. Therefore, looking at these properties, MWCNTs are used in photocatalytic activity as a co-catalyst [14 - 17]. Due to the unique properties of MWCNTs adsorbed on ZnS and CdS NCs surface and efficiency of photocatalyst improving by increasing charge separation and retardation for the recombination of charge carriers (electron/holes). Therefore, we concluded that MWCNTs prevent photo corrosion in photocatalyst and enhanced photocatalytic activity in doped nanocomposites [18 - 21].

To improve the efficiency of hydrogen production, many researchers have developed several efficient photocatalysts such as titanium dioxide (TiO_2); nitrides, phosphides, metal-based sulphides; nanocomposites; and core-shell catalysts (Christoforidis and Fornasiero, 2017; Madhumitha *et al*., 2018) [22, 23]. We will briefly review the recent progress made in photocatalytic decomposition for the generation of Hydrogen. Pérez *et al*. (2012) synthesized TiO_2–ZnO mixed oxides for hydrogen generation from a water-ethanol mixture. Hybrid photocatalysts were successfully splitting the water and the result TiO_2–ZnO is more active than the TiO_2 semiconductor [24]. Wang *et al*. (2014) tested the photocatalytic performance of Cd/CdS using a photoreactor and Xe lamp (300 W). Doping of Cd nanocrystals into CdS improved photocatalytic activity [25]. Synthesis $Cd_{0.4}Zn_{0.6}S/TiO_2$ was investigated using a sealed quartz reactor with 450 nm LED radiation source (Kozlova *et al*., 2015). TiO_2 increases charge separation by low charge carrier recombination [26]. Ruban and Sellappa (2016) reported the synthesis of (Cds-ZnS)/TiO_2 core-shell NPs and tested the photocatalytic activity using Plexiglas tubular reactor. The composite photocatalyst showed a high hydrogen production compare to the individual catalysts [27]. The photocatalytic activity investigated with different parameters such as catalyst dosage, pH value and the physical properties was studied using Pt-modified commercial TiO_2 photocatalyst (Melián *et al*., 2016) [28]. The composite photocatalyst synthesis by CdS/TiO_2-WO_3 ternary hybrid from formic acid was studied (Y.L. Chen *et al.*, 2016). The lifetime of e^--h^+ pairs in the ternary hybrid is extended further when compared with the binary hybrid [29]. Liu *et al.* (2016) synthesized N-TiO_2−$x@MoS_2$ and tested the photocatalytic H_2 generation performance. The magnificent photocatalytic performance of photocatalyst was ascribed to the composite effect, which is possible due to the narrow bandgap and finally restrained the recombination of charge carriers (e^--h^+) [30]. Xie *et al.* (2017) prepared Pt-doped TiO_2 −ZnO photocatalyst and studied its performance using a Pyrex glass reactor. The result indicates that Pt could promote charge separation [31]. Li *et al.* (2019) compared the photocatalytic hydrogen production efficiency of CdS nanoparticles and nanosheets using Pt as cocatalyst. The small sizes of crystalline NPs show a high photocatalytic activity [32].

Recent work in progress on dopant concentration of MWCNTs in photocatalyst is following: Composting TiO_2 with CNTs and graphene shows a high photocatalytic activity because CNTs increase charge separation of charge carriers and also CNTs improved the stability of photocatalysts. In environmental development applications, CNTs are used because of their owning property and high surface area (Yao *et al.*, 2008) [33]. Synthesis CNTs/TiO_2 composites enhanced the photocatalytic performance of TiO_2 photocatalyst (Cooke *et al.*, 2010) [34]. Wongaree *et al.* (2015) have studied the promotion of the photocatalytic activity of TiO_2 by the incorporation of CNTs. It was also reported that the properties of CNTs are superior to the other catalyst supporting materials like activated carbon or graphite. The degradation efficiency of dye is higher in CNTs/ TiO_2 composites material because CNTs have a lower value of bandgap that improved the visible light absorption in comparison to TiO_2 nanocrystals [35]. Ashkarran *et al.* (2015) have studied the enhancement of visible light photo-induced activity using TiO_2 nanoparticles immobilized on CNTs. The decolourization of Rhodamine B is high in composite material CNT-TiO_2 because CNTs increase the visible light absorption [36].

In addition, enhancement in photocatalytic activity and dye degradation study under visible light is also an important part of our study. Dye degradation study of photocatalytic activity revealed that MWCNTs doped nanocomposites MWCNTS/ZnS and MWCNTs/CdS nanocomposites exhibited high photocatalytic performance in comparison to ZnS and CdS NCs [37 - 39].

EXPERIMENTAL

Materials and Methods

Synthesis MWCNTs were purchased from helix company, China. Methylene blue (aqueous) solution was purchased from Merck chemicals. Cadmium Sulfide (CdS) orange 99% and Zinc Sulfide (ZnS) white 99% extra pure was purchased from Loba Chemie Pvt. Ltd. All synthesis chemicals were provided by credible suppliers and the purity (99%) of chemicals were verified by chemical reactions.

Synthesis

MWCNTs/ZnS and MWCNTs/CdS composites were synthesized by chemical route. MWCNTs were intercalated with ZnS and CdS NCs with different wt% ratios. 1PPM (concentration 1mg/l) aqueous solution of methylene blue (MB) was prepared to observe the photocatalytic response. MWCNTs/ZnS and MWCNTs/CdS composites were added into 1PPM aqueous solution of MB and stirred for 1 hour at room temperature for complete mixing.

Characterization

Structural characterization of MWCNTs/ZnS and MWCNTs/CdS nanocomposites studied by XRD (X-ray diffraction) on Bruker-AXS D8 Advance with Cu-Kα high energy beam (λ=0.1542nm) in the range of 20° to 70° with a scanning rate of 0.01 steps/s. The morphology of prepared samples was studied by SEM (scanning electron microscope) on MIRA II LMH, TESCAN with the 25KV accelerated voltage. The bonding behaviour of composites was investigated by Fourier transform infrared spectrometer (Bruker) with 500-4000 cm^{-1} range in KBr mode. Optical bandgap and absorption were studied under a UV-visible spectrophotometer (Shimadzu, Japan) over a wavelength 200-800 nm range. The luminescence of composite material spectra was investigated on a photoluminescence (PL) spectrometer (Horiba Scientific, Japan) operating with a 200-800nm Xenon flash lamp as a light source. The characteristic absorption peak of methylene blue 664 nm was used for the photocatalytic degradation process and the monitored parameter.

RESULTS AND DISCUSSION

Structural Analysis (XRD)

Fig. (**1**) shows XRD spectra of Synthesis MWCNTs, CdS, ZnS, MWCNTs/CdS, and MWCNTs/ZnS nanocomposites. The MWCNTs have crystalline planes (002) and (100) with diffraction peaks at 2 of 25.07and 43.19 respectively (JCPDS card no. 89-8487). The crystalline planes (100), (002), (101), (102), (110), (103) and (112) of CdS and ZnS with the hexagonal structure were observed. The seven predominant characteristic diffraction peaks observed for CdS at equal to 27.0, 28.6, 30.6, 39.7, 47.6, 51.8, 55.5 (JCPDS card no. 65-3414), and ZnS NCs equal to 24.9, 26.6, 28.3, 36.7, 43.8, 47.9, 51.15 (JCPDS card no.36-1450), respectively. All the observed peaks confirm that composites consisting of MWCNTs/ZnS and MWCNTs/CdS were successfully observed. The characteristic peaks in MWCNTs/ZnS and MWCNTs/CdS nanocomposites were shifting due to the overlapping of MWCNTs with ZnS and CdS NCs. Scherer's equation is used to calculate the average particle size of the composites [40 - 43].

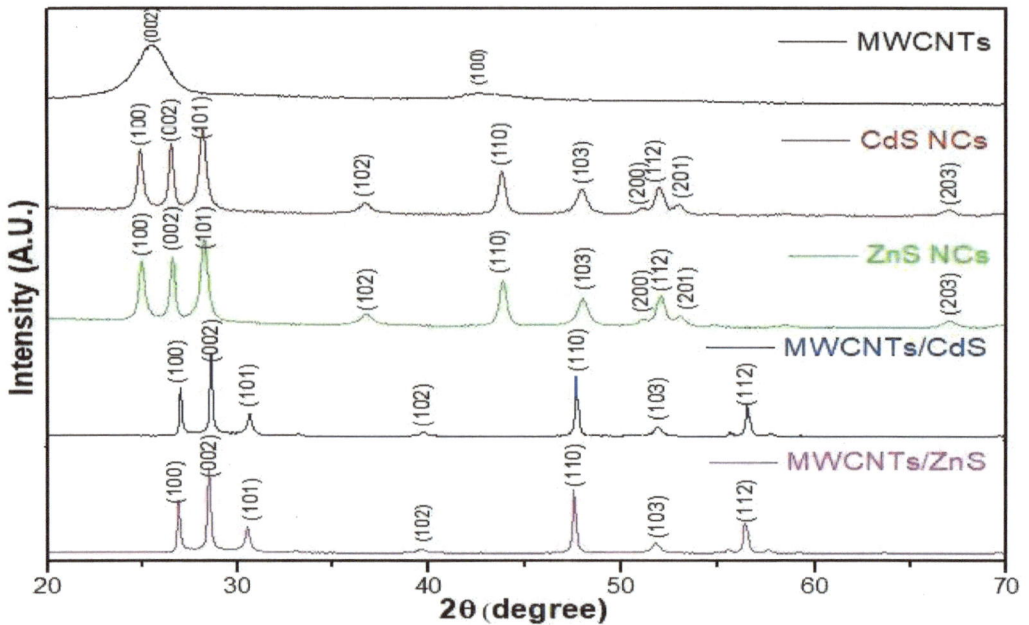

Fig. (1). XRD spectra of (**a**) MWCNTs, (**b**) CdS, (**c**) ZnS, (**d**) MWCNTs/CdS nanocomposites, and (**e**) MWCNTs/ZnS nanocomposites.

$$D = \frac{K\lambda}{\beta Cos\theta} \qquad (1)$$

Where k = 0.94 (shape factor constant), β = FWHM value of the peak. The crystalline size observed for ZnS was 38.52nm and for MWCNTs/ZnS nanocomposites was 13.18nm respectively. Similarly, for CdS, the observed size was 32.57nm and for MWCNTs/CdS nanocomposites was 8.48nm. Therefore intercalation of MWCNTs in ZnS and CdS nanocrystals reduces the particle size of the sub-nanometre range. A good photocatalyst would have a smaller particle size, therefore according to the result, MWCNTs is an excellent co-catalyst and effectively could enhance the photocatalytic activity.

Bonding Analysis (FT-IR)

Fig. (2) represents FT-IR spectra of MWCNTs/ZnS and MWCNTs/CdS nano-composites. It was used to analyze composite's existence of oxygen accommodate functionalities and change after compensating. FTIR spectra show two characteristic curve bands at 1637 and 3313 cm$^{-1.}$ The intense band at 1637 cm^{-1} was due to the stretching vibrations of the carbonyl group and the other intense band at 3313cm^{-1} was assigned to the vibration of O-H stretching.

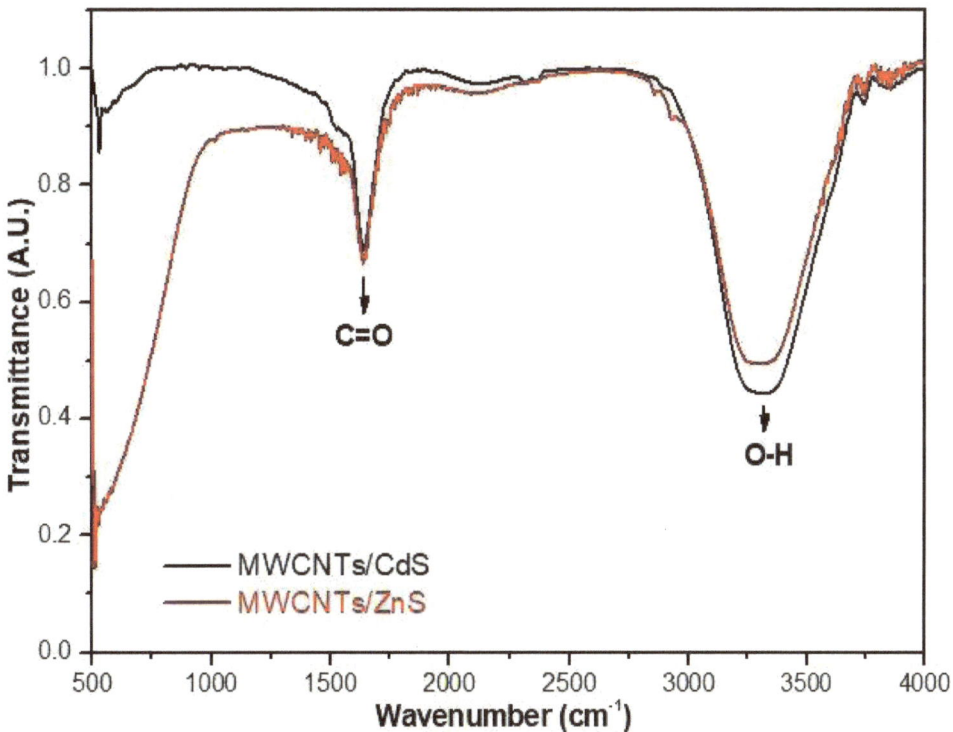

Fig. (2). The FT-IR spectra of MWCNTs/ZnS and MWCNTs/CdS nanocomposites.

Surface Morphology Analysis (SEM)

Fig. **3(a-c)** represents the SEM images of synthesized MWCNTs, MWCNTs/CdS, and MWCNTs/ZnS nanocomposites, respectively. SEM images show the shape of MWCNTs/CdS and MWCNTs/ZnS nanocomposites as quasi-spherical with uniform distribution of NCs. SEM images confirm that MWCNTs are closely in contact with several CdS and ZnS NCs. It is evident existence of strong interaction between MWCNTs and sulfide semiconductors. This strong interaction increases the production of electron-hole (e-h) pairs.

Fig. (3). SEM image of **(a)** Synthesized MWCNTs, **(b)** MWCNTs/CdS nanocomposites, and **(c)** MWCNTs/ZnS nanocomposites.

Fig. (**4**) represents the EDX (energy-dispersive X-ray spectroscopy) image of the chemical composition percentage of MWCNTs/CdS nanocomposites. The weight percentage and atomic percentage of MWCNTs/CdS nanocomposites have been shown in EDX Table **1**. EDX spectroscopy was used to analyze the purity of nanocomposites, EDX Table **1** represents that no other chemical composition was found that indicates the purity of photocatalyst.

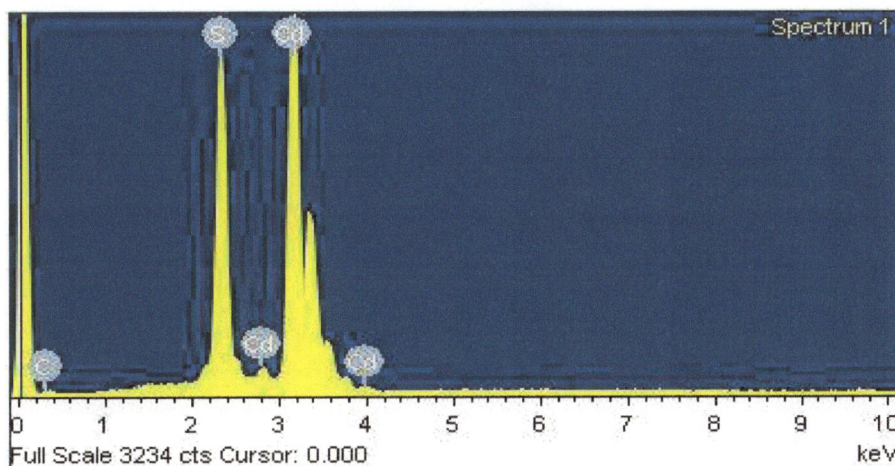

Fig. (4). EDX image of MWCNTs/CdS nanocomposites.

Table 1. The elemental composition percentage of MWCNTs/CdS nanocomposites by EDX technique.

MWCNTs/CdS Nanocomposites Elemental Composition		
Element	**Weight Percentage**	**Atomic Percentage**
Carbon (C)	4.39	21.34
Sulfur (S)	22.33	40.62
Cadmium (Cd)	73.28	38.03
Total	**100**	**100**

Fig. (**5**) represents the EDX (energy-dispersive X-ray spectroscopy) image of chemical composition percentage of MWCNTs/ZnS nanocomposites. The weight percentage and atomic percentage of MWCNTs/ZnS nanocomposites has been shown in EDX Table **2**. EDX Table **2** represents that no other chemical composition was found that indicate that purity of photocatalyst.

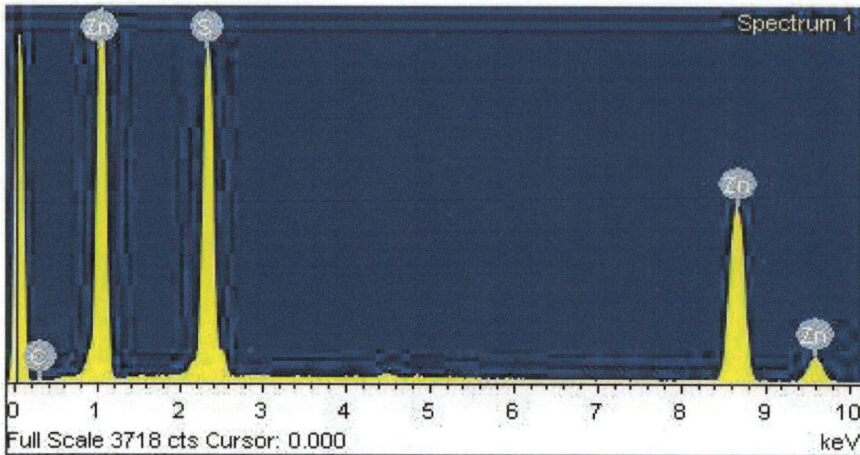

Fig. (5). EDX image of MWCNTs/ZnS nanocomposites.

Table 2. The elemental composition percentage of MWCNTs/ZnS nanocomposites by EDX technique.

MWCNTs/ZnS Nanocomposites Elemental Composition		
Element	**Weight Percentage**	**Atomic Percentage**
Carbon (C)	7.41	24.48
Sulfur (S)	30.68	37.95
Zinc (Zn)	61.91	37.57
Total	**100**	**100**

Optical Absorption Analysis (UV-Vis)

To study the enhancement in photocatalytic activity and decomposition study of dye in photocatalyst UV-visible spectroscopy is a very important tool. Fig. **6(a,b)** show optical absorption spectra of ZnS, MWCNTs/ZnS, CdS, and MWCNTs/CdS respectively.

The energy bandgap (Eg) of nanocomposites can be evaluated by drawing a Tauc plot between $(\alpha h\upsilon)^2$ versus $(h\upsilon)$. According to Tauc relation [44 - 46]:

$$(\alpha h\upsilon)^2 = A\,(h\upsilon - Eg)^n \qquad\qquad (2)$$

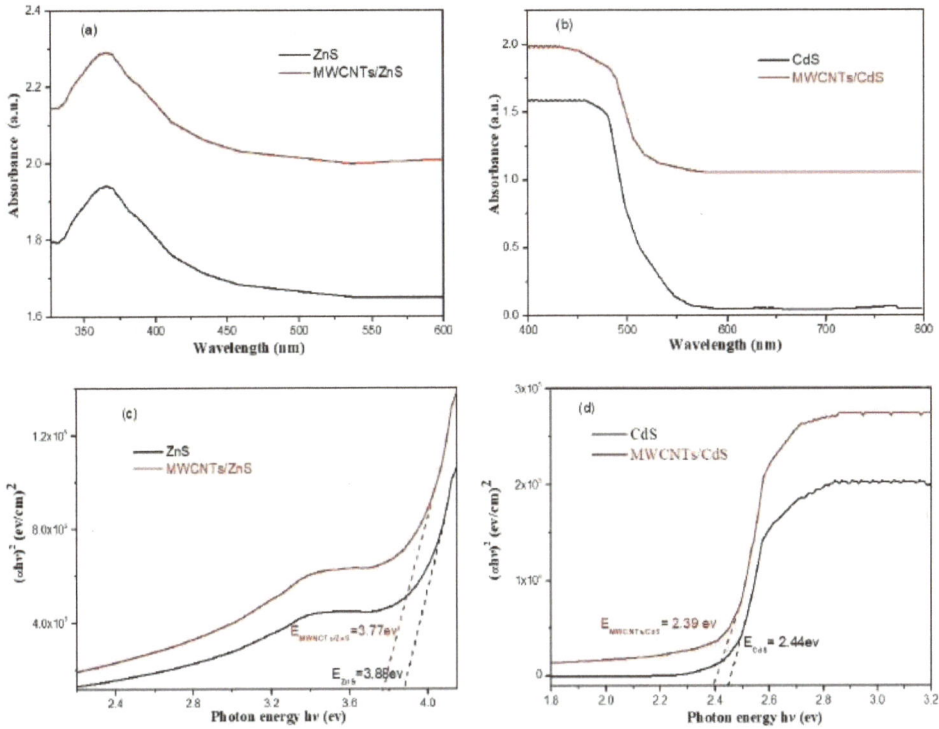

Fig. (6). (a) Absorption spectrum of ZnS and MWCNTS/ZnS nanocomposites, (b) absorption spectrum of CdS and MWCNTs/CdS nanocomposites, (c) Tauc plot of ZnS and MWCNTS/ZnS nanocomposites, (d) Tauc plot of CdS and MWCNTs/CdS nanocomposites.

Where A = constant, $h\upsilon$ = photon energy, and n = 0.5 (direct allowed transition). The energy bandgap as shown in Fig. (6c) can be calculated by abrupt drop curve and intercept tangent line given the value of energy bandgap 3.77 eV and 3.88 eV for MWCNTs/ZnS nanocomposites and ZnS NCs, respectively. Similarly, Fig. (6d) represents the energy bandgap using Tauc relation calculated for MWCNTs/CdS nanocomposites and CdS are 2.39 eV and 2.44 eV, respectively. The reduction of energy bandgap in MWCNTs/ZnS and MWCNTs/CdS nanocomposites increase visible light absorption and produces more photo-charge carriers.

Photocatalytic Activity Analysis

Fig. 7(a and b) shows the photocatalytic degradation spectra of methylene blue under visible light for a different time duration in the existence of MWCNTs/ZnS nanocomposites and MWCNTs/CdS nanocomposites, respectively. The absorption peaks of MWCNTs/ZnS and MWCNTs/CdS nanocomposites corresponding to methylene blue rapidly decrease and finally vanish with increasing exposure time. No other absorption peaks emerging out in the visible region pointed out that methylene blue is degraded into inorganic ions, without the creation of new organic particles. Therefore, the result indicates that the adsorption of MWCNTs in photocatalyst increases the rate of photo-reaction.

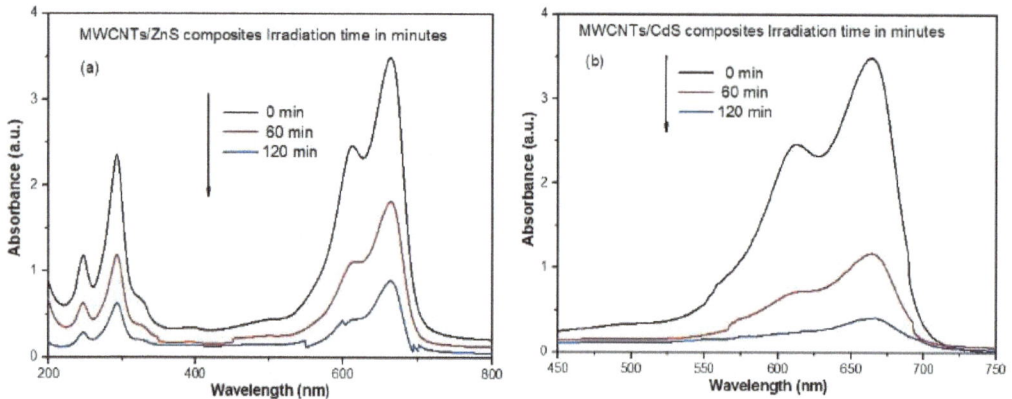

Fig. (7). The UV- visible spectrum of methylene blue under visible light **(a)** MWCNTs/ZnS composites and **(b)** MWCNTs/CdS composites.

The percentage degradation of methylene blue can be determined using the Beer-Lambert law [47 - 50]:

$$R = \frac{C_o - C}{C_o} \times 100\% \tag{3}$$

Where C_o is the initial concentration of methylene blue and after a time t concentration is C.

Fig. **(8)** shows the comparative study of percentage degradation for MWCNTs/ZnS and MWCNTs/CdS nanocomposites with time. According to the plot, MWCNTs/CdS nanocomposites show rapid and higher percentage degradation in comparison to ZnS NCs, CdS NCs, and MWCNTs/ZnS nanocomposites. The sequence of rapid and higher percentage degradation of photocatalyst is MWCNTs/CdS nanocomposites greater > MWCNTs/ZnS nanocomposites > CdS NCs > ZnS NCs. These results show that MWCNTs could successfully promote the degradation rate of ZnS and CdS NCs.

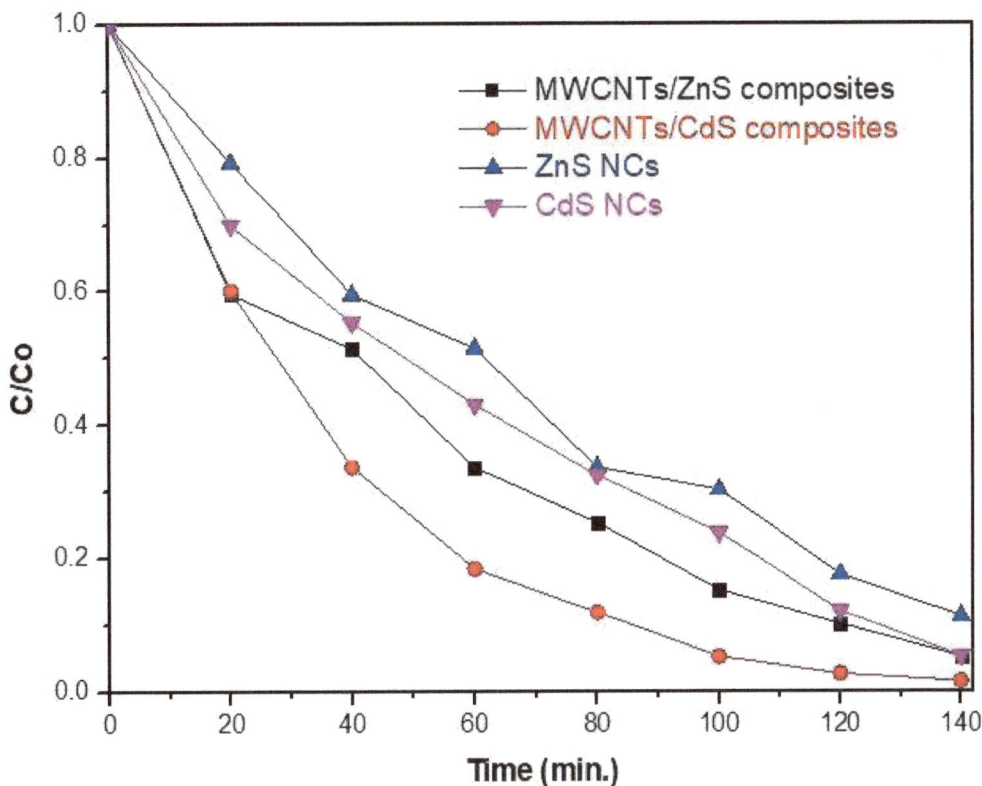

Fig. (8). The plot of C/C_o versus visible light exposure time for MWCNTs/ZnS nanocomposites, MWCNTs/CdS nanocomposites, ZnS NCs and CdS NCs.

Fig. (9) shows plots in $\ln(C/C_o)$ versus 't' of pseudo first-order-kinetics with R^2 = 0.9681, 0.9269, 0.9790, 0.9951 and value of k = 0.0134, 0.0089, 0.0082 and 0.0065 minute^{-1} corresponds to the ZnS, CdS, MWCNTs/ZnS and MWCNTs/ CdS nanocomposites, respectively. The value of R^2 revealed that reactions follow first-order kinetics under visible light. The values of reaction rate constants 'k' in indicating the higher degradation efficiency of MWCNTs/CdS nanocomposites. According to Table **3** on the basis of the value of reaction rate constant K (min^{-1}) the order of degradation efficiency is given below:

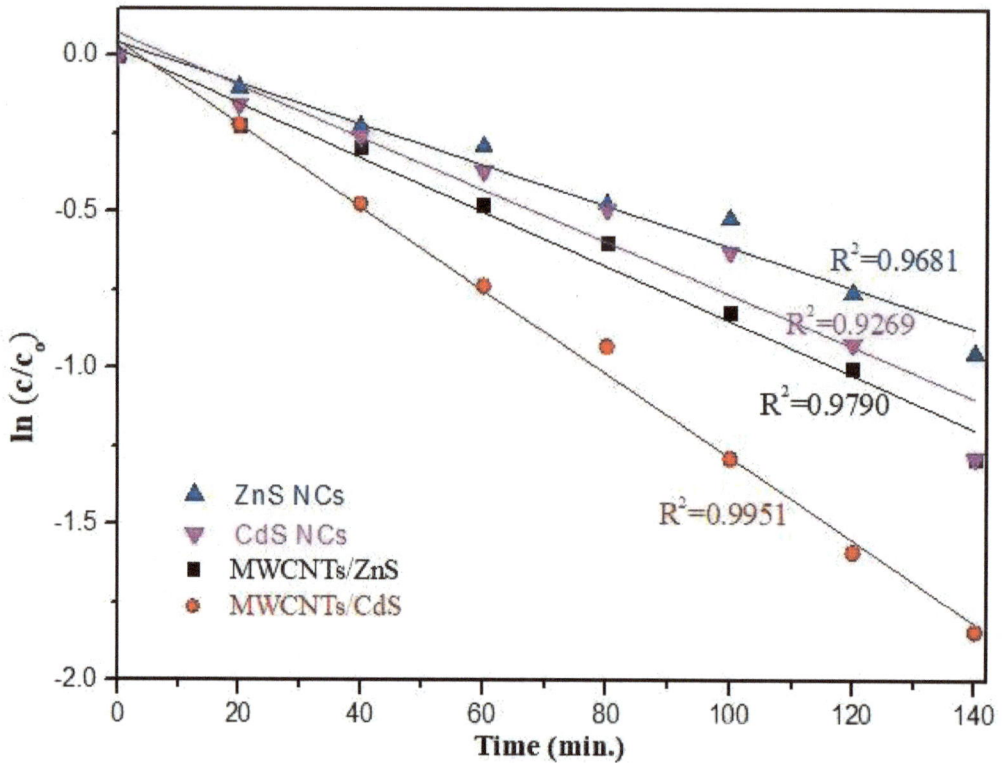

Fig. (9). The plot of ln C/C_o versus visible light exposure time for ZnS, CdS, MWCNTs/ZnS, and MWCNTs/CdS nanocomposites.

Table 3. The different values of R^2 and K by pseudo-first-order-kinetics of photocatalysts.

S.No.	Material	ln(C/Co)	
		R^2	K (min^{-1})
1.	ZnS	0.9681	0.0134
2.	CdS	0.9269	0.0089
3.	MWCNTs/ZnS	0.9790	0.0082
4.	MWCNTs/CdS	0.9951	0.0065

Fig. **(10)** represents the photoluminescence spectra of photocatalyst composites. The photoluminescence spectra of ZnS NCs and MWCNTs/ZnS composites have emission peaks obtained at 335.02nm with higher intensity in Fig. **10(a)**. In a similar way, the photoluminescence spectra of CdS NCs and MWCNTs/CdS composites have emission peaks obtained at 684.4 nm with higher intensity in Fig. **10(b)**. The reduction in PL intensity after doping of MWCNTs indicates that charge recombination is effectively restrained in nanocomposites. Therefore, the photoluminescence spectrum of photocatalyst verifies the result of UV visible spectroscopy.

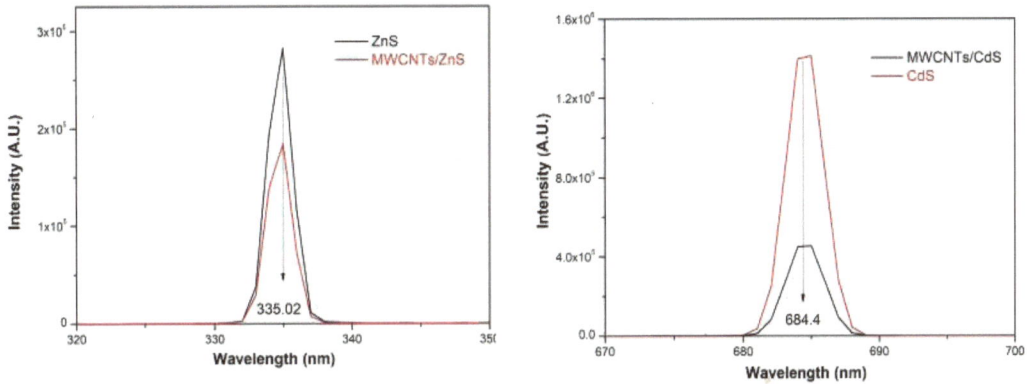

Fig. (10). Photoluminescence (PL) spectra of a solution of dye (MB) in the existence of (a) ZnS NCs & MWCNTs/ZnS nanocomposites, and (b) CdS NCs & MWCNTs/CdS nanocomposites.

Possible Photocatalytic Mechanism

Fig. (11) shows possible reaction mechanism of photocatalytic activity under visible light. For the photocatalytic process, the energy of the photon (hυ) absorbed by the photocatalyst must be higher than its bandgap energy (Eg) **(Step 1)**. On the photocatalyst surface excited electron moves in the conduction band and vacant space of valance band is filled by holes after absorption of photon energy by the semiconductor. **Step 2** represents excited electrons and holes produced in bulk, and they separate and migrate on the semiconductor surface. For redox reaction with water, CO_2 or organic contaminants at the surface of semiconductor electron acts as strong reductants and holes act as an oxidizer **(step 3)**. A few photo charge carriers do not participate in redox reactions because they emitted light in the form of heat. Therefore pair of charge carriers electron and holes recombine in the semiconductor surface in **step 2**, to reduce the recombination of charge carriers used MWCNTs as a co-catalyst. MWCNTs trapped and store excited electrons from the semiconductor surface and forward it to a direct redox reaction, as a result, reduce the recombination of charge carriers and enhance the photocatalytic activity. Following reactions are possible for photocatalytic activity [51 - 54]:

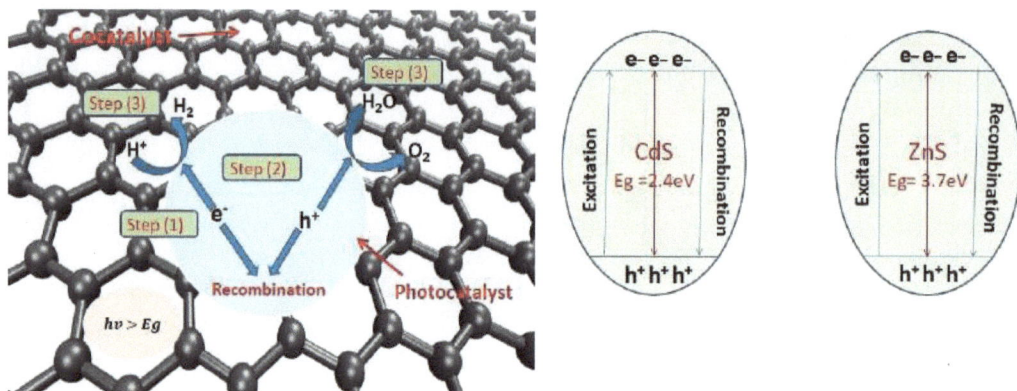

Fig. (11). The possible reaction of photocatalytic activity in the presence of visible light.

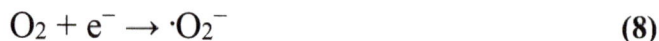

$$MWCNTs/ZnS + h\upsilon \rightarrow MWCNTs/ZnS + e^- + h^+ \qquad (4)$$

$$MWCNTs/CdS + h\upsilon \rightarrow MWCNTs/CdS + e^- + h^+ \qquad (5)$$

$$h^+ + H_2O \rightarrow \cdot OH + H^+ \qquad (6)$$

$$OH^- + h^+ \rightarrow \cdot OH \qquad (7)$$

$$O_2 + e^- \rightarrow \cdot O_2^- \qquad (8)$$

In the end, the photocatalytic reaction obtained CO_2 and H_2O [55, 56].

$$MB + \cdot OH/\cdot O_2^- \rightarrow CO_2 + H_2O + NH_4^+ + NO_3^- + SO_2^{4-} + Cl^- \qquad (9)$$

CONCLUSION

This work demonstrates the important photocatalytic responses of MWCNTS/ZnS and MWCNTs/CdS nanocomposites by using methylene blue dye as a degradation agent. XRD spectra ensure the formation of MWCNTS/ZnS and MWCNTs/CdS nanocomposites with a hexagonal crystal structure and distributed particle size. Optical absorption spectra confirm reduced bandgap due to intercalation of MWCNTs in CdS and ZnS nanocrystals. The photodegradation efficiency of MWCNTs/CdS nanocomposites was found to be higher than MWCNTs/ZnS nanocomposites, which could absorb visible light more efficiently. Therefore, MWCNTs intercalation with CdS and ZnS photocatalyst revealed high photocatalytic performance and decomposition of MB under visible light.

CONSENT FOR PUBLICATION

Not applicable.

CONFLICT OF INTEREST

The authors declare no conflict of interest, financial or otherwise.

ACKNOWLEDGMENTS

The authors especially thank Central Analytical Facilities, Manipal University Jaipur, Jaipur for characterization facilities for this work.

REFERENCES

[1]   Lin, C.C.; Chiang, Y.J. Feasibility of using a rotating packed bed in preparing coupled ZnO/SnO$_2$ photocatalysts. *J. Ind. Eng. Chem.,* **2012**, *18*(4), 1233-1236.
        [http://dx.doi.org/10.1016/j.jiec.2011.11.152]

[2]   Hu, J.S.; Ren, L.L.; Guo, Y.G.; Liang, H.P.; Cao, A.M.; Wan, L.J.; Bai, C.L. Mass production and high photocatalytic activity of ZnS nanoporous nanoparticles. *Angew. Chem. Int. Ed.,* **2005**, *44*(8), 1269-1273.
        [http://dx.doi.org/10.1002/anie.200462057] [PMID: 15651014]

[3]   Kudo, A.; Sekizawa, M. Photocatalytic H$_2$ evolution under visible light irradiation of Ni- doped ZnS photocatalyst. In: *chem. commun*; , **2000**; 15, pp. 1371-1372.
        [http://dx.doi.org/10.1039/b003297m]

[4]   Feng, S.; Zhao, J.; Zhu, Z. The manufacture of carbon nanotubes decorated with ZnS to enhance the ZnS photocatalytic activity. *N. Carbon Mater.,* **2008**, *23*(3), 228-234.
        [http://dx.doi.org/10.1016/S1872-5805(08)60025-6]

[5]   Fujiwara, H.; Hosokawa, H.; Murakoshi, K.; Wada, Y.; Yanagida, S. Surface characteristics of ZnS Nano crystals relating to their photo catalysis for CO$_2$ reduction. *Langmuir,* **1998**, *14*(18), 5154-5159.
        [http://dx.doi.org/10.1021/la9801561]

[6]   Fang, X.; Zhai, T.; Gautam, U.K.; Li, L.; Wu, L.; Bando, Y.; Golberg, D. ZnS nanostructures: From synthesis to applications. *Prog. Mater. Sci.,* **2011**, *56*(2), 175-287.
        [http://dx.doi.org/10.1016/j.pmatsci.2010.10.001]

[7]   Yaamaguchi, T.; Yamamoto, Y.; Tanka, T.; Yoshida, A. Preparation and characterization of (Cd,Zn)S thin films by chemical bath deposition for photovoltaic devices. In: *Thin solid film*; , **1999**; 344, pp. 516-519.
        [http://dx.doi.org/10.1016/S0040-6090(98)01665-4]

[8]   Ahmed, M.; Rasool, K.; Imran, Z.; Ratiq, M.A.; Hasan, M.M. *Structural and electrical properties of ZnS Nano particles*; IEEE, **2011**, Vol. 1, pp. 4577-0066.

[9]   Tao, H.; Jin, Z.; Wang, W.; Yang, J.; Hong, Z. Preparation and characteristics of CdS thin films by dip-coating method using its nanocrystal ink. *Mater. Lett.,* **2011**, *65*(9), 1340-1343.
        [http://dx.doi.org/10.1016/j.matlet.2011.01.077]

[10]   Pandya, S.G. Preparation and characterization of CdS Nanocrystalline thin film grown by chemical Method. *IJRSR,* **2016**, *7*, 14887-14890.

[11]   Wenyi, L.; Xun, C.; Qiulong, C.; Zhibin, Z. Influence of growth process on the structural, optical and electrical properties of CBD-CdS films. *Mater. Lett.,* **2005**, *59*(1), 1-5.
        [http://dx.doi.org/10.1016/j.matlet.2004.04.008]

[12]   Chate, P.A.; Patil, S.S.; Patil, J.S.; Sathe, D.J.; Hankare, P.P. Synthesis, optoelectronic properties and photoelectrochemical performance of CdS thin films. *Physica B,* **2013**, *411*, 118-121.
[http://dx.doi.org/10.1016/j.physb.2012.11.032]

[13]   Abdulwahab, S.; Lahewil, Z.; Al-Douri, Y.; Hashim, U. *Sol. Energy,* **2012**, *86*, 3234-3240.
[http://dx.doi.org/10.1016/j.solener.2012.08.013]

[14]   Robel, I.; Bunker, B.A.; Kamat, P.V. Single-Walled Carbon Nanotube-CdS Nanocomposites as Light-Harvesting Assemblies: Photoinduced Charge-Transfer Interactions. *Adv. Mater.,* **2005**, *17*(20), 2458-2463.
[http://dx.doi.org/10.1002/adma.200500418]

[15]   Kim, Y.K.; Park, H. Light-harvesting multi-walled carbon nanotubes and CdS hybrids: Application to photocatalytic hydrogen production from water. *Energy Environ. Sci.,* **2011**, *4*(3), 685-694.
[http://dx.doi.org/10.1039/C0EE00330A]

[16]   Woan, K.; Pyrgiotakis, G.; Sigmund, W. Photocatalytic Carbon-Nanotube-TiO$_2$ Composites. *Adv. Mater.,* **2009**, *21*(21), 2233-2239.
[http://dx.doi.org/10.1002/adma.200802738]

[17]   Silva, C.G.; Faria, J.L. Photocatalytic oxidation of phenolic compounds by using a carbon nanotube-titanium dioxide composite catalyst. *ChemSusChem,* **2010**, *3*(5), 609-618.
[http://dx.doi.org/10.1002/cssc.200900262] [PMID: 20437451]

[18]   Chen, X.; Shen, S.; Guo, L.; Mao, S.S. Semiconductor-based photocatalytic hydrogen generation. In: *chemical reviews*; , **2010**; 110, pp. 6503-6570.
[http://dx.doi.org/10.1021/cr1001645]

[19]   Jang, J.; Kim, H.; Joshi, U.; Jang, J.; Lee, J. Fabrication of CdS nanowires decorated with TiO$_2$ nanoparticles for photocatalytic hydrogen production under visible light irradiation. *Int. J. Hydrogen Energy,* **2008**, *33*(21), 5975-5980.
[http://dx.doi.org/10.1016/j.ijhydene.2008.07.105]

[20]   Qutub, N.; Pirzada, B.M.; Umar, K.; Mehraj, O.; Muneer, M.; Sabir, S. Synthesis, characterization and visible-light driven photocatalysis by differently structured CdS/ZnS sandwich and core–shell nanocomposites. *Physica E,* **2015**, *74*, 74-86.
[http://dx.doi.org/10.1016/j.physe.2015.06.023]

[21]   Soltani, N.; Saion, E.; Hussein, M.Z.; Erfani, M.; Abedini, A.; Bahmanrokh, G.; Navasery, M.; Vaziri, P. Visible light-induced degradation of methylene blue in the presence of photocatalytic ZnS and CdS nanoparticles. *Int. J. Mol. Sci.,* **2012**, *13*(12), 12242-12258.
[http://dx.doi.org/10.3390/ijms131012242] [PMID: 23202896]

[22]   Christoforidis, K.C.; Fornasiero, P. Photocatalytic Hydrogen Production: A Rift into the Future Energy Supply. *ChemCatChem,* **2017**, *9*(9), 1523-1544.
[http://dx.doi.org/10.1002/cctc.201601659]

[23]   Madhumitha, A.; Preethi, V.; Kanmani, S. Photocatalytic hydrogen production using TiO$_2$ coated iron-oxide core shell particles. *Int. J. Hydrogen Energy,* **2018**, *43*(8), 3946-3956.
[http://dx.doi.org/10.1016/j.ijhydene.2017.12.127]

[24]   Pérez-Larios, A.; Lopez, R.; Hernández-Gordillo, A.; Tzompantzi, F.; Gómez, R.; Torres-Guerra, L.M. Improved hydrogen production from water splitting using TiO$_2$–ZnO mixed oxides photocatalysts. *Fuel,* **2012**, *100*, 139-143.
[http://dx.doi.org/10.1016/j.fuel.2012.02.026]

[25]   Wang, Q.; Lian, J.; Li, J.; Wang, R.; Huang, H.; Su, B.; Lei, Z. Highly Efficient Photocatalytic Hydrogen Production of Flower-like Cadmium Sulfide Decorated by Histidine. *Sci. Rep.,* **2015**, *5*(1), 13593.
[http://dx.doi.org/10.1038/srep13593] [PMID: 26337119]

[26]   Kozlova, E.A.; Kurenkova, A.Y.; Semeykina, V.S.; Parkhomchuk, E.V.; Cherepanova, S.V.;

Gerasimov, E.Y.; Saraev, A.A.; Kaichev, V.V.; Parmon, V.N. Effect of Titania Regular Macroporosity on the Photocatalytic Hydrogen Evolution on $Cd_{1-x}Zn_x S/TiO_2$ Catalysts under Visible Light. *ChemCatChem,* **2015**, *7*(24), 4108-4117.
[http://dx.doi.org/10.1002/cctc.201500897]

[27]   Ruban, P.; Sellappa, K. Concurrent Hydrogen Production and Hydrogen Sulfide Decomposition by Solar Photocatalysis. *Clean (Weinh.),* **2016**, *44*(8), 1023-1035.
[http://dx.doi.org/10.1002/clen.201400563]

[28]   Melián, E.P.; López, C.R.; Santiago, D.E.; Quesada-Cabrera, R.; Méndez, J.A.O.; Rodríguez, J.M.D.; Díaz, O.G. Study of the photocatalytic activity of Pt-modified commercial TiO_2 for hydrogen production in the presence of common organic sacrificial agents. *Appl. Catal. A Gen.,* **2016**, *518*, 189-197.
[http://dx.doi.org/10.1016/j.apcata.2015.09.033]

[29]   Chen, Y.L.; Lo, S.L.; Chang, H.L.; Yeh, H.M.; Sun, L.; Oiu, C. Photocatalytic hydrogen production of the CdS/TiO_2-WO_3 ternary hybrid under visible light irradiation. *Water Sci. Technol.,* **2016**, *73*(7), 1667-1672.
[http://dx.doi.org/10.2166/wst.2015.639] [PMID: 27054739]

[30]   Liu, M.; Chen, Y.; Su, J.; Shi, J.; Wang, X.; Guo, L. Photocatalytic hydrogen production using twinned nanocrystals and an unanchored NiSx co-catalyst. *Nat. Energy,* **2016**, *1*(11), 16151.
[http://dx.doi.org/10.1038/nenergy.2016.151]

[31]   Xie, M.Y.; Su, K.Y.; Peng, X.Y.; Wu, R.J.; Chavali, M.; Chang, W.C. Hydrogen production by photocatalytic water-splitting on Pt-doped TiO_2–ZnO under visible light. *J. Taiwan Inst. Chem. Eng.,* **2017**, *70*, 161-167.
[http://dx.doi.org/10.1016/j.jtice.2016.10.034]

[32]   Li, X.; Deng, Y.; Jiang, Z.; Shen, R.; Xie, J.; Liu, W.; Chen, X. Photocatalytic Hydrogen Production over CdS Nanomaterials: An Interdisciplinary Experiment for Introducing Undergraduate Students to Photocatalysis and Analytical Chemistry. *J. Chem. Educ.,* **2019**, *96*(6), 1224-1229.
[http://dx.doi.org/10.1021/acs.jchemed.9b00087]

[33]   Yao, Y.; Li, G.; Ciston, S.; Lueptow, R.M.; Gray, K.A. Photoreactive TiO_2/carbon nanotube composites: synthesis and reactivity. *Environ. Sci. Technol.,* **2008**, *42*(13), 4952-4957.
[http://dx.doi.org/10.1021/es800191n] [PMID: 18678032]

[34]   Cooke, D.J.; Eder, D.; Elliott, J.A. Role of Benzyl Alcohol in Controlling the Growth of TiO_2 on Carbon Nanotubes. *J. Phys. Chem. C,* **2010**, *114*(6), 2462-2470.
[http://dx.doi.org/10.1021/jp909117x]

[35]   Wongaree, M.; Chiarakorn, S.; Chuangchote, S. Photocatalytic Improvement under Visible Light in TiO_2 Nanoparticles by Carbon Nanotube Incorporation. *J. Nanomater.,* **2015**, *2015*, 1-10.
[http://dx.doi.org/10.1155/2015/689306]

[36]   Ashkarran, A.A.; Fakhari, M.; Hamidinezhad, H.; Haddadi, H.; Nourani, M.R. TiO_2 nanoparticles immobilized on carbon nanotubes for enhanced visible-light photo-induced activity. *J. Mater. Res. Technol.,* **2015**, *4*(2), 126-132.
[http://dx.doi.org/10.1016/j.jmrt.2014.10.005]

[37]   Lucas, M.S.; Peres, J.A. Degradation of Reactive Black 5 by Fenton/UV-C and ferrioxalate/H_2O_2/solar light processes. *Dyes Pigments,* **2007**, *74*(3), 622-629.
[http://dx.doi.org/10.1016/j.dyepig.2006.04.005]

[38]   Aksu, Z. Application of biosorption for the removal of organic pollutants: a review. *Process Biochem.,* **2005**, *40*(3-4), 997-1026.
[http://dx.doi.org/10.1016/j.procbio.2004.04.008]

[39]   Somasiri, W.; Li, X.F.; Ruan, W.Q.; Jian, C. Evaluation of the efficacy of upflow anaerobic sludge blanket reactor in removal of colour and reduction of COD in real textile wastewater. *Bioresour. Technol.,* **2008**, *99*(9), 3692-3699.

[http://dx.doi.org/10.1016/j.biortech.2007.07.024] [PMID: 17719776]

[40]   Al-Tabbakh, A.A.; Karatepe, N.; Al-Zubaidi, A.B.; Benchaabane, A.; Mahmood, N.B. Crystallite size and lattice strain of lithiated spinel material for rechargeable battery by X-ray diffraction peak-broadening analysis. *Int. J. Energy Res.,* **2019**, *43*(5), 1903-1911.
       [http://dx.doi.org/10.1002/er.4390]

[41]   Rajesh Kumar, B.; Hymavathi, B. X-ray peak profile analysis of solid-state sintered alumina doped zinc oxide ceramics by Williamson–Hall and size-strain plot methods. *Journal of Asian Ceramic Societies,* **2017**, *5*(2), 94-103.
       [http://dx.doi.org/10.1016/j.jascer.2017.02.001]

[42]   Yaremiy, V. X-ray Analysis of NiCrxFe2-xO4 Nanoparticles Using Debye-Scherrer, Williamson-Hall and Size-strain Plot Methods. *Journal of nano and electronic physics,* **2019**, *11*, 04020.

[43]   Ahmadipour, M.; Abu, M.J.; Ab Rahman, M.F.; Ain, M.F.; Ahmad, Z.A. Assessment of crystallite size and strain of CaCu $_3$ Ti $_4$ O $_{12}$ prepared *via* conventional solid-state reaction. *Micro & Nano Lett.,* **2016**, *11*(3), 147-150.
       [http://dx.doi.org/10.1049/mnl.2015.0562]

[44]   Tauc, J.; Grigorovici, R.; Vancu, A. Optical Properties and Electronic Structure of Amorphous Germanium. *Phys. Status Solidi, B Basic Res.,* **1966**, *15*(2), 627-637.
       [http://dx.doi.org/10.1002/pssb.19660150224]

[45]   Davis, E.; Mott, N. Conduction in non-crystalline systems V. Conductivity, optical absorption and photoconductivity in amorphous semiconductors. *Philos. Mag,* **1970**, *22*, 0903-0922.
       [http://dx.doi.org/10.1080/14786437008221061]

[46]   Raciti, R.; Bahariqushchi, R.; Summonte, C.; Aydinli, A.; Terrasi, A.; Mirabella, S. Optical bandgap of semiconductor nanostructures: Methods for experimental data analysis. *J. Appl. Phys.,* **2017**, *121*(23), 234304.
       [http://dx.doi.org/10.1063/1.4986436]

[47]   Mitschele, J. Beer-Lambert Law. *J. Chem. Educ.,* **1996**, *73*(11), A260.
       [http://dx.doi.org/10.1021/ed073pA260.3]

[48]   Aarthi, T.; Narahari, P.; Madras, G. Photocatalytic degradation of Azure and Sudan dyes using nano TiO$_2$. *J. Hazard. Mater.,* **2007**, *149*(3), 725-734.
       [http://dx.doi.org/10.1016/j.jhazmat.2007.04.038] [PMID: 17540499]

[49]   Wang, K.; Yu, L.; Yin, S.; Li, H.; Li, H. Photocatalytic degradation of methylene blue on magnetically separable FePc/Fe$_3$O$_4$ nanocomposite under visible irradiation. *Pure Appl. Chem.,* **2009**, *81*(12), 2327-2335.
       [http://dx.doi.org/10.1351/PAC-CON-08-11-23]

[50]   Abdollahi, Y.; Abdullah, A.H.; Zainal, Z.; Yusof, N.A. Photocatalytic degradation of p-cresol by zinc oxide under UV irradiation. *Int. J. Mol. Sci.,* **2011**, *13*(1), 302-315.
       [http://dx.doi.org/10.3390/ijms13010302] [PMID: 22312253]

[51]   Pawar, R.C.; Kang, S.; Park, J.H.; Kim, J.; Ahn, S.; Lee, C.S. Evaluation of a multi-dimensional hybrid photocatalyst for enrichment of H $_2$ evolution and elimination of dye/non-dye pollutants. *Catal. Sci. Technol.,* **2017**, *7*(12), 2579-2590.
       [http://dx.doi.org/10.1039/C7CY00466D]

[52]   Zhang, Y.C.; Du, Z.N.; Li, K.W.; Zhang, M.; Dionysiou, D.D. High-performance visible-light-driven SnS$_2$/SnO$_2$ nanocomposite photocatalyst prepared *via* in situ hydrothermal oxidation of SnS$_2$ nanoparticles. *ACS Appl. Mater. Interfaces,* **2011**, *3*(5), 1528-1537.
       [http://dx.doi.org/10.1021/am200102y] [PMID: 21476553]

[53]   Rafatullah, M.; Sulaiman, O.; Hashim, R.; Ahmad, A. Adsorption of methylene blue on low-cost adsorbents: A review. *J. Hazard. Mater.,* **2010**, *177*(1-3), 70-80.
       [http://dx.doi.org/10.1016/j.jhazmat.2009.12.047] [PMID: 20044207]

[54] Chen, C.; Ma, W.; Zhao, J. Semiconductor-mediated photodegradation of pollutants under visible-light irradiation. *Chem. Soc. Rev.,* **2010**, *39*(11), 4206-4219.
[http://dx.doi.org/10.1039/b921692h] [PMID: 20852775]

[55] Zhao, J.; Yang, X. Photocatalytic oxidation for indoor air purification: a literature review. *Build. Environ.,* **2003**, *38*(5), 645-654.
[http://dx.doi.org/10.1016/S0360-1323(02)00212-3]

[56] Das, D.P.; Biswal, N.; Martha, S.; Parida, K.M. Solar-light induced photodegradation of organic pollutants over CdS-pillared zirconium–titanium phosphate (ZTP). *J. Mol. Catal. Chem.,* **2011**, *349*(1-2), 36-41.
[http://dx.doi.org/10.1016/j.molcata.2011.08.012]

CHAPTER 10

Organic Solar Cells: Fundamentals, Working Principle and Device Structures

Shyam Sunder Sharma[1,*], **Atul Kumar Dadhich**[2] and **Subodh Srivastava**[3]

[1] *Department of Physics, Government Women Engineering College, Ajmer 305002, India*

[2] *Department of Electrical Engineering, Vivekananda Global University, Jaipur 302012, India*

[3] *Department of Physics, Vivekananda Global University, Jaipur 302012, India*

Abstract: New photovoltaic energy technologies are helping to provide ecologically acceptable renewable energy sources while also lowering carbon dioxide emissions from fossil fuels and biomass. Organic photovoltaic (OPV) technology is a novel type of solar technology based on conjugated polymers and small molecules. These solar cells have enticed triable attention in recent years due to their potential of providing mechanical flexible, light weight, low cost and environmental friendly solar cells with highly tunable electrical and chemical properties. In particular, bulk-heterojunction organic solar cells (OSCs) made up of a blend of a p-type conjugated polymer as a donor and an n-type semiconductor as an acceptor is thought to be a viable method. The fundamental physics of OSCs, their operating mechanism, novel materials used and device architectures are discussed in this chapter. The technological development for large-area fabrication and the studies on stability issues of the flexible OSCs will be the main focus of the researchers in the next step. The chapter also reviews the present state of OSC production and the problems that it faces, as well as issues of stability and deterioration.

Keywords: Bulk-heterojunction, Conjugated polymer, Device architecture, Fullerene, Organic solar cells.

INTRODUCTION

Renewable energy production has surpassed all predictions in the last decade. Global installed capacity and production from all renewable technologies have risen significantly, and policy support has continued to spread in all parts of the globe. Renewable energy and energy efficiency have long been recognized as having many societal advantages, including reduced energy prices and carbon emission, improved air quality and public health, and increased job creation and

* **Corresponding author Shyam Sunder Sharma:** Department of Physics, Govt. Women Engineering College, Ajmer-305002, India; Tel: 9414778985; E-mail: shyam@gweca.ac.in

Dibya Prakash Rai (Ed.)

economic growth. The energy crises of the 1970s, as well as later economic downturns, were major drivers in the fast expansion of renewable energy sources. The momentum has been aided by a growing focus on preventing climate change and adjusting to its consequences. Due to low operating costs and privileged access to power networks, renewable energy will have the largest percentage of the global electricity mix ever in 2020, with an estimated 29 percent [1]. The contribution of renewable energy to the global heat, electricity, and transportation sectors has continuously grown. In 2019, contemporary renewable energy was expected to contribute 11.2 percent of total final energy consumption (TFEC). Renewable energy accounted for the most TFEC (6.0 percent), followed by renewable heat (4.2 percent) and transportation biofuels (1.0 percent) [2].

Solar PV is becoming the most cost-effective alternative for power generation, and its demand is continuously growing. In recent years, the cost of producing power from solar energy has decreased substantially. Since 2010, the global weighted average levelised cost of energy generated by utility-scale solar photovoltaics (PV) has decreased by 85% through 2020 [3]. Solar PV experienced another record-breaking year, with about 139 gigawatts (GW) of renewable energy capacity built globally throughout the year, bringing the total global capacity (both on-grid and off-grid) to an estimated 760 GW [4]. Now, the idea of achieving high shares of renewable energy is considered feasible. However, the renewable energy sector still faces numerous challenges. Furthermore, if the increase in global temperature is to be limited to two degrees Celsius above pre-industrial levels as specified in the Paris Agreement, developments and investment in renewable energy, as well as enhancement in energy efficiency, must continue [5].

To meet the climate objectives, a rapid decarbonisation of the energy sector using renewable energy technology is necessary. The shift has been set in motion during the last decade, but it will need a deliberate and persistent effort to complete. Renewables will continue to exceed expectations and promote a cleaner energy future as ambitious objectives, creative policies, and technical advancements are implemented. More Renewable energy generation is critical to averting a global energy crisis, and photovoltaic (PV) devices have been highlighted as a key technology for this purpose. There has always been a search for more efficient, reliable, and cost-effective PV technology, which has given rise to new generations of solar cell.

Organic solar cells (OSCs) have attracted the attention of global scientists as they promise to be a potential means of cost-effective energy production. The concept of bulk-heterojunction (BHJ) solar cells using a conjugated polymer donor and a fullerene acceptor was first introduced by Yu *et al.* in 1995 [6]. Due to benefits such as lightweight, flexibility, and roll-to-roll manufacturing *via* printing methods, BHJ organic solar cells (OSCs) have piqued academic and industry attention since then [7 - 10]. The amazing success of OSCs is due to the constant development of novel materials and device manufacturing processes, as well as advances in film morphology and device physics [11, 12]. Because of substantial progress in the development of novel electron-donor and electron-acceptor materials, power conversion efficiencies (PCEs) for OSCs have recently been increased to above 18%, indicating great application potential [13 - 15]. However, it should be emphasized that top-performing devices often have very tiny surfaces (0.04 cm^2) and are manufactured using very hazardous halogenated solvents, which represents one of the major roadblocks to commercializing OSCs [10, 16, 17]. Highly efficient materials and the most frequently used device processing approaches for lab-scale devices have been demonstrated to be incompatible with industrial-scale large area OSC modules [18].

OSCs are still missing the numbers that would allow them to be economically successful. OSCs will not be as stable and efficient as traditional Si solar cells, but they can still be more economic and competitive because of their low cost, and they still need reasonable efficiency and lifetime. Traditional solar cells have struggled for decades to provide an efficient solution to the energy crisis, but the problems with OSCs are different and it is believed that these problems can be easily resolved. This technology would soon serve the society in a better way. Compared to traditional Si solar cells, OSCs will not only be economical but also more fascinating.

SOLAR ENERGY

The sun has been the source of energy flow for survival in the eco-system for billions of years. The usage of solar energy began with focusing the sun's heat with glass and mirrors to light fires, and today solar-powered automobiles, homes, and technological devices are commonplace. The energy from the sun is free, but the devices used to transform it are not. Although there has been a lot of research and development into numerous methods to harness the sun's energy, the current technology only meets approximately a tenth of one percent of world energy consumption. The sun emits energy in the form of electromagnetic radiation into space, with the quantity reaching the earth varying based on latitude and location, as well as weather patterns. The sun's inner and outer temperatures are around 2×10^7 K and 6000 K, respectively. The sun generates a large quantity of heat and limitless energy through a fusion process or proton-proton reaction that takes place in its core. The fusion process of the sun is seen in Fig. (**1**).

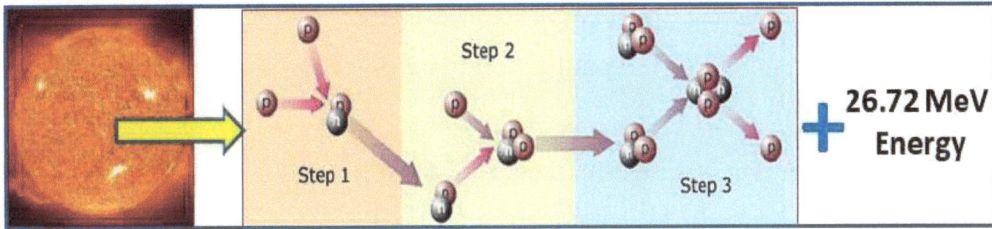

Fig. (1). The fusion Process in Sun.

The proton-proton chain reaction in the sun releases net energy of 26.72 MeV including 2.04 MeV energy due to the annihilation of two positrons with two electrons. The steps of fusion reactions are as follows:

$$2{}_1^1H + 2{}_1^1H \rightarrow 2{}_1^2H + 2{}_1^0e^+ + 2v + 0.84 \text{ Mev} \qquad \text{(Step-1)}$$

$$2{}_1^2H + 2{}_1^1H \rightarrow 2{}_2^3He + 2\gamma + 11.04 \text{ Mev} \qquad \text{(Step-2)}$$

$${}_2^3He + {}_2^3He \rightarrow {}_2^4He^+ + 2{}_1^1H + 12.8 \text{ Mev} \qquad \text{(Step-3)}$$

$$4{}_1^1H \rightarrow {}_2^4He^+ + 2{}_1^0e^+ + 2v + 2\gamma + 24.68 \text{ Mev} \qquad \text{(Net reaction)}$$

The reaction involves fusing of four protons (hydrogen atoms) resulting into the formation of the Helium nucleus with a huge amount of energy. This energy is a result of the change in the total mass of the atoms according to Einstein's mass and energy relation $E = mc^2$ [19].

Solar Spectrum

The sun emits a huge amount of radiant energy into the solar system on a regular basis. Although the Earth gets just a small portion of this energy, each square meter (m^2) of the outer edge of the Earth's atmosphere receives an average of 1367 watts (W). Some of this radiation, such as X-rays and ultraviolet rays, is absorbed and reflected by the atmosphere. Nonetheless, the quantity of solar energy reaching the Earth's surface every hour is higher than the whole amount of energy consumed by the world's human population in a year. The spectral power distribution of solar radiation is described by the concept of Air Mass. Just above the atmoshere, the spectral power distribution of solar radiation is known as Air Mass 0 (AM 0) and equal to 1367 W/m^2 . The radiation that reaches sea level in a clear sky at high noon (at zenith) is known as "air mass 1" (AM1) radiation. As the sun moves lower in the sky, more energy is lost as light travels a longer path through the atmosphere. When the sun is at an angle θ to the zenith, the Air Mass is the ratio of the path length of the sun's rays through the atmosphere when the sun is at a given angle θ to the path length when the sun is at its zenith. AM 1.5 corresponds to the spectral power distribution observed when the sun light is coming from an angle of 48.2° to the zenith. The typical spectrum of sunlight at the Earth's surface has been expressed by AM1.5G (where G stands for "global" and includes both direct and diffuse radiation).

The 30% sun light gets reflected back and 70% sun light is available in earth's atmosphere. Thus, the standard AM1.5 G (both direct and indirect radiation named as global) spectrum is "normalized" to 1000 W/m² and known as the terrestrial spectrum [20]. The solar irradiance spectrum is shown in Fig. (2). When compared to the space solar spectrum (extraterrestrial), there are some wavelengths where the amount of photons is considerably reduced. Photons are absorbed by atmospheric gases, the most well-known of which is ozone (O_3), which absorbs higher-energy (shorter-wavelength) UV radiation below 400 nm. Water vapor in the atmosphere absorbs photons with wavelengths about 900, 1100 and 1400 nm [21].

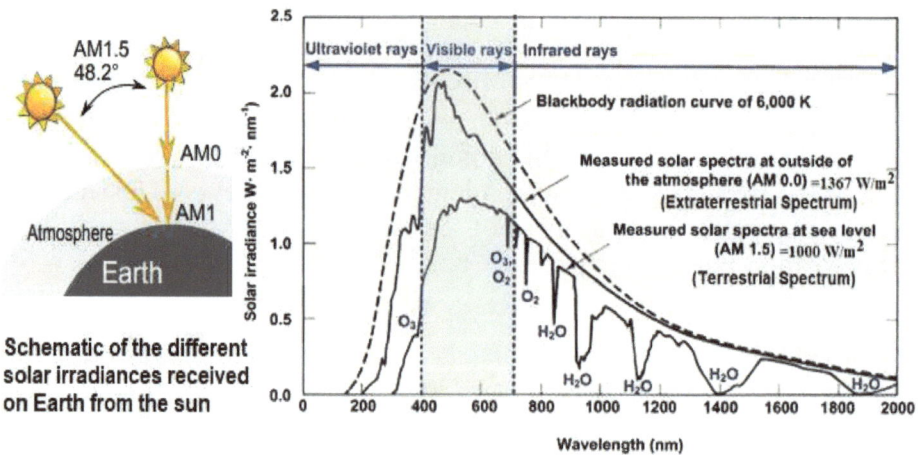

Fig. (2). Solar power spectrum with respect to wavelength.

SOLAR CELL

Background of Solar Photovoltaic Technology

Solar cells use the photovoltaic phenomenon discovered by French physicist Edmond Becquerel in 1839 [22] to convert sunlight directly into electrical energy [23]. When photons contact a semiconductor surface, some of them are absorbed by the valence band electrons, who are subsequently excited into the conduction band, leaving a hole behind. The built-in electric field in a typical photovoltaic device pulls these excited electrons and holes away and collects them before they can relax. Charles Fritts created the first large-area solar cell in 1883 by sandwiching a layer of selenium between iron plate and semi-transparent gold top [24]. Although these devices had very low efficiency (under 1%), they were the beginning stage of one of these days' most dynamically developing areas of engineering [25]. Because of the necessity for power production on satellites, a significant development in the field of photovoltaic devices began in the early 1960s. Researchers began working on silicon electronics in the same period, which was primarily based on the concept of silicon p-n junctions. The first silicon solar cell demonstrated at Bell Laboratories had a 6% efficiency at a cost per Watt of $200 [26]. At the time, the most apparent application was space, where the cost of the cell was immaterial. However, the 1973 oil crisis in oil-dependent western countries prompted a fresh interest in photovoltaics for purposes other than space, and the first practical solar modules for terrestrial use were built in 1976 [27]. As a result, photovoltaics began to compete in markets where traditional energy was prohibitively expensive, such as isolated regions or rural areas in poor nations. The potential of solar cells for terrestrial power generation was soon realized, resulting in the rapid growth of the photovoltaic sector. Solar cell efficiency began to improve in the 1980s and reached a 20% efficiency milestone in 1985 for Silicon solar cells [28]. Over the next decade, solar cell output will rise by 15-30% each year, lowering costs largely owing to economies of scale. Further developments in recent years have reduced the cost of solar cells to approximately $1 per Watt, making their usage on the ground more realistic. In addition, innovative photovoltaic materials and device design have piqued the interest of numerous scientists. The accomplishment of both high power conversion efficiency and low cost in the same device technology is the holy grail of solar cell development. Green *et al.* [29] provided a detailed evaluation of the present state of PV technology.

Today, silicon is the most widely used technology in photovoltaic (PV) cells across the world. Commercially available solar panels have a power conversion efficiency of approximately 15-20%, but state-of-the-art inorganic solar cells have a record power conversion efficiency of close to 39 percent [30]. They are tough to make and need high temperatures in several manufacturing processes, resulting in high prices for the mass market, and the idea is not appropriate for large-scale applications. As a result, new types of materials and concepts for photovoltaic devices must be developed in order to reduce the cost of solar cells. Organic semiconductors are one of them, and they are being looked at as a possible option for the next generation of solar cells. As illustrated in Fig. (**3**), the evolution of solar cells may be divided into three generations.

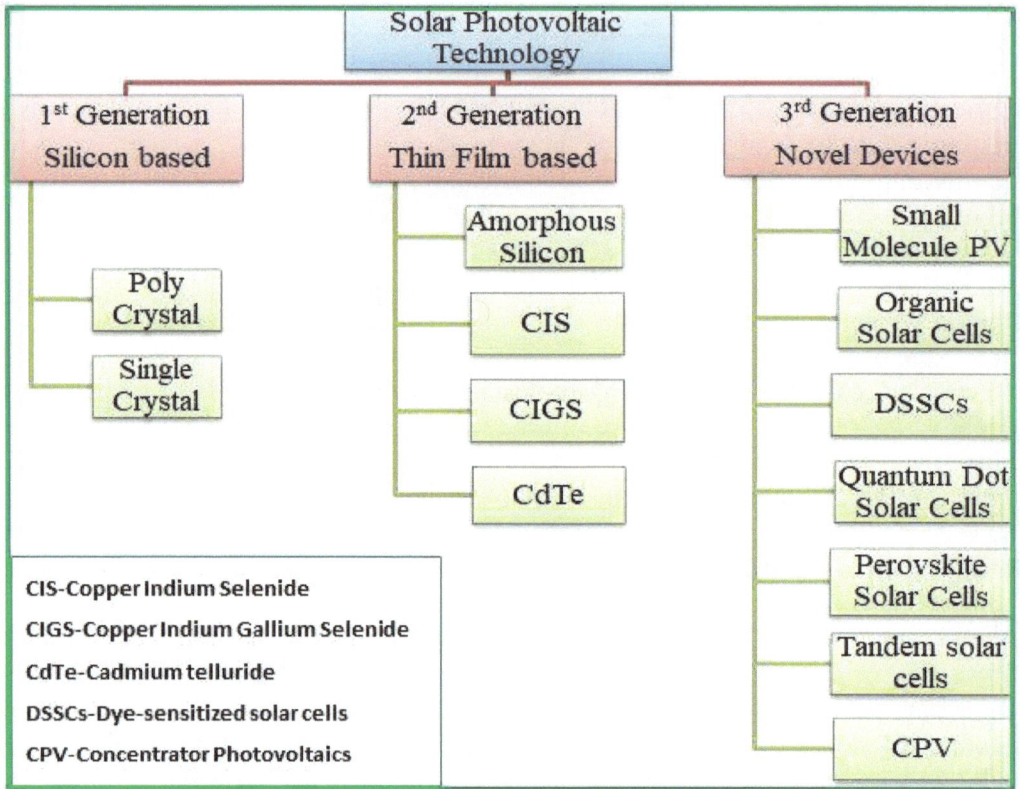

Fig. (3). Types of the generation of solar cells.

Photovoltaic Effect

The photovoltaic effect is the generation of a voltage or electric current in a substance or device when light is illuminated. When French physicist A.H. Becquerel discovered that illuminating light on an electrode submerged in a conductive fluid created an electric current in 1839, he discovered the photovoltaic effect. W. Smith discovered photoconductivity in selenium in 1873, which led to the development of photovoltaic technology. Photovoltaic devices, often known as solar cells, are devices that have a photovoltaic effect. In general, this phenomenon occurs in semiconductor devices when photons from light are absorbed in the semiconducting material, exciting electrons from the valance band to the conduction band. Photogenerated electrons are electrons in the conduction band that leave equivalent holes in the valance band. To generate energy, these electrons and holes must be removed. Because electrons (-ve) and holes (+ve) have opposing charges, they have a tendency to recombine, which should be avoided. The solar cells are constructed and manufactured in such a manner that the photogenerated electrons and holes drift and diffuse in opposing directions and accumulate there. The light-absorbing material is placed between two electrodes of a solar cell. An electromotive force, or photo voltage, develops across the device due to the collection of photogenerated charge carriers on the electrodes on opposing sides of the semiconductor. If the device is linked to an electrical circuit, an electric current will flow across the circuit, converting the light energy into electricity. The built-in electric field provides drift, whilst the concentration gradient of photogenerated charge carriers causes diffusion. If the incoming photons' energy is smaller than the semiconductor's band gap, the photons will not be absorbed and no photovoltaic effect will be detected. In thermoelectric materials, when light absorption causes heating, the photovoltaic effect may be seen. The heating would raise the temperature, and the Seebeck effect would create an electromotive force owing to the temperature gradient in thermoelectric materials. The electric current would pass through if it was linked to an electric circuit.

Transformation from Inorganic to Organic

Polymers, often known as plastics, have largely displaced conventional materials such as wood, leather, metal, glass, and ceramics due to their superior qualities, low cost, and simplicity of manufacturing. Polymers have been utilized as electrically insulating materials in electrical switches and other specialty goods since the discovery of Bakelite. As a result, polymers were often regarded as just insulating materials. The belief that polymers were insulators was transformed in 1976, when Alan J. Heeger, Alan G. MacDiarmid, and Hideki Shirakawa discovered that by exposing polyacetylene to iodine vapor, the conductivity of the organic polymer "polyacetylene" could be changed by several orders of magnitude, approaching that of metals [31]. "For the invention and development of conductive polymers," they were awarded the Nobel Prize in Chemistry in the year 2000. Conducting polymers are a "new family of materials" having electrical and optical characteristics similar to semiconductors and metals, as well as manufacturing benefits and mechanical qualities similar to plastics.

The discovery of these materials not only opened the door to a vast library of materials but also to futuristic technologies based on organic semiconducting and conducting materials serving as the active material in electronic applications. This finding has also spawned a new study field known as organic electronics, which lies at the intersection of chemistry, physics and engineering. Based on the components that make up a substance, it might be categorized as organic or inorganic. Organic compounds are usually described as any molecular substance that contains the element carbon (C) in conjunction with other atoms, which may now be produced in a huge variety of ways. Since the discovery of conducting polymers 30 years ago, there has been significant progress in the creation of organic semiconductors [7, 32]. Many of these organic compounds, particularly polymers, may be made into inks from a solution. This enables the production of organic electrical devices in large quantities at a reasonable cost on a variety of flexible substrates.

Conjugated polymers and small molecules are semiconducting materials with a high absorption rate that might be utilized to generate energy. On the basis of sp^2-hybridized carbon atoms, conducting polymers typically have an alternating single bond–double bond structure (conjugation). This results in a highly delocalized π-electron system with a high propensity to polarize electronically. This permits both visible light absorption due to π–π^* transitions between the bonding and antibonding p_z orbitals, and electrical charge tranport, these two conditions that semiconductors must meet for power production in solar cells. Because of its simplicity of processing, light weight, mechanical flexibility, and potential for low-cost manufacturing in vast areas, organic solar cells have a lot of commercial promise. They have high absorption coefficients, and their material characteristics may be modified by changing their chemical composition. They can also be treated using a variety of low-cost coating methods such spin coating, doctor blading, and ink jet printing. One more benefit is the possibility to change the colour of an organic solar cell for architectural purposes [33]. The cost of these materials is low, and the manufacturing procedures are far simpler and less expensive than inorganic materials, which have highly easy and cost-effective purification and production steps. Organic solar cells (OSC) may become a viable alternative to inorganic solar cells in the near future due to these benefits. Simultaneously, scientific effort and industrial interest in technology are rapidly increasing. The general trend of nanotechnology has recently emerged in the field of solar photovoltaic. New photovoltaic materials and systems have emerged as a result of nanoscale material engineering, which might lead to the development of low-cost solar cells in the future.

ORGANIC SOLAR CELLS

Fundamental of Organic Solar Cell

Polymer solar cells have a number of inherent benefits, including inexpensive material and production costs, light weight, and flexibility. The conversion of sunlight into electrical energy is the subject of photovoltaics. Since the initial realization of a silicon sun cell in 1954 by Chapin, Fuller, and Pearson in the Bell labs [26], traditional photovoltaic solar cells based on inorganic semiconductors have progressed significantly [34]. Today, silicon remains the most widely used photovoltaic solar cell technology, with mono-crystalline devices achieving power conversion efficiency of 15–20 percent. Despite the fact that the solar energy sector has been extensively supported for many years, the pricing of silicon solar cell-based power plants or panels is still not competitive with other traditional combustion processes, with the exception of a few niche items. Organic materials that can be treated under less demanding circumstances are one way to reduce solar cell production costs. Organic photovoltaics have been developed for more than 30 years, but the research area has accelerated significantly in the recent two decades [35 - 37]. Every year, the quantity of solar energy used to light up Earth's landmass is almost 3,000 times the entire amount of energy used by human beings. Photovoltaic devices, on the other hand, must convert sunlight to electricity with a particular degree of efficiency in order to compete with energy from fossil fuels. Scientists have long thought that the purity of the donor and acceptor materials for polymer/organic cells, is the key to high efficiency organic photovoltaic cells, which are considerably less expensive to manufacture than silicon-based solar cells. Organic solar cells are a relatively recent path to a low-cost, large-area energy source. Organic materials that can be solution processed enable low-cost deposition processes including spin coating, doctor blading, inkjet printing, and eventually roll-to-roll manufacturing. It is feasible to modify the chemical and physical characteristics of materials because of the virtually unlimited change of the molecular structure, allowing for tremendous design flexibility [38]. Organic solar cells can absorb most of the light in extremely thin layers (less than 100 nm) due to their high absorption coefficient, resulting in considerable material savings. Organic materials, on the other hand, are frequently extremely disordered, which limits charge carrier transit and device performance. Organic solar cells can be distinguished by their manufacturing method, material properties, and device design. Single layer, bi-layer hetero junction [39], and bulk hetero junction [40] device designs exist, with the diffuse bi-layer hetero junction serving as a bridge between the bi layer and the bulk hetero junction. The single layer architecture uses only one active material, whereas the other structures use two types of materials: electron donors (D) and electron acceptors (A). The charge production method differs between these architectures: single layer devices often require a Scotty barrier at one contact, which allows light excitations to be separated in the barrier field. To separate the electron from the hole, the DA solar cells use light induced electron transfer [7]. The photo-induced electron transfer happens from the excited state of the donor lowest unoccupied molecular orbital (LUMO) to the LUMO of acceptor, therefore the acceptor must be a good electron acceptor with a higher electron affinity. Both the electron and the hole must reach the opposing electrodes, the cathode and the anode, for charge separation. An outside circuit can therefore receive a direct current.

Despite a decade of study, the device efficiency of solar devices based on these materials is still almost an order of magnitude lower than that of inorganic materials. These organic devices are restricted by poor exciton dissociation and charge transport due to high rates of exciton recombination and low charge carrier mobility, respectively. Although it is well understood that important photophysical processes like efficient exciton dissociation take place at the interface between donor and acceptor materials, the short exciton diffusion length in organic semiconductors in comparison to the film thickness required for efficient light absorption is a significant limiting factor in achieving high device efficiency [41]. Blend [42, 43] and stacked layer hetero-junctions [44 - 46] are two methods for overcoming this restriction. These geometries provide multiple exciton dissociation sites with the large interfacial areas between the donor and acceptor species, as well as distinct charge routes, thus limiting charge carrier recombination and allowing for more thicker absorbent organic films. Due to closely spaced dissociation interfaces and distinct charge routes for electrons and holes, blends of electron acceptor and donor materials result in substantially improved device performances and high degrees of photoluminescence quenching when compared to pure film counterparts [47 - 52]. Because the diffusion length of excitons in organic semiconductors is short, it's critical to have dissociating centers close to the exciton generation sites. This has been accomplished using several interfaces in the form of bulk hetero-junction, particularly in photovoltaic systems [53 - 55]. Blending techniques are frequently employed in polymeric systems with great effort in order to generate a nano-structured interface with a high interfacial area between the D and A species by guiding and optimizing blend shape and percolation route between D and A domains [49, 56]. Currently, the most efficient organic photovoltaic devices use the bulk-heterojunction principle [42, 43, 53, 55, 57, 58], in which the intermixing of electron-donating (p-type) and electron-accepting (n-type) molecules results in the formation of microscopic hetero junctions between domains of the two materials throughout the bulk film. Early promising results have sparked a lot of interest in creating solar cells with this bulk heterojunction structure because of its simplicity of processing, device flexibility and potential for low-cost production [6, 42, 59].

Organic Semiconductors

Organic materials have the potential to build a long-term solution for large-scale power generation that is economically viable and based on ecologically acceptable resources that are infinitely available. Organic semiconductors are a less costly alternative to inorganic semiconductors such as Si, and they can have exceptionally high optical absorption coefficients, allowing for the manufacture of ultra-thin solar cells. The ability to build thin flexible devices utilizing high throughput, low temperature methods that utilize well-established printing techniques in a roll-to-roll process [60] is another appealing characteristic of organic PVs. Because flexible plastic substrates may be used in an easily scaled high-speed printing process, the balance of system cost for organic PVs can be reduced, leading to a faster energy pay-back time. All organic semiconductors have a conjugated π-electrons electrical structure. Some of the organic semiconductors used in OSCs are shown in Fig. (**4**).

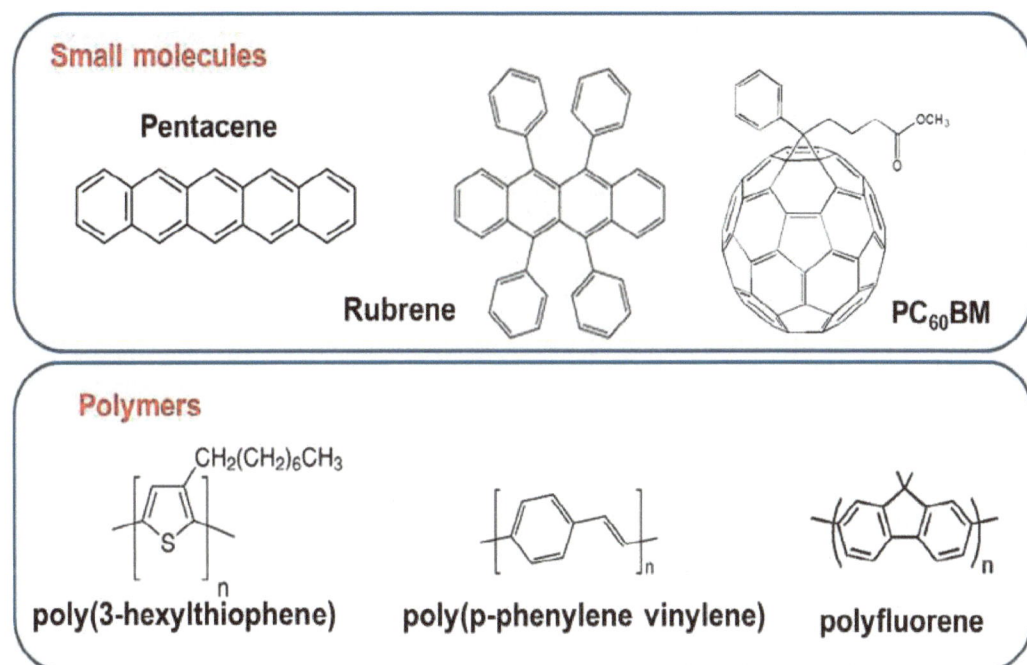

Fig. (4). Conjugated polymers and small molecules used in OSCs.

A conjugated organic system is made up of alternating single and double carbon-carbon bonds. Single bonds are known as σ-bonds and are connected with localized electrons, whereas double bonds have two bonds namely σ-bond and π-bond. The π-electrons are considerably more mobile than the σ-electrons; they can move from site to site between carbon atoms owing to the mutual overlap of π orbitals along the conjugation path, which causes the wave functions to delocalize along the conjugated backbone. The π-bands are either vacant (known as the

Lowest Unoccupied Molecular Orbital - LUMO) or loaded with electrons known as the (called the Highest Occupied Molecular Orbital - HOMO) [61]. The band gap of conjugated polymers is determined by the gap of HOMO–LUMO [8]. These materials have a band gap ranging from 1 to 4 eV [62]. This π-electron system possesses all of the electrical characteristics of organic materials, including light absorption and emission, charge production and transport [63].

The charge transport of Organic semiconductor is fundamentally distinct from standard inorganic physics, which uses the concepts of band conduction and free charge carriers. Instead, carriers jump from one localized state to the next across the material, a phenomenon is known as hopping [64]. This hopping mechanism leads to a comparatively limited charge carrier mobility due to the energetic and spatial disorder in the materials. As a result, organic semiconductor devices are not designed to compete with high-speed applications like silicon computer chips, but rather with applications that combine ease of processing (and hence low cost) with moderate performance. Field effect transistors (FETs) for identifying tags or sensors, light emitting diodes (LEDs) for large-area lighting or displays, memory devices, and solar cells are just a few examples. The structure of an organic semiconductor is shown in Fig. (5).

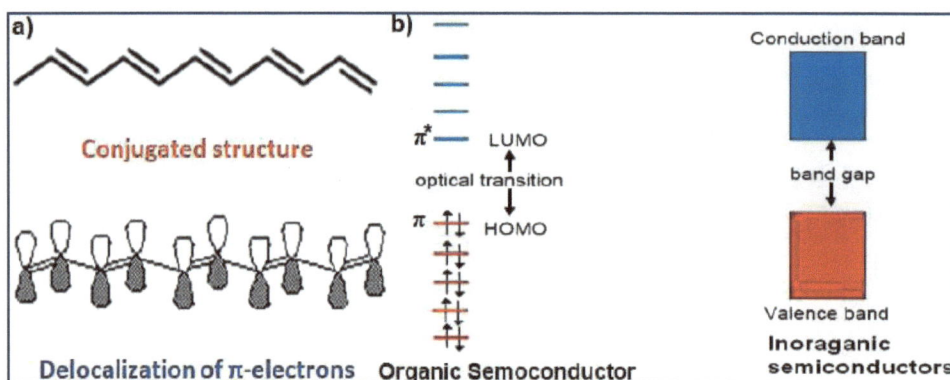

Fig. (5). Organic Semiconductor a) Conjugated structure and b) band diagram.

Molecular Bonds in Conducting Semiconductor

An atom is the fundamental unit of matter, consisting of a compact, positively charged nucleus and a cloud of negatively charged electrons around it. Quantum mechanics describes such microscopic systems, in which each elementary particle is linked with a wave function $\Psi(r, t)$. The density function $|\Psi(r, t)|^2$, which reflects the probability of detecting the particle at the location r at time t, is the square of the modulus of the wave function. The electrons of atom can only exist in specific quantum states. Only wave functions that are Schrodinger equation solutions are permitted [65]. The shape and energy of these wave functions are determined by three quantum numbers namely the principal quantum number n, the orbital angular momentum quantum number l and the magnetic quantum number m_l. Each orbital can only hold two electrons at a time, one from each spin (up or down). The shells and subshells are also designated K, L, M, N,... and s, p, d, f,..., respectively. The s orbitals, which are sphere-shaped and nonzero at the center of the nucleus, and the p orbitals, which resemble dumbbells with their two ellipsoid-shaped lobes separated by a nodal plane at the nucleus, are the two most intriguing types of atomic orbitals in organic electronics as shown in Fig. (6).

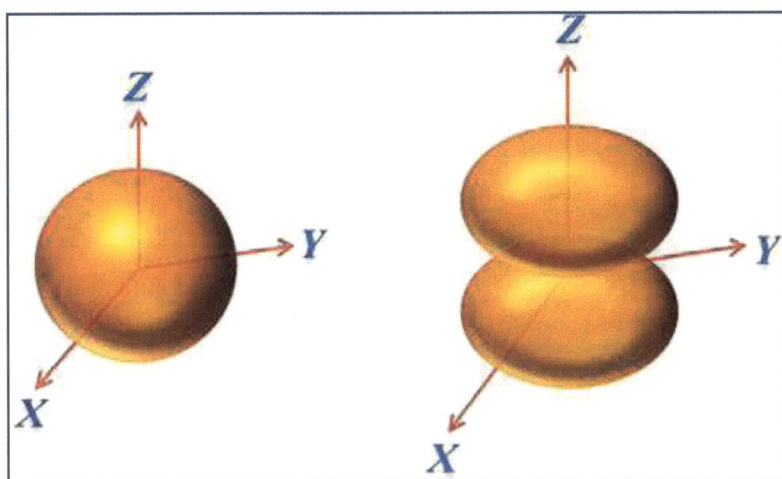

Fig. (6). Illustration of *s* orbital *p* orbital in Nodal Plane at the Nucleus [66].

The chemical, electrical and optical characteristics of materials are determined by the electrons in the outermost shell, known as valence electrons. They also play a role in forming chemical bonds with other atoms. When two atoms are placed close together, valence electron contact causes their atomic orbitals to overlap. These atomic orbitals are combined to produce molecular orbitals, which are represented by linear combinations of atomic orbitals. As illustrated in Fig. (8), the interactions between the atomic orbitals are either constructive ($\Psi+$) or destructive ($\Psi-$), resulting in an increase or reduction in the electronic density between the nuclei. The former is a bonding molecular orbital, whereas the latter is antibonding. The bonding orbital stabilizes the molecule and has lower energy than the initial atomic orbitals, whereas the antibonding orbital destabilizes the molecule and has greater energy than the original atomic orbitals. The bonding and antibonding orbitals of a conjugated polymer are shown in Fig. (7).

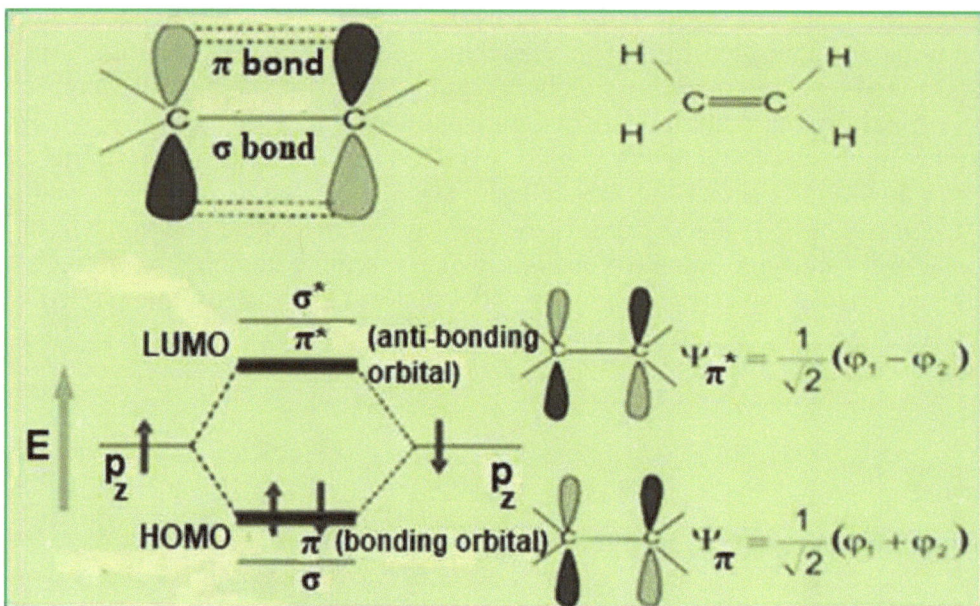

Fig. (7). Molecular bonding in Conjugated Polymer.

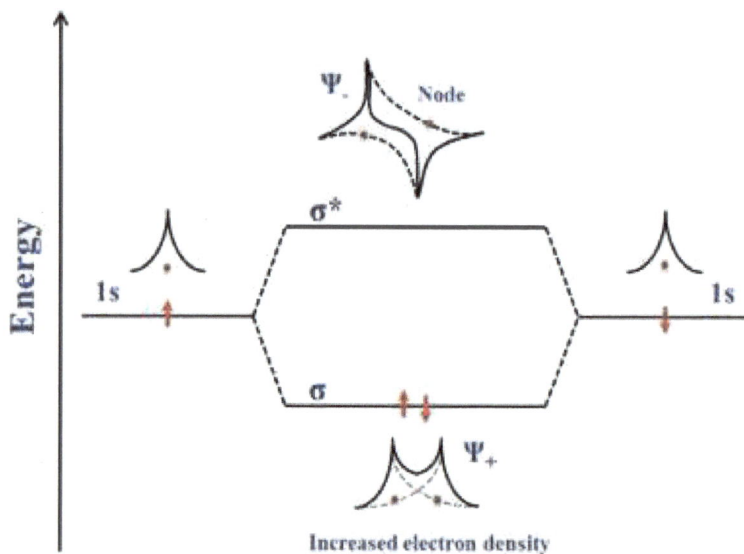

Fig. (8). The formation of bonding and anti-bonding molecular orbitals and the splitting of energy levels for a dehydrogenate molecule [66].

As a result, the original energy levels are separated, and the strength of the atomic interactions is determined by the spacing of the energy levels. The electrons will occupy the lower energy orbitals first, and the atoms will form a stable bond if the overall energy of the system is less than that of the two isolated atoms. The highest occupied molecular orbital (HOMO) is the orbital with the highest energy and is occupied with electrons, whereas the lowest unoccupied molecular orbital (LUMO) is the orbital with the lowest energy and is empty (LUMO). If the bond is symmetrical with regard to rotation around the bond axis, it is termed a σ bond, and if it is not, it is called a π bond. In π bond, the orbitals have a nodal plane that passes across the nuclei [66]. The equivalent bonds of antibonding orbitals are designated by the letters σ* and π*. Because the atomic orbitals overlap more, the σ bonds are typically stronger than the π bonds. As a result, π orbitals have more energy than σ orbitals. Covalent bonds are intra molecule bonds in which the atoms share electrons. Due to a difference in electro negativity, the electrons in a bond may be divided unequally between the atoms, resulting in the development of an electric dipole moment along the bond axis. Polar bonds are the name for these types of bonds. The electron pair is almost entirely found at the more electronegative atom due to a substantial difference in electro negativity between the atoms participating in the connection. Such type bond is known as ionic bond. The van der Waals bond formed by interactions between permanent and/or induced dipoles and the hydrogen bond formed between hydrogen atoms and electronegative atoms carrying an electron lone pair, such as oxygen, nitrogen, and so on, are intermolecular bonds, both of which are significantly weaker than intramolecular bonds where as the van der Waals bonds are weaker than the hydrogen bonds. The Formation of bonding and antibonding molecular orbitals, as well as the splitting of energy levels for a dehydrogenate molecule, are depicted in Fig. (**8**).

Working Mechanism of Organic Solar Cells

In general, there are four important processes that have to be optimized to obtain high power conversion efficiency from an organic photovoltaic cell. Organic materials are responsible for generating free charge carriers from sunlight in organic solar cells. The operational mechanism of OPVs can be summarized as follows:

Photon Absorption and Exciton Generation

An exciton (bound electron hole pair) is created when an electron in HOMO of organic material absorbs a photon and is excited into the lowest unoccupied molecular orbital (LUMO). The absorption spectra of the photoactive organic layer should match the solar emission spectrum for effective photon collecting, and the layer should be thick enough to absorb all incident light. When an incoming photon has an energy of $h\nu \geq E_g$, where E_g is the band gap of semiconductor, an electron (e) in the HOMO of donor is excited to the LUMO, leaving a hole (h) in the HOMO level. Singlet exciton with opposing spin is the name given to this e–h pair [67]. Fig. (**9**) depicts the working mechanism of an organic photovoltaic cells.

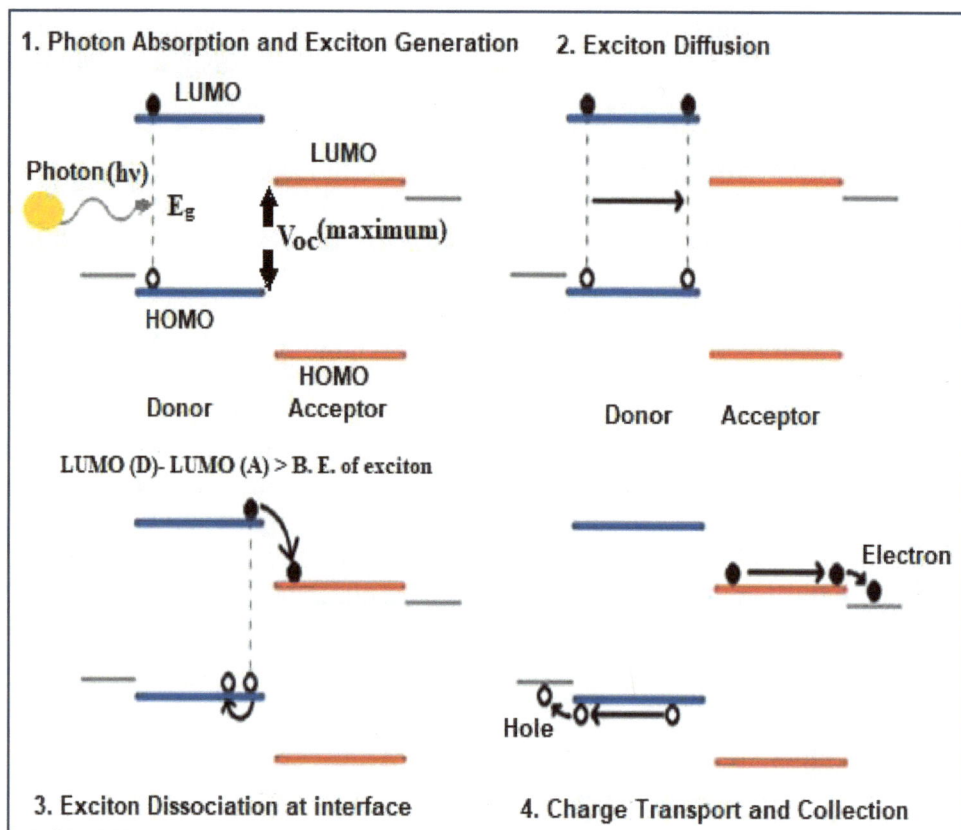

Fig. (9). Operational Mechanism of OPV.

Exciton Diffusion

The exciton diffuses to the interface of electron donor and electron acceptor materials. The thermal energy at ambient temperature is insufficient to dissociate a photo-generated exciton into free charge carriers (polarons) due to the high exciton binding energy (small relative permittivity $\varepsilon_r \approx 3-4$) in conjugated polymers [68]. As a result, PV devices based on organic semiconductors differ considerably from those based on inorganic materials in terms of design and operating principles. To dissolve the firmly bound exciton into free charge carriers, an efficient electron acceptor is often employed in OSCs.

Exciton Dissociation

Organic semiconductors usually exhibit high excitonic binding energies of the order of 0.2–0.5 eV [69]. Therefore, the bulk-heterojunction design is used to make the most efficient organic solar cells, in which the photogenerated excitons only dissociate when the potential difference between the donor and acceptor interfaces is greater than the exciton binding energy (Fig. **10**). An electron can hop from the LUMO of the donor to the LUMO of the acceptor after photo-excitation from the HOMO to the LUMO of donor [70]. However, this photo-induced charge transfer process may lead to free charges only if the hole remains on the donor due to its higher HOMO level. If HOMO of the acceptor is greater, the exciton transfers itself entirely to the lower-band gap material, resulting in energy loss [71].

Charge Transport and Collection

Now, the separated charge carriers are transported and collected by the opposite electrodes of the cell. Thus, electrons move towards the cathode and holes move towards the anode (Fig. **9**). The movement of electrons and holes depends upon the work function of the electrodes. Work function is the amount of energy required to remove an electron from a material [72]. Therefore, a lower work function material will accept an electron easily. Aluminum, which is generally used as the cathode for organic solar cells is a low work function metal. It will easily accept an electron from the LUMO level of an acceptor. On the other hand, Indium tin oxide (ITO) is a widely used high work function material in organic solar cells. It will extract holes from donor molecules easily. Thus, the liberated electrons and holes must then be delivered to the electrodes *via* percolated donor and acceptor channels to create the photocurrent [70]. The carriers must move through the active materials to the electrodes in order to collect the photogenerated charges. Spin-coating is commonly used to deposit the active layer in polymer solar cells. The polymer chains are organized in a disorderly way in such a spin-coated film. Charge carriers will be limited to small segments due to conformational and chemical flaws in polymer chains and molecules. As a result, the delocalization length of charge carriers is restricted to almost molecular dimensions. The energy of the localized states available to charge carriers is distributed according to the π-conjugation lengths of the polymer segments [66]. Thus, the charge transport in organic semiconductors is based on carriers hopping between localized states associated with organic molecules. Because of significant scattering for electrons in this process, the electron mobility in organic semiconductors is very low.

Parameters of Organic Solar Cell

Fig. (**10a**) depicts the calculated I–V curve of a solar cell in dark and illumination. It mostly resides in the fourth quadrant, implying that it generates energy. As a power supply, the illuminated characteristic may be presented inverted like in Fig. (**10b**). The maximum power rectangle is a rectangle that corresponds to the values of V and I at which the power generated ($P_{max}=V_{max}I_{max}$) reaches its maximum. Basically, there are five fundamental electrical parameters to characterise an organic solar cell namely Short-circuit Current (I_{sc}), Open-circuit Voltage (V_{oc}), Fill Factor (FF), Efficiency(η) and External Quantum Efficiency (EQE) or Incident Photon to Current Conversion Efficiency (IPCE).

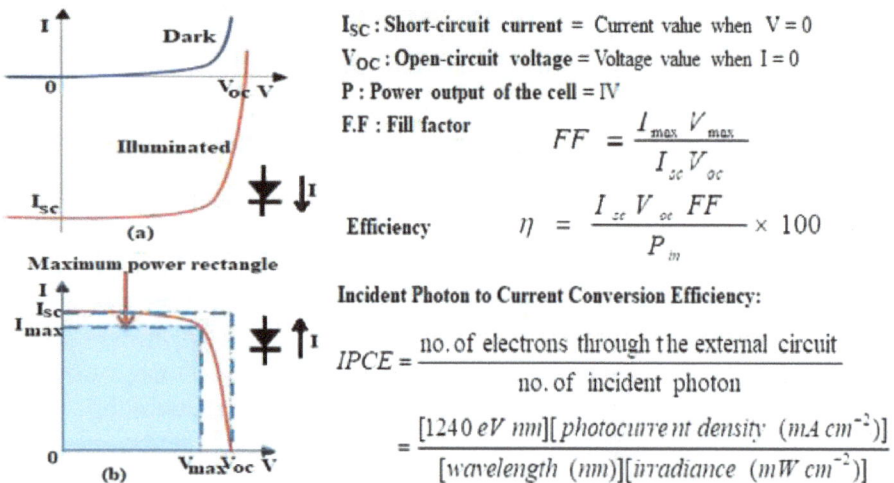

I_{SC} : Short-circuit current = Current value when V = 0

V_{OC} : Open-circuit voltage = Voltage value when I = 0

P : Power output of the cell = IV

F.F : Fill factor

$$FF = \frac{I_{max} V_{max}}{I_{sc} V_{oc}}$$

Efficiency

$$\eta = \frac{I_{sc} V_{oc} FF}{P_{in}} \times 100$$

Incident Photon to Current Conversion Efficiency:

$$IPCE = \frac{no.\,of\ electrons\ through\ the\ external\ circuit}{no.\,of\ incident\ photon}$$

$$= \frac{[1240\,eV\ nm][photocurrent\ density\ (mA\ cm^{-2})]}{[wavelength\ (nm)][irradiance\ (mW\ cm^{-2})]}$$

Fig. (10). I-V Characteristic curve of the Solar cell.

The quantum efficiency (QE) of a solar cell is defined as the ratio of the number of electrons in the external circuit produced by an incident photon of a particular wavelength. It is classified as internal and external quantum efficiency, denoted by IQE(λ) and EQE(λ), respectively. The key difference between internal and external quantum efficiency is that internal quantum efficiency is calculated using the only photons that are absorbed by the solar cell, whereas the external quantum efficiency is calculated using all the photons that are incident on the cell surface. Moreover, internal quantum efficiency always has a higher value than external quantum efficiency. If the internal quantum efficiency is known, the total photo generated current is given by

$$I_{ph} = q \int \phi(\lambda)\{1 - R(\lambda)\}IQE(\lambda)d\lambda$$

where $\phi(\lambda)$ is the photon flux incident on the cell at wavelength and $R(\lambda)$ is the reflection coefficient from the top surface [73].

Device Architectures of Organic Solar Cell

Organic solar cells (OSCs) have been considered as a potential option in comparision to their inorganic counterpart due to lightweight, flexible, lowcost, easily processed and have less environmental impact. Organic solar cells are generally made of three layers namely an active layer, an anode layer of high work function and a cathode layer of low work function. Organic material is referred as an active layer and kept in between the anode and cathode layers as shown in Fig. (**11**). As one of the electrode should be transparent to light, ordinarily, indium tin oxide has been used as a transparent anode layer and a metal electrode such as aluminum or calcium or magnesium has been used as a cathode layer.

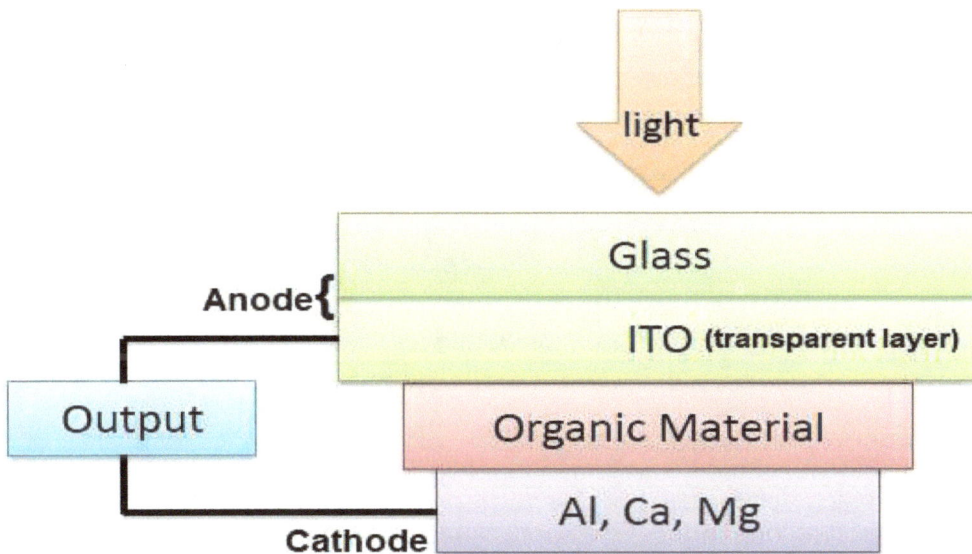

Fig. (11). Schematic lay-out of an organic solar cell.

A breakthrough made in a laboratory at UC Santa Barbara in 1992 led to the development of the technology to manufacture plastic solar cells [74]. The possible interaction of our semiconducting polymers with the well-known fullerene molecules piqued our curiosity. Low diffusion length of electron/ hole and high binding energy are two main problems related to OSCs. Thus, a lot of research work has been carried out on device architectures of organic solar cells to improve efficient charge transfer across the cell [75]. The different device architectures of organic solar cells are discussed below.

Monolayer Solar Cell

The initial attempts to build organic solar cells were based on single layer (homojunction cell) of organic materials between two conducting contacts. The organic material can absorb photons with enough energy to promote an electron from the HOMO to the LUMO when the device is exposed to light. The exciton is formed when the excited electron and the remaining positively charged hole in the HOMO form a coulombically bound pair. The charge carriers are split when the binding energy of exciton is overcome, and they drift towards their respective electrodes due to the built-in electric field. The device architecture of a monolayer solar cell is shown in Fig. (**12**).

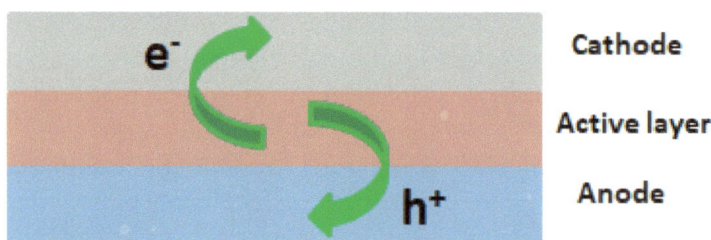

Fig. (12). Schematic device structure of a single layer Solar Cell.

However, the maximum power conversion efficiency (PCE) achieved with these single-layer OSCs is 0.1% [76]. The explanation for this is because organic semiconductors have an lower dielectric constant than their inorganic counterparts, resulting in greater exciton binding energies in the 0.2-0.5 eV range [69]. As a result, at typical working circumstances, the exciton binding energy will be an order of magnitude more than the thermal energy $k_B T$, and thermal excitations will not be sufficient to dissociate the excitons. Thus, the recombination of hole and electron is the main constraint in this device architecture. Hence, an extra driving force will be needed to break apart photogenerated excitons and increase the power conversion efficiency of solar cell. For this, a bilayer device structure of OPV was developed by introducing an acceptor layer between a semiconducting polymer (donor) and a cathode.

Bilayer Solar Cell

This architecture consists of two layers as p-type electron donor and n-type electron acceptor sandwiched between anode and cathode to overcome the problem of high exciton binding energy [77] as shown in Fig. (**13**).

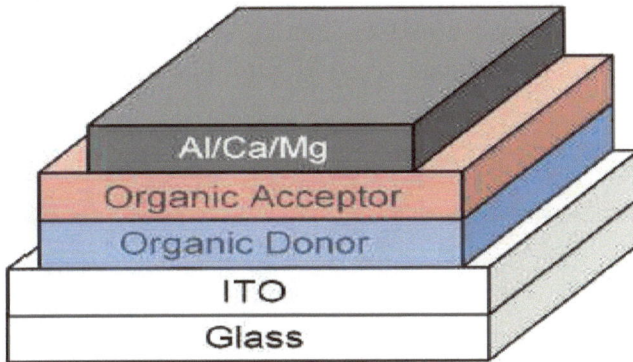

Fig. (13). Schematic structure of a Bi-layer organic Solar Cell.

In 1986, Tang introduced this structure using phthalocyanines as a donor layer and perylene derivatives as an acceptor layer [39]. Photon absorption occurs primarily in the so-called donor material in most of the investigated bilayer systems. The excitons are then dispersed throughout the donor material, eventually reaching the planar donor/acceptor (D/A) interface. The difference in electro negativity between the donor and acceptor acts as a driving force for the excitons to dissociate, and the excited electron is transported to the acceptor LUMO. Fig. (**14**) depicts the energy level diagram of a donor-acceptor system.

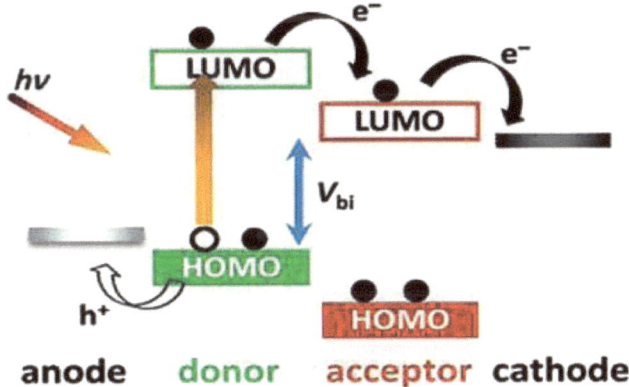

Fig. (14). Energy level diagram of a donor-acceptor system.

It is worth noting that the converse can also happen, with excitons generated in the acceptor material being separated at the D/A interface through hole transfer from the acceptor to the donor. The maximum power conversion efficiency achieved with these bilayer OSCs is 1% [39], which is still significantly lower than that of inorganic solar cells. The main reason for this is the intrinsically short exciton diffusion length (typically in the range of 10–20 nm) of excitons in organic semiconductors, so only those excitons created near the interface between donor and acceptor contribute to the photocurrent as shown in Fig. (**15**). This limits the photocurrent and hence overall PCE of the bi-layer OSCs is less [7].

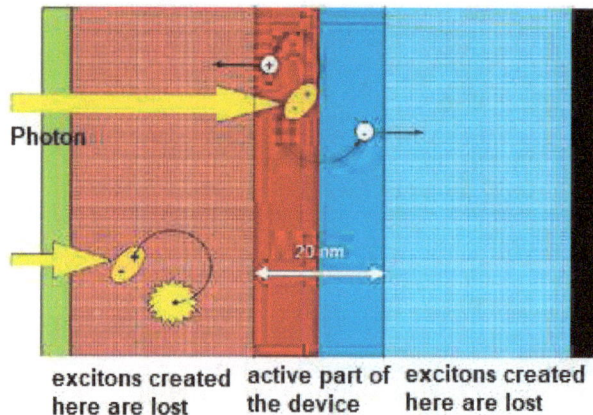

Fig. (15). Exciton diffusion in a bi-layer OSC.

Bulk Heterojunction Solar Cell

In this device structure, active layer of bulk heterojunction (BHJ) solar cell is made up of a mix of a donor and acceptor materials, and known as polymer blend. Both components create an interconnected network, allowing D/A interfaces to spread over the active layer. When the phase separation length scale is on the order of the exciton diffusion length, a perfect BHJ morphology is achieved. As a result, all excitons generated in the donor will be able to reach a D/A interface, resulting in significantly higher photocurrents than the bilayer idea could provide. To enhance hole transport to the anode and smooth the ITO surface, a layer of poly (styrenesulfonate) doped poly (3,4-ethylenedioxythiophene) or PEDOT:PSS is frequently placed on top of the ITO electrode. The active layer, which is made up of a bulk heterojunction of donor and acceptor materials, is deposited either by spin coating or doctor blading or printing techniques. The metallic top electrodes are then thermally vaporized in a high vacuum setup as a final step. The basic structure of BHJ solar cell is illustrated in Fig. (**16**).

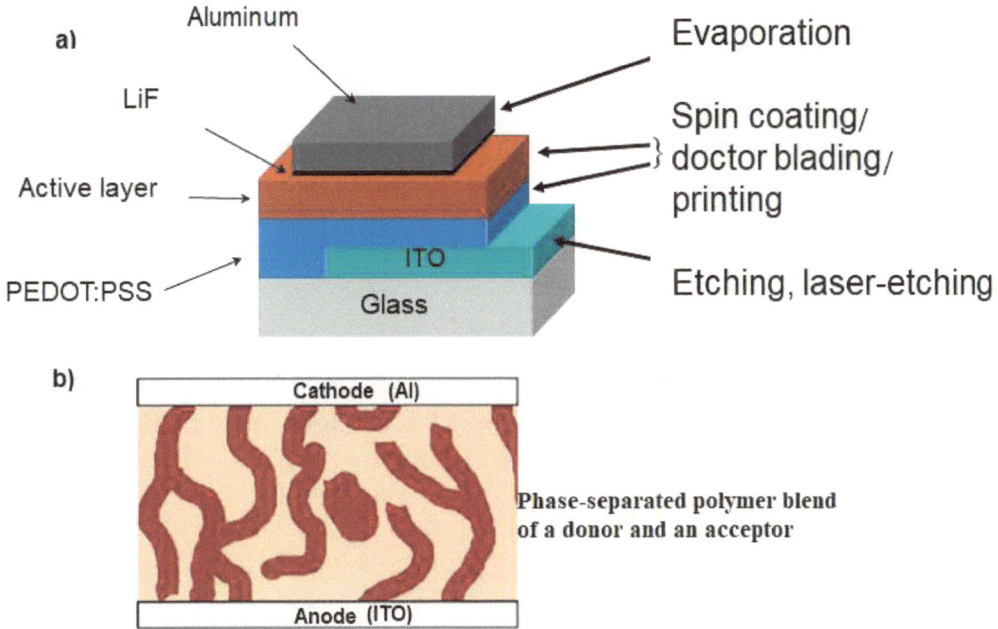

Fig. (16). a) Basic device structure of BHJ organic solar cell and b) Typical BHJ morphology of a solution processed device.

In 1995, the concept of bulk heterojunction solar cell was proposed by Alan Heeger *et al.* [6]. They investigated a phase-separated polymer blend of poly [2-methoxy-5-(20 -ethyl-hexyloxy)-l, 4-phenylene vinylene], MEH-PPV, as a donor and C_{60}, as an acceptor and reported that the carrier collection efficiency (η_c) and energy conversion efficiency (η_e) of polymer photovoltaic cells were improved by blending of the semiconducting polymer with C_{60} or its functionalized derivatives due to the efficient charge separation through a bicontinuous network of internal donor-acceptor heterojunctions.

The next significant improvement in PCE of blend OPV cells was achieved using poly (3-hexylthiophene) (P3HT) as a donor polymer. Sharma *et al.* reported improvement in the power conversion efficiency (PCE) of the bulk heterojunction organic photovoltaic device based on poly (3-hexylthiophene): [6, 6]-phenyl–C61–butyric acid methyl ester (P3HT:PCBM) blend by incorporating a small molecule SM having absorption band in the longer wavelength region. The overall PCE of the device based on thermally annealed blend is improved up to 4.1% due to the increased crystallinity of the blend through thermal annealing [78]. Gang Li *et al.* reported 4.4% PCE from the optimized BHJ solar cell fabricated using P3HT and soluble derivative of C_{60} known as PCBM [79]. John A. Mikroyannidis *et al.* fabricated the organic solar cells based on the blending of

P3HT as a donor and modified fullerene (F) with 4-nitro-4'-hydroxy-α-cyano-stilbene as an acceptor. This device shows a higher V_{oc} of 0.86 V and a short circuit current (J_{sc}) of 8.5 mA cm^{-2}, resulting in a power conversion efficiency (PCE) of 4.23% and it further improved up to 5.25% with the OSC based on the P3HT:F blend deposited from a mixture of solvents (chloroform/acetone) and subsequent thermal annealing at 120 °C [80]. Fullerene-based OSCs have shown ~12% PCE by using low bandgap polymer donors such as PffBT4T-C$_9$C$_{13}$ [81].

A simplified band diagram for a BHJ OSC device is shown in Fig. (**17**). To remove holes from the organic layer, the anode is made of a material with a high work function. To extract the electrons, a metal with a low work function is used as the cathode. An energetic mismatch between the work function of the electrode and the donor HOMO or acceptor LUMO energy level might cause injection barriers for holes and electrons. As a result, thin charge transport layers between the metallic contact and the organic layer are frequently employed to decrease the energy barrier, allowing the charge to flow more freely from the organic material to the electrodes. In any event, for a good device layout, careful consideration of the cathode and anode materials, as well as the usage of electron and hole transport layers, is critical. Indium tin oxide (ITO) is the most widely utilized transparent conductor in OPV due to its great optical transparency [76].

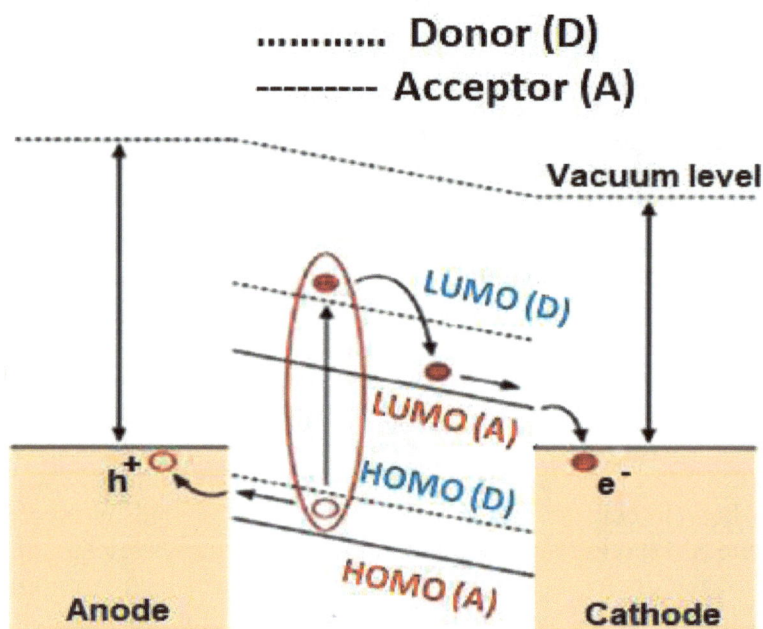

Fig. (17). Energy band diagram of an BHJ organic solar.

Inverted BHJ Solar Cell

The stability and lifespan of BHJ solar cells in the open air are one of the major problems [6, 82]. In BHJ solar cells, a low-work-function metal (e.g., Al) is used as an cathode, which is sensitive to air and moisture and leads to the oxidization of the cathode quickly. Also, the acidic PEDOT:PSS interfacial layer directly contacts the ITO glass and can etch the ITO resulting in a degradation of the device performance [83]. The inverted BHJ solar cell is a significant improvement over a single BHJ solar cell in terms of stability and endurance. In comparison to the normal BHJ cell, this novel design has an inverted geometry [84]. When compared to a single BHJ solar cell, the charge carriers travel in the other direction due to the device's shape. Aluminum is employed as a cathode in a standard BHJ solar cell, however, because it has a poor work function, it oxidizes in air. The high work function metal, such as silver, is utilized to extract holes from the active layer, whereas the transparent ITO electrode is used to remove electrons. To prevent charge recombination, a hole blocking layer made of Zinc oxide (ZnO) is often put between the active layer and the transparent cathode, ITO [85]. The device architecture of the inverted BHJ is shown in Fig. (**18**).

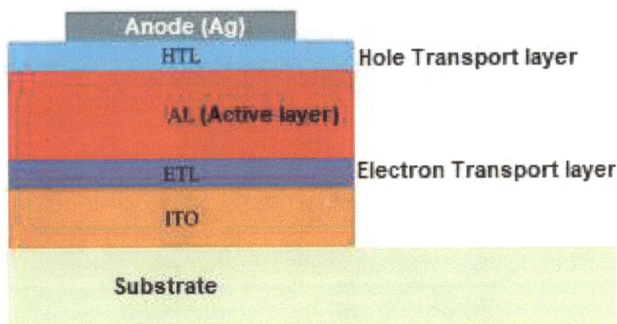

Fig. (18). Schematic device structure of inverted bulk heterojunction OSC.

Wang *et al.* showed that the stability of inverted small-molecule OSCs (CuPc/C_{60}) could be greatly improved to about 950 h in the air [86]. Norrman *et al.* fabricated a device structure of ITO/ZnO/ PCBM:P3HT/PEDOT:PSS/Ag, and found that oxygen causes more significant degradation of encapsulated OSCs, whereas both water and oxygen cause serious degradation of unencapsulated OSCs. The top PEDOT:PSS layer protects the intrusion of air more efficiently than the aluminum cathodes used in conventional OSCs [87]. Tan *et al.* fabricated inverted OSCs with a structure of ITO/ PDMAEMA/PCBM:P3HT/MoO3/Ag, in which the polymer Poly (2- N,N-dimethylaminoethyl methacrylate) (PDMAEMA) reduces the work function of ITO from 4.38 eV to 3.94 eV. These devices exhibit an average PCE of 3.2% and super high stability in the air [88].

Tandem BHJ Solar Cell

Owing to the thermalization of heated charge carriers and restricted light absorption, the efficiency of normal and inverted BHJ organic solar cells has been recommended to be restricted to 10% power conversion efficiency [89]. As a result, devices with a tandem design have recently attracted attention [90, 91]. The basic device design of a tandem organic solar cell is shown in Fig. (**19**). The red layer depicts a BHJ active layer that absorbs red wavelengths, while the blue layer depicts a BHJ active layer that absorbs blue wavelengths. Three distinct cells can be stacked together in the tandem arrangement, with a third layer of near infrared (NIR) absorbing molecules [92].

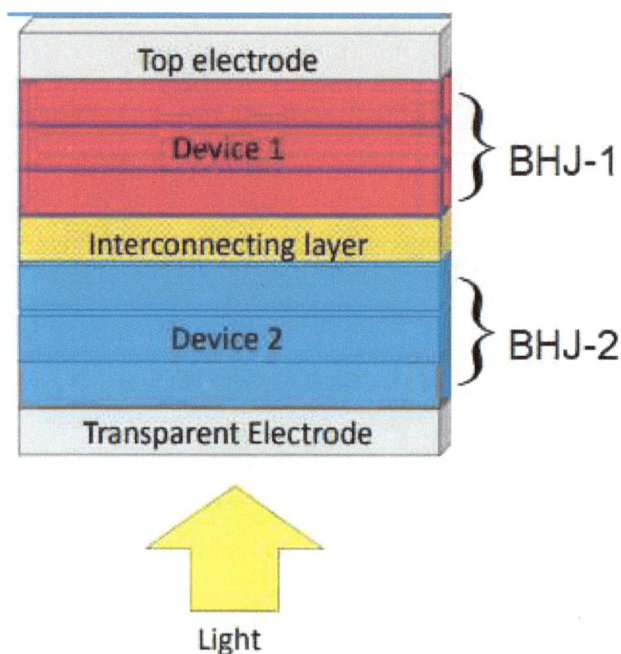

Fig. (19). Basic device architecture of Tandem organic solar cell.

Two BHJ solar cells are stacked together and linked by a charge transfer intermediate layer in this design. Tandem cells may absorb both long and short wavelength parts of the sun spectrum to produce electrons, as distinct chemical systems can absorb various wavelength regions of the spectrum. The whole sun spectrum may be used from this tandem arrangement with correct adjustment of energy collecting chemical devices. Li, M. *et al.* reported the single-junction devices based on DR3TSBDT and DPPEZnP-TBO with an architecture of ITO/PEDOT:PSS/active layer/PFN/Al with a power conversion efficiency 12.5%

[93]. Xiaozhou Che *et al.* demonstrated a tandem cell with an efficiency of 15% (for 2 mm^2 cells) that combines a solution-processed non-fullerene-acceptor-based infrared absorbing subcell on a visible-absorbing fullerene-based subcell grown by vacuum thermal evaporation [94]. With the ternary OPV of PBDB-T-2F:Tf-F-4FIC in the front cell with PTB7-Th:O6T-4F:PC71BM in the back cell as tandem arrangement, the maximum efficiency achievable is 18.6% [95]. However, fabrication difficulties exist, such as the wetting behavior of connecting layers, the use of various solvents for different layers, and complicated band gap matching [96].

Energy Losses in Organic Solar Cells

Optical and electronics losses are two main types of losses that influence the performance of organic solar cells (OSCs), and both losses must be avoided as much as possible to increase the PCE of these solar cells. Reflection losses, inadequate light absorption in the active layer and optical distribution density are all examples of optical losses. These losses are occurred during the absorption of photons by the active layer of the OSCs and control the corresponding value of J_{sc} of the devices. It is necessary to develop materials (donors and acceptors) with larger absorption profiles in order to absorb the majority of photons and enhance photon harvesting efficiency. Exciton recombination losses, thermalization losses in the bulk of the active layer and energy or voltage loss are all examples of electronics losses and occurred during exciton diffusion and exciton dissociation. Therefore, the donor and acceptor materials with proper device architecture must be carefully selected for efficient charge transfer.

In a typical binary fullerene based OSC, the highest occupied molecular orbital (HOMO) and lowest unoccupied molecular orbital (LUMO) energy levels of donor (D) and acceptor (A) should align as shown in Fig. (**20**), where the energy level offsets between D and A (*i.e.* ΔHOMO and ΔLUMO) need to be larger than 0.3 eV to provide enough driving force to dissociate the exciton. However, recent investigations have validated that this limitation can be overcome in NFA-based devices, where the energy level offsets can be minimized [97 - 99].

Practically, the V_{oc} value of BHJ organic solar cells is proportional to the difference between the LUMO energy level of A and HOMO energy level of D (*i.e.* (LUMO]$_A$-[HOMO]$_D$). Therefore, the trade-off between the J_{sc} and V_{oc} should be considered, particularly in the pursuit of acceptors with strong absorption in near infra-red (NIR). Improvement of J_{sc} is possible using narrow bandgap acceptors but would cause reduced V_{oc} values, when employing same donor due to the deeper LUMO energy levels for the acceptor. In order to understand the origin of voltage loss and increase the value of V_{oc} in different photovoltaic devices, energy loss (E_{loss}) is defined as follows according to the Shockley-Queisser (SQ) theory [100]:

$$E_{loss} = qV_{loss} = E_g - qV_{OC}$$

$$E_{loss} = \left(E_g - qV_{OC}^{SQ}\right) + \left(qV_{OC}^{SQ} - qV_{OC}^{rad}\right) + \left(qV_{OC}^{rad} - qV_{OC}\right)$$

$$E_{loss} = \left(q\Delta V_{OC}^{SQ}\right) + \left(q\Delta V_{OC}^{rad}\right) + \left(q\Delta V_{OC}^{non-rad}\right)$$

$$E_{loss} = \Delta E1 + \Delta E2 + \Delta E3$$

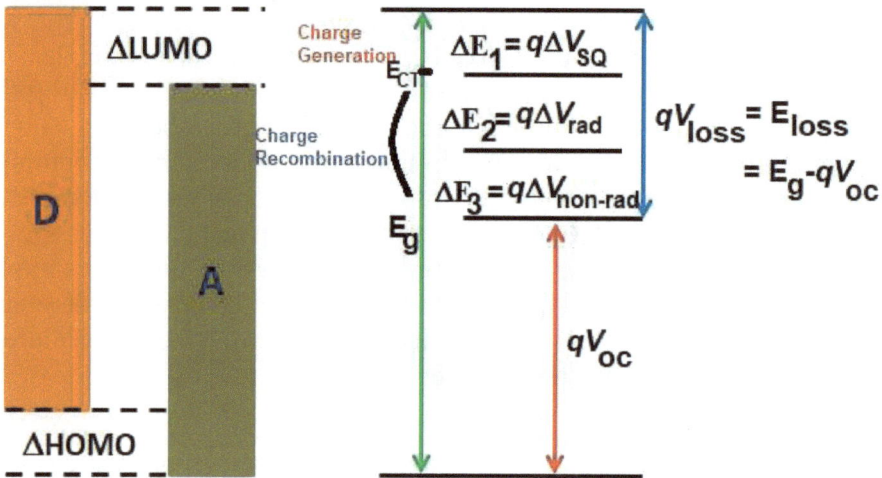

Fig. (20). Energy levels of donor and acceptor bilayer BHJ organic Solar cell with various components of Energy loss.

where, E_g is the band gap of the blend film, q represents to elementary charge, is the maximum voltage based on the Shockley– Queisser (SQ) limit, where several assumptions are made including the EQE being a step-function and the absence of non-radiative recombination. The is the voltage when only radiative recombination exists *i.e.* no non-radiative recombination.

ΔE_1 originated above the optical bandgap is unavoidable (the radiative loss and excess kinetic energy loss). ΔE_2 can be caused by the additional radiative recombination stemming from the absorption below the bandgap *i.e.* deviation from a step-function like EQE. Compared to other solar cells, the larger ΔE_2 in

OSCs mainly results from the charge transfer (CT) state absorption for the existence of energy offset between HOMO and LUMO (ΔE_{offset}). ΔE_3 is a result of non-radiative recombination and can be originated from the vibronic coupling traps and impurities in the solar cells [6, 101]. It can be estimated by measuring the EQE_{EL} of the devices and using the following equation: $\Delta E_3 = -T \ln(EQE_{EL})$, where EQE_{EL} is the electroluminescence external quantum efficiency of a photovoltaic device. E_3 can be minimized when the luminescence property of the device is improved [99]. Therefore, decreasing the $\Delta_{Eoffset}$ and enhancing the EQE_{EL} are two effective methods to reduce the energy loss and hence increase the Voc of the OSCs.

Thus, the relatively large non-radiative recombination losses in OSCs due to extremely low electroluminescence quantum efficiency (EQE_{EL}) of polymer blends (typically in the range of $10^{-6} - 10^{-8}$) and the existence of a significant energy offset between the LUMO level of the donor/acceptor materials are responsible for the large voltage loss in high-efficiency OSCs [102, 103]. The energy offset is the driving force of charge separation in OSCs. A significant driving force is necessary to achieve fast and efficient charge separation in OSCs, which leads to the voltage loss in state-of-the-art OSCs [104 - 107]. The existence of a significant driving force in state-of-the-art OSCs creates a problematic trade-off between the V_{oc} and short-circuit current density (J_{sc}) of the OSCs, which limits the maximum achievable efficiency for OSCs. Therefore, it is important to demonstrate an OSC with a minimal driving force for efficient charge separation. Small non-radiative recombination is another way to further increase the V_{oc}, which will improve the EQE_{EL} of the solar cells. If both of these (small driving force and high EQE_{EL}) can be achieved, it will be possible to achieve efficient OSCs with a small voltage loss, which then could remove the negative trade-off between V_{oc} and J_{sc}. Despite the early success in OSCs, the intrinsic weak light-harvesting capability of fullerenes limits their further application. In this context, non-fullerene acceptors based on acceptor-donor-acceptor (A-D-A) small molecules have been developed for OSCs. These non-fullerene acceptors have the strong visible-near infra-red light-harvesting capability and good electron mobility, delivering higher short-circuit current density (J_{sc}) and tuned the energy offset between the donor and acceptor to enhance PCE in solar cells than fullerene acceptors [108, 109]. Qunping Fan *et al.* reported that the the PF5-Y5 based all organic solar cells achieved a high PCE of up to 14.5% [110]. Polymer solar cells (PSCs) are based on a ternary bulk heterojunction (BHJ) active layer consisting of the conjugated polymer PTB7-Th as a donor and two non-fullerene acceptors, that is, a wide bandgap acceptor based on perylene-diimide (PDI) unit, TPhEPDI-4, and narrow bandgap A-D-A acceptor denoted as BThIND-Cl showed an efficiency of 16.40% and a low energy loss of 0.53 eV [111].

CONCLUSION

OSCs are considered a low cost technology for solar energy conversion due to their attractive properties of light weight, mechanical flexibility and suitability for large-scale solution processing manufacturing. Recently, the power conversion efficiencies (PCEs) for OSCs have increased to over 18% because of significant progress in the development of new electron-donor and electron-acceptor materials. However, these devices usually have very small areas and are processed from highly toxic solvents. Stability, relatively low efficiency and large-scale production of OSCs compared to silicon-based solar cells are major issues where much attention is still required from the researchers to go ahead to a next step. Present research efforts are focused on resolving these issues through the development of novel materials, device architectures and processing techniques. The outcome of these research efforts is expected to open the potential of emerging solar cell technologies and lead to their future commercialization.

CONSENT FOR PUBLICATION

Not applicable.

CONFLICT OF INTEREST

The authors declare no conflict of interest, financial or otherwise.

ACKNOWLEDGEMENTS

One of the authors (S. S. sharma) would like to acknowledge the Department of Science & Technology, SERB Division, Govt. of India (Award # SR/FTP/PS-112/2012), Material Science lab developed at GWEC, Ajmer under TEQIP-II and RTU (ATU) CRS project (Award # TEQIP-III/ RTU (ATU) /CRS/2019-20/63) under TEQIP-III for providing research facilities.

REFERENCES

[1]   Global Electricity Review. , **2021**.

[2]   IEA. *World Energy Balances,* **2020**.

[3]   International Renewable Energy Agency (IRENA). , **2021**.

[4]   International Renewable Energy Agency (IRENA), Renewable Capacity Statistics 2021 (Abu Dhabi: March 2021). **2021**.

[5]   IEA,. World Energy Model Documentation on Sustainable Development Scenario (2020).

[6]   Yu, G.; Gao, J.; Hummelen, J.C.; Wudl, F.; Heeger, A.J. Polymer photovoltaic cells: enhanced efficiencies *via* a network of internal donor-acceptor heterojunctions. *Science,* **1995**, *270*(5243), 1789-1791.
        [http://dx.doi.org/10.1126/science.270.5243.1789]

[7]   Vijay, Y.K.; Sharma, S.S. *Synthesis and Characterization of Organic Photovoltaic Cells: The Solar Energy Harvesting at Lower Cost Photovoltaic is a Challenging Task. Efforts in the Organic Photovoltaic*; VDM: Germany, **2010**.

[8]   Mikroyannidis, J.A.; Sharma, G.D.; Sharma, S.S.; Vijay, Y.K. Novel Low Band Gap Phenylenevinylene Copolymer with BF $_2$ −Azopyrrole Complex Units: Synthesis and Use for Efficient Bulk Heterojunction Solar Cells. *J. Phys. Chem. C,* **2010**, *114*(3), 1520-1527.
[http://dx.doi.org/10.1021/jp910467c]

[9]   Shyam, S. Sharma, Khushboo Sharma and G.D. Sharma, Efficient bulk heterojunction photovoltaic devices based on modified PCBM. *Nanotechnol. Rev.,* **2015**, *4*(5), 419-428.
[http://dx.doi.org/10.1515/ntrev-2014-0041]

[10]   Inganäs, O. Organic photovoltaics over three decades. *Adv. Mater.,* **2018**, *30*(35)1800388
[http://dx.doi.org/10.1002/adma.201800388] [PMID: 29938847]

[11]   Lu, L.; Zheng, T.; Wu, Q.; Schneider, A.M.; Zhao, D.; Yu, L. Recent advances in bulk heterojunction polymer solar cells. *Chem. Rev.,* **2015**, *115*(23), 12666-12731.
[http://dx.doi.org/10.1021/acs.chemrev.5b00098] [PMID: 26252903]

[12]   Hou, J.; Inganäs, O.; Friend, R.H.; Gao, F. Organic solar cells based on non-fullerene acceptors. *Nat. Mater.,* **2018**, *17*(2), 119-128.
[http://dx.doi.org/10.1038/nmat5063] [PMID: 29358765]

[13]   Liu, Q.; Jiang, Y.; Jin, K.; Qin, J.; Xu, J.; Li, W.; Xiong, J.; Liu, J.; Xiao, Z.; Sun, K.; Yang, S.; Zhang, X.; Ding, L. 18% Efficiency organic solar cells. *Sci. Bull. (Beijing),* **2020**, *65*(4), 272-275.
[http://dx.doi.org/10.1016/j.scib.2020.01.001]

[14]   Li, S.; Li, C.Z.; Shi, M.; Chen, H. New phase for organic solar cell research: Emergence of Y-series electron acceptors and their perspectives. *ACS Energy Lett.,* **2020**, *5*(5), 1554-1567.
[http://dx.doi.org/10.1021/acsenergylett.0c00537]

[15]   Cai, Y.; Huo, L.; Sun, Y. Recent advances in wide-bandgap photovoltaic polymers. *Adv. Mater.,* **2017**, *29*(22)1605437
[http://dx.doi.org/10.1002/adma.201605437] [PMID: 28370466]

[16]   Wang, K.; Li, Y.; Li, Y. Challenges to the stability of active layer materials in organic solar cells. *Macromol. Rapid Commun.,* **2020**, *41*(4)1900437
[http://dx.doi.org/10.1002/marc.201900437] [PMID: 31894897]

[17]   Cui, H.Q.; Peng, R.X.; Song, W.; Zhang, J.F.; Huang, J.M.; Zhu, L.Q.; Ge, Z.Y. Optimization of ethylene glycol doped PEDOT:PSS transparent electrodes for flexible organic solar cells by drop-coating method. *Chin. J. Polym. Sci.,* **2019**, *37*(8), 760-766.
[http://dx.doi.org/10.1007/s10118-019-2257-5]

[18]   Wang, G.; Adil, M.A.; Zhang, J.; Wei, Z. Large-area organic solar cells: Material requirements, modular designs, and printing methods. *Adv. Mater.,* **2019**, *31*(45)1805089
[http://dx.doi.org/10.1002/adma.201805089] [PMID: 30506830]

[19]   Rainville, S.; Thompson, J.K.; Myers, E.G.; Brown, J.M.; Dewey, M.S.; Kessler, E.G., Jr; Deslattes, R.D.; Börner, H.G.; Jentschel, M.; Mutti, P.; Pritchard, D.E. A direct test of E = mc2. *Nature.,* **2005**.

[20]   Green, M.A. Solar cells—Operating principles, technology and system applications. *Sol. Energy,* **1982**, *28*(5), 447.
[http://dx.doi.org/10.1016/0038-092X(82)90265-1]

[21]   Gueymard, C.A.; Myers, D.; Emery, K. Proposed reference irradiance spectra for solar energy systems testing. *Sol. Energy,* **2002**, *73*(6), 443-467.
[http://dx.doi.org/10.1016/S0038-092X(03)00005-7]

[22]   Becquerel, A.E. Memoire sur les effects d'electriques produits sous l'influence des rayons solaires. *Comptes Rendus de L'Academie des Sciences,* **1839**, *9*, 561-567.

[23]   Cummerow, R.L. Photovoltaic Effect in p − n Junctions. *Phys. Rev.,* **1954**, *95*(1), 16-21.
[http://dx.doi.org/10.1103/PhysRev.95.16]

[24]   Fritts, C.E. On a new form of selenium photocell. *Am. J. Sci.,* **1883**, *26*, 465.
[http://dx.doi.org/10.2475/ajs.s3-26.156.465]

[25]   *J. Nelson, The Physics of Solar Cells*; Imperial College Press: UK, **2003**.

[26]   Chapin, D.M.; Fuller, C.S.; Pearson, G.L. A New Silicon *p-n* Junction Photocell for Converting Solar Radiation into Electrical Power. *J. Appl. Phys.,* **1954**, *25*(5), 676-677.
[http://dx.doi.org/10.1063/1.1721711]

[27]   Green, M.A. Silicon photovoltaic modules: a brief history of the first 50 years. *Prog. Photovolt. Res. Appl.,* **2005**, *13*(5), 447-455.
[http://dx.doi.org/10.1002/pip.612]

[28]   Gregory, M. Wilson The 2020 photovoltaic technologies roadmap. *J. Phys. D: Appl. Phys.,* **2020**, *53*, 493001.

[29]   Green, M.A.; Dunlop, E.D.; Hohl-Ebinger, J.; Yoshita, M.; Kopidakis, N.; Ho-Baillie, A.W.Y. Solar cell efficiency tables (Version 55). *Prog. Photovolt. Res. Appl.,* **2020**, *28*(1), 3-15.
[http://dx.doi.org/10.1002/pip.3228]

[30]   Masson, G.; Latour, M.; Rekinger, M.; Theologitis, I-T.; Papoutsi, M. *Global Market Outlook for Photovoltaics 2013-2017*; European Photovoltaic Industry Association, **2017**.

[31]   Hall, N. Twenty-five years of conducting polymers. *Chem. Commun. (Camb.),* **2003**, *7*(1), 1-4.
[PMID: 12610942]

[32]   Anthony, J.E.; Facchetti, A.; Heeney, M.; Marder, S.R.; Zhan, X. n-Type organic semiconductors in organic electronics. *Adv. Mater.,* **2010**, *22*(34), 3876-3892.
[http://dx.doi.org/10.1002/adma.200903628] [PMID: 20715063]

[33]   Scharber, M.; Hinsch, A.; Fostiropoulos, K. Photovoltaik - Neue Horizonte: Jahrestagung des ForschungsVerbunds Sonnenenergie. *Berlin,* **2003**, 111-115.

[34]   Foley, T.; Thornton, K.; Hinrichs-rahlwes, R.; Sawyer, S.; Sander, M.; Taylor, R.; Teske, S.; Lehmann, H.; Alers, M.; Hales, D. REN12- GSR2015_Onlinebook_low1.pdf.. **2015**.

[35]   Mikroyannidis, J.A.; Sharma, S.S.; Vijay, Y.K.; Sharma, G.D. Novel Low Band Gap Small Molecule and Phenylenevinylene Copolymer with Cyanovinylene 4-Nitrophenyl Segments: Synthesis and Application for Efficient Bulk Heterojunction Solar Cells. *ACS Appl. Mater. Interfaces,* **2010**, *2*(1), 270-278.
[http://dx.doi.org/10.1021/am9006897]

[36]   Singh, M.; Kurchania, R.; Mikroyannidis, J.A.; Sharma, S.S.; Sharma, G.D. An A–D–A small molecule based on the 3,6-dithienylcarbazole electron donor (D) unit and nitrophenyl acrylonitrileelectron acceptor (A) units for solution processed organic solar cells. *J. Mater. Chem. A Mater. Energy Sustain.,* **2013**, *1*(6), 2297-2306.
[http://dx.doi.org/10.1039/C2TA00749E]

[37]   Martin, G.; Keith, E.; Yoshihiro, H.; Wilhelm, W. D. D. Solar Cells Utilizing Small Molecular Weight Organic Semiconductors. *Prog. Photovolt. Res. Appl.,* **2015**, *15*, 659-676.

[38]   Scharber, M.C.; Mühlbacher, D.; Koppe, M.; Denk, P.; Waldauf, C.; Heeger, A.J.; Brabec, C.J. Design rules for donors in bulk-heterojunction solar cells—towards 10% energy-conversion efficiency. *Adv. Mater.,* **2006**, *18*(6), 789-794.
[http://dx.doi.org/10.1002/adma.200501717]

[39]   Tang, C.W. Two-layer organic photovoltaic cell. *Appl. Phys. Lett.,* **1986**, *48*(2), 183-185.
[http://dx.doi.org/10.1063/1.96937]

[40]   Heeger, A.J. 25th anniversary article: Bulk heterojunction solar cells: understanding the mechanism of

operation. *Adv. Mater.,* **2014**, *26*(1), 10-28.
[http://dx.doi.org/10.1002/adma.201304373] [PMID: 24311015]

[41]   Peumans, P.; Yakimov, A.; Forrest, S.R. Small molecular weight organic thin-film photodetectors and solar cells. *J. Appl. Phys.,* **2003**, *93*(7), 3693-3723.
[http://dx.doi.org/10.1063/1.1534621]

[42]   Brabec, C.J.; Sariciftci, N.S.; Hummelen, J.C. Plastic Solar Cells. *Adv. Funct. Mater.,* **2001**, *11*(1), 15-26.
[http://dx.doi.org/10.1002/1616-3028(200102)11:1<15::AID-ADFM15>3.0.CO;2-A]

[43]   Shaheen, S.E.; Brabec, C.J.; Sariciftci, N.S.; Padinger, F.; Fromherz, T.; Hummelen, J.C. 2.5% efficient organic plastic solar cells. *Appl. Phys. Lett.,* **2001**, *78*(6), 841-843.
[http://dx.doi.org/10.1063/1.1345834]

[44]   Yakimov, A.; Forrest, S.R. High photovoltage multiple-heterojunction organic solar cells incorporating interfacial metallic nanoclusters. *Appl. Phys. Lett.,* **2002**, *80*(9), 1667-1669.
[http://dx.doi.org/10.1063/1.1457531]

[45]   Chasteen, S.V.; Härter, J.O.; Rumbles, G.; Scott, J.C.; Nakazawa, Y.; Jones, M.; Hörhold, H-H.; Tillman, H.; Carter, S.A. Comparison of blended versus layered structures for poly(p-phenylene vinylene)-based polymer photovoltaics. *J. Appl. Phys.,* **2006**, *99*(3)033709
[http://dx.doi.org/10.1063/1.2168046]

[46]   Schultes, S.M.; Sullivan, P.; Heutz, S.; Sanderson, B.M.; Jones, T.S. The role of molecular architecture and layer composition on the properties and performance of CuPc-C_{60} photovoltaic devices. *Mater. Sci. Eng. C,* **2005**, *25*(5-8), 858-865.
[http://dx.doi.org/10.1016/j.msec.2005.06.039]

[47]   Dittmer, J.J.; Petritsch, K.; Marseglia, E.A.; Friend, R.H.; Rost, H.; Holmes, A.B. Photovoltaic properties of MEH-PPV/PPEI blend devices. *Synth. Met.,* **1999**, *102*(1-3), 879-880.
[http://dx.doi.org/10.1016/S0379-6779(98)00852-2]

[48]   Arango, A.C.; Carter, S.A.; Brock, P.J. Charge transfer in photovoltaics consisting of interpenetrating networks of conjugated polymer and TiO2 nanoparticles. *Appl. Phys. Lett.,* **1999**, *74*(12), 1698-1700.
[http://dx.doi.org/10.1063/1.123659]

[49]   Geens, W.; Shaheen, S.E.; Wessling, B.; Brabec, C.J.; Poortmans, J.; Serdar Sariciftci, N. Dependence of field-effect hole mobility of PPV-based polymer films on the spin-casting solvent. *Org. Electron.,* **2002**, *3*(3-4), 105-110.
[http://dx.doi.org/10.1016/S1566-1199(02)00039-3]

[50]   Ma, W.; Yang, C.; Gong, X.; Lee, K.; Heeger, A.J. Thermally Stable, Efficient Polymer Solar Cells with Nanoscale Control of the Interpenetrating Network Morphology. *Adv. Funct. Mater.,* **2005**, *15*(10), 1617-1622.
[http://dx.doi.org/10.1002/adfm.200500211]

[51]   Reyes-Reyes, M.; Kim, K.; Carroll, D.L. High-efficiency photovoltaic devices based on annealed poly(3-hexylthiophene) and 1-(3-methoxycarbonyl)-propyl-1- phenyl-(6,6)C61 blends. *Appl. Phys. Lett.,* **2005**, *87*(8)083506
[http://dx.doi.org/10.1063/1.2006986]

[52]   Li, G.; Shirotriya, V.; Huang, J.; Yao, Y.; Mariarty, T.; Emery, K.; Yang, Y. High-efficiency solution processable polymer photovoltaic cells by self-organization of polymer blends"Nat. *Mater.,* **2005**, *4*, 864-868.

[53]   Bucher, Léo; Tanguy, Loïc; Fortin, Daniel; Desbois, Nicolas; Pierre, D Harvey; Ganesh, D Sharma; Claude, P Gros A Very Low Band Gap Diketopyrrolopyrrole–Porphyrin Conjugated Polymer. *Chem Plus Chem.,* **2017**, *82*(4) Special Issue: Biofest.

[54]   Oh, S.W.; Woo Rhee, H.; Lee, C.; Chul Kim, Y.; Kyeong Kim, J.; Yu, J.W. The photovoltaic effect of the p–n heterojunction organic photovoltaic device using a nano template method. *Curr. Appl. Phys.,*

2005, *5*(1), 55-58.
[http://dx.doi.org/10.1016/j.cap.2003.11.079]

[55]   Schilinsky, P.; Waldauf, C.; Brabec, C.J. Recombination and loss analysis in polythiophene based bulk heterojunction photodetectors. *Appl. Phys. Lett.,* **2002**, *81*(20), 3885-3887.
[http://dx.doi.org/10.1063/1.1521244]

[56]   Martens, T.; D'Haen, J.; Munters, T.; Beelen, Z.; Goris, L.; Manca, J.; D'Olieslaeger, M.; Vanderzande, D.; De Schepper, L.; Andriessen, R. Disclosure of the nanostructure of MDMO-PPV:PCBM bulk hetero-junction organic solar cells by a combination of SPM and TEM. *Synth. Met.,* **2003**, *138*(1-2), 243-247.
[http://dx.doi.org/10.1016/S0379-6779(02)01311-5]

[57]   Halls, J.J.M.; Walsh, C.A.; Greenham, N.C.; Marseglia, E.A.; Friend, R.H.; Moratti, S.C.; Holmes, A.B. Efficient photodiodes from interpenetrating polymer networks. *Nature,* **1995**, *376*(6540), 498-500.
[http://dx.doi.org/10.1038/376498a0]

[58]   Padinger, F.; Rittberger, R.S.; Sariciftci, N.S. Effects of Postproduction Treatment on Plastic Solar Cells. *Adv. Funct. Mater.,* **2003**, *13*(1), 85-88.
[http://dx.doi.org/10.1002/adfm.200390011]

[59]   Nelson, J. Organic photovoltaic films. *Curr. Opin. Solid State Mater. Sci.,* **2002**, *6*(1), 87-95.
[http://dx.doi.org/10.1016/S1359-0286(02)00006-2]

[60]   Vilkman, M.; Hassinen, T.; Kerunen, M.; Pretot, R.; Van Der Schaaf, P.; Ruotsalainen, T.; Sandberg, H.G.O. Fully roll-to-roll processed organic top gate transistors using a printable etchant for bottom electrode patterning. Org. Electron. Physics. *Mater. Appl.,* **2015**, *20*(February), 8-14.

[61]   Roncali, J. Molecular engineering of the band gap of π-conjugated systems: facing technological applications. *Macromol. Rapid Commun.,* **2007**, *28*(17), 1761-1775.
[http://dx.doi.org/10.1002/marc.200700345]

[62]   Tzamalis, G.; Lemaur, V.; Karlsson, F.; Holtz, P.O.; Andersson, M.; Crispin, X.; Cornil, J.; Berggren, M. Fluorescence light emission at 1eV from a conjugated polymer. *Chem. Phys. Lett.,* **2010**, *489*(1-3), 92-95.
[http://dx.doi.org/10.1016/j.cplett.2010.02.049]

[63]   Le, T-H.; Yoon, H. Fundamentals of Conjugated Polymer Nanostructures, Conjugated Polymer Nanostructures for Energy Conversion and Storage Applications. WILEY-VCH GmbH, **2021**; pp. 3-42.

[64]   Liu, C.; Huang, K.; Park, W.T.; Li, M.; Yang, T.; Liu, X.; Liang, L.; Minari, T.; Noh, Y-Y. A unified understanding of charge transport in organic semiconductors: the importance of attenuated delocalization for the carriers. *Mater. Horiz.,* **2017**, *4*(4), 608-618.
[http://dx.doi.org/10.1039/C7MH00091J]

[65]   Landau, L.D.; Lifshitz, E.M. *Quantum Mechanics,* 3rd ed; , **1977**.

[66]   Dey, Anamika; Singh, Ashish; Das, Dipjyoti; Iyer, Parameswar Krishnan; Semiconductors, Organic A New Future of Nanodevices and Applications. **2015**.
[http://dx.doi.org/10.1007/978-3-319-14774-1_4]

[67]   Gregg, B.A. Excitonic solar cells. *J. Phys. Chem. B,* **2003**, *107*(20), 4688-4698.
[http://dx.doi.org/10.1021/jp022507x]

[68]   Nelson, J. Polymer:fullerene bulk heterojunction solar cells. *Mater. Today,* **2011**, *14*(10), 462-470.
[http://dx.doi.org/10.1016/S1369-7021(11)70210-3]

[69]   Alvarado, S.; Seidler, P.; Lidzey, D.; Bradley, D. Direct determination of the exciton binding energy ofconjugated polymers using a scanning tunneling microscope. *Phys. Rev. Lett.,* **1998**, *81*(5), 1082-1085.
[http://dx.doi.org/10.1103/PhysRevLett.81.1082]

[70]   Dimitrov, S.D.; Durrant, J.R. Materials design considerations for charge generation in organic solar cells. *Chem. Mater.*, **2014**, *26*(1), 616-630.
[http://dx.doi.org/10.1021/cm402403z]

[71]   Menke, S.M.; Ran, N.A.; Bazan, G.C.; Friend, R.H. Understanding energy loss in organic solar cells: toward a new efficiency regime. *Joule*, **2018**, *2*(1), 25-35.
[http://dx.doi.org/10.1016/j.joule.2017.09.020]

[72]   Brabec, C.J.; Cravino, A.; Meissner, D.; Sariciftci, N.S.; Rispens, M.T.; Sanchez, L.; Hummelen, J.C.; Fromherz, T. The influence of materials work function on the open circuit voltage of plastic solar cells. *Thin Solid Films*, **2002**, *403-404*, 368-372.
[http://dx.doi.org/10.1016/S0040-6090(01)01586-3]

[73]   *Tom Markvart and Luis Castañer*, 2nd ed; Practical Handbook of Photovoltaics Fundamentals and Applications, **2012**.

[74]   Sariciftci, N.S.; Smilowitz, L.; Heeger, A.J.; Wudl, F. Photoinduced electron transfer from a conducting polymer to buckminsterfullerene. *Science*, **1992**, *258*(5087), 1474-1476.
[http://dx.doi.org/10.1126/science.258.5087.1474] [PMID: 17755110]

[75]   Ragoussi, M.E.; Torres, T. New generation solar cells: concepts, trends and perspectives. *Chem. Commun. (Camb.)*, **2015**, *51*(19), 3957-3972.
[http://dx.doi.org/10.1039/C4CC09888A] [PMID: 25616149]

[76]   Frank, J. Kampas, M. Gouterman, Photovoltaic Properties of octaethylporphyrin and tetraethylporphyrin. *J. Phys. Chem.*, **1977**, *81*, 690-695.

[77]   Tiwari, Sanjay; Tiwari, Tanya; Carter, Sue A.; Scott, J. Campbell; Yakhmi, J.V. Advances in Polymer-Based Photovoltaic Cells: Review of Pioneering Materials, Design, and Device Physics. *Handbook of Ecomaterials*, **2017**,

[78]   Sharma, S.S.; Sharma, G.D.; Mikroyannidis, J.A. Improved power conversion efficiency of bulk heterojunction poly(3-hexylthiophene):PCBM photovoltaic devices using small molecule additive. *Sol. Energy Mater. Sol. Cells*, **2011**, *95*(4), 1219-1223.
[http://dx.doi.org/10.1016/j.solmat.2010.12.013]

[79]   Li, G.; Shrotriya, V.; Huang, J.; Yao, Y.; Moriarty, T.; Emery, K.; Yang, Y. High-efficiency solution processable polymer photovoltaic cells by self-organization of polymer blends. *Nat. Mater.*, **2005**, *4*(11), 864-868.
[http://dx.doi.org/10.1038/nmat1500]

[80]   John, A. Mikroyannidis; Antonis, N. Kabanakis; Sharma, S.S.; Ganesh, D.Sharma. A Simple and Effective Modification of PCBM for Use as an Electron Acceptor in Efficient Bulk Heterojunction Solar Cells. **2011**, *21*(4), 746-755.

[81]   Zhao, J.; Li, Y.; Yang, G.; Jiang, K.; Lin, H.; Ade, H.; Ma, W.; Yan, H. Efficient organic solar cells processed from hydrocarbon solvents. *Nat. Energy*, **2016**, *1*(2), 15027.
[http://dx.doi.org/10.1038/nenergy.2015.27]

[82]   Sun, Y.; Seo, J.H.; Takacs, C.J.; Seifter, J.; Heeger, A.J. Inverted polymer solar cells integrated with a low-temperature-annealed sol-gel-derived ZnO Film as an electron transport layer. *Adv. Mater.*, **2011**, *23*(14), 1679-1683.
[http://dx.doi.org/10.1002/adma.201004301] [PMID: 21472797]

[83]   Cao, Huanqi; Weidong, He; Yiwu, Mao; Xiao, Lin; Ken, Ishikawa; James, H. Dickerson; Wayne, P. Hess Recent progress in degradation and stabilization of organic solar cells. *Journal of Power Sources.*, **2014**, *264*, 168e183.

[84]   Sean, E. Shaheen, Nikos Kopidakis, David S. Ginley, Matthew S. White, and Dana C. Olson, Inverted bulk-heterojunction plastic solar cells. *SPIE Newsroom*, **2007**.

[85]   Liang, Z.; Zhang, Q.; Jiang, L.; Cao, G. ZnO cathode buffer layers for inverted polymer solar cells.

Energy Environ. Sci., **2015**, *8*(12), 3442-3476.
[http://dx.doi.org/10.1039/C5EE02510A]

[86]   Wang, M.L.; Song, Q.L.; Wu, H.R.; Ding, B.F.; Gao, X.D.; Sun, X.Y.; Ding, X.M.; Hou, X.Y. Small-molecular organic solar cells with C_{60}/Al composite anode. *Org. Electron.,* **2007**, *8*(4), 445-449.
[http://dx.doi.org/10.1016/j.orgel.2007.03.001]

[87]   Norrman, K.; Madsen, M.V.; Gevorgyan, S.A.; Krebs, F.C. Degradation patterns in water and oxygen of an inverted polymer solar cell. *J. Am. Chem. Soc.,* **2010**, *132*(47), 16883-16892.
[http://dx.doi.org/10.1021/ja106299g] [PMID: 21053947]

[88]   Jin Tan, M.; Zhong, S.; Wang, R.; Zhang, Z.; Chellappan, V.; Chen, W. Biopolymer as an electron selective layer for inverted polymer solar cells. *Appl. Phys. Lett.,* **2013**, *103*(6)063303
[http://dx.doi.org/10.1063/1.4817931]

[89]   Jørgensen, M.; Norrman, K.; Gevorgyan, S.A.; Tromholt, T.; Andreasen, B.; Krebs, F.C. Stability of polymer solar cells. *Adv. Mater.,* **2012**, *24*(5), 580-612.
[http://dx.doi.org/10.1002/adma.201104187] [PMID: 22213056]

[90]   Gilot, J.; Wienk, M.M.; Janssen, R.A.J. Optimizing polymer tandem solar cells. *Adv. Mater.,* **2010**, *22*(8), E67-E71.
[http://dx.doi.org/10.1002/adma.200902398] [PMID: 20217802]

[91]   You, J.; Dou, L.; Yoshimura, K.; Kato, T.; Ohya, K.; Moriarty, T.; Emery, K.; Chen, C.C.; Gao, J.; Li, G.; Yang, Y. A polymer tandem solar cell with 10.6% power conversion efficiency. *Nat. Commun.,* **2013**, *4*(1), 1446.
[http://dx.doi.org/10.1038/ncomms2411] [PMID: 23385590]

[92]   Chen, C.C.; Dou, L.; Gao, J.; Chang, W.H.; Li, G.; Yang, Y. High-performance semi-transparent polymer solar cells possessing tandem structures. *Energy Environ. Sci.,* **2013**, *6*(9), 2714-2720.
[http://dx.doi.org/10.1039/c3ee40860d]

[93]   Li, M.; Gao, K.; Wan, X.; Zhang, Q.; Kan, B.; Xia, R.; Liu, F.; Yang, X.; Feng, H.; Ni, W.; Wang, Y.; Peng, J.; Zhang, H.; Liang, Z.; Yip, H-L.; Peng, X.; Cao, Y.; Chen, Y. Solution-processed organic tandem solar cells with power conversion efficiencies >12%. *Nat. Photonics,* **2017**, *11*(2), 85-90.
[http://dx.doi.org/10.1038/nphoton.2016.240]

[94]   Che, X.; Li, Y.; Qu, Y.; Forrest, S.R. High fabrication yield organic tandem photovoltaics combining vacuum- and solution-processed subcells with 15% efficiency. *Nat. Energy,* **2018**, *3*(5), 422-427.
[http://dx.doi.org/10.1038/s41560-018-0134-z]

[95]   RezaNekoveib, R.Jeyakumarc, Organic tandem solar cells with 18.6% efficiency. *Sol. Energy,* **2020**, *198*(1), 160-166.

[96]   Albrecht, S.; Yilmaz, S.; Dumsch, I.; Allard, S.; Scherf, U.; Beaupré, S.; Leclerc, M.; Neher, D. Solution Processed Organic Tandem Solar Cells. *Energy Procedia,* **2012**, *31*, 159-166.
[http://dx.doi.org/10.1016/j.egypro.2012.11.178]

[97]   Liu, J.; Chen, S.; Qian, D.; Gautam, B.; Yang, G.; Zhao, J.; Bergqvist, J.; Zhang, F.; Ma, W.; Ade, H.; Inganäs, O.; Gundogdu, K.; Gao, F.; Yan, H. Fast charge separation in a non-fullerene organic solar cell with a small driving force. *Nat. Energy,* **2016**, *1*(7), 16089.
[http://dx.doi.org/10.1038/nenergy.2016.89]

[98]   Li, S.; Zhan, L.; Sun, C.; Zhu, H.; Zhou, G.; Yang, W.; Shi, M.; Li, C.Z.; Hou, J.; Li, Y.; Chen, H. Highly efficient fullerene free organic solar cells operated at near zero highest occupied molecular orbital offsets. *J. Am. Chem. Soc.,* **2019**, *141*(7), 3073-3082.
[http://dx.doi.org/10.1021/jacs.8b12126] [PMID: 30685975]

[99]   Fu, H.; Wang, Y.; Meng, D.; Ma, Z.; Li, Y.; Gao, F.; Wang, Z.; Sun, Y. Suppression of recombination energy losses by decreasing the energetic offsets in perylene Diimide-based non-fullerene organic solar cells. *ACS Energy Lett.,* **2018**, *3*(11), 2729-2735.
[http://dx.doi.org/10.1021/acsenergylett.8b01665]

[100] Shockley, W.; Queisser, H.J. Detailed Balance Limit of Efficiency of *p-n* Junction Solar Cells. *J. Appl. Phys.,* **1961**, *32*(3), 510-519.
[http://dx.doi.org/10.1063/1.1736034]

[101] Benduhn, J.; Tvingstedt, K.; Piersimoni, F.; Ullbrich, S.; Fan, Y.; Tropiano, M.; McGarry, K.A.; Zeika, O.; Riede, M.K.; Douglas, C.J.; Barlow, S.; Marder, S.R.; Neher, D.; Spoltore, D.; Vandewal, K. Intrinsic non-radiative voltage losses in fullerene-based organic solar cells. *Nat. Energy,* **2017**, *2*(6), 17053.
[http://dx.doi.org/10.1038/nenergy.2017.53]

[102] Vandewal, K.; Tvingstedt, K.; Gadisa, A.; Inganäs, O.; Manca, J.V. Relating the open-circuit voltage to interface molecular properties of donor:acceptor bulk heterojunction solar cells. *Phys. Rev. B Condens. Matter Mater. Phys.,* **2010**, *81*(12)125204
[http://dx.doi.org/10.1103/PhysRevB.81.125204]

[103] Burke, T.M.; Sweetnam, S.; Vandewal, K.; McGehee, M.D. Beyond Langevin recombination: how equilibrium between free carriers and charge transfer states determines the open-circuit voltage of organic solar cells. *Adv. Energy Mater.,* **2015**, *5*(11)1500123
[http://dx.doi.org/10.1002/aenm.201500123]

[104] Peng, Q.; Huang, Q.; Hou, X.; Chang, P.; Xu, J.; Deng, S. Enhanced solar cell performance by replacing benzodithiophene with naphthodithiophene in diketopyrrolopyrrole-based copolymers. *Chem. Commun. (Camb.),* **2012**, *48*(93), 11452-11454.
[http://dx.doi.org/10.1039/c2cc36324k] [PMID: 23086539]

[105] Ran, N.A.; Love, J.A.; Takacs, C.J.; Sadhanala, A.; Beavers, J.K.; Collins, S.D.; Huang, Y.; Wang, M.; Friend, R.H.; Bazan, G.C.; Nguyen, T.Q. Harvesting the full potential of photons with organic solar cells. *Adv. Mater.,* **2016**, *28*(7), 1482-1488.
[http://dx.doi.org/10.1002/adma.201504417] [PMID: 26663421]

[106] Ma, Z.; Wang, E.; Vandewal, K.; Andersson, M.R.; Zhang, F. Enhance performance of organic solar cells based on an isoindigo-based copolymer by balancing absorption and miscibility of electron acceptor. *Appl. Phys. Lett.,* **2011**, *99*(14)143302
[http://dx.doi.org/10.1063/1.3645622]

[107] Kawashima, K.; Tamai, Y.; Ohkita, H.; Osaka, I.; Takimiya, K. High-efficiency polymer solar cells with small photon energy loss. *Nat. Commun.,* **2015**, *6*(1), 10085.
[http://dx.doi.org/10.1038/ncomms10085] [PMID: 26626042]

[108] Zhao, J.; Li, Y.; Lin, H.; Liu, Y.; Jiang, K.; Mu, C.; Ma, T.; Lin Lai, J.Y.; Hu, H.; Yu, D.; Yan, H. High-efficiency non-fullerene organic solar cells enabled by a difluorobenzothiadiazole-based donor polymer combined with a properly matched small molecule acceptor. *Energy Environ. Sci.,* **2015**, *8*(2), 520-525.
[http://dx.doi.org/10.1039/C4EE02990A]

[109] Lin, H.; Chen, S.; Li, Z.; Lai, J.Y.L.; Yang, G.; McAfee, T.; Jiang, K.; Li, Y.; Liu, Y.; Hu, H.; Zhao, J.; Ma, W.; Ade, H.; Yan, H. High-performance non-fullerene polymer solar cells based on a pair of donor–acceptor materials with complementary absorption properties. *Adv. Mater.,* **2015**, *27*(45), 7299-7304.
[http://dx.doi.org/10.1002/adma.201502775] [PMID: 26462030]

[110] Fan, Q.; An, Q.; Lin, Y.; Xia, Y.; Li, Q.; Zhang, M.; Su, W.; Peng, W.; Zhang, C.; Liu, F.; Hou, L.; Zhu, W.; Yu, D.; Xiao, M.; Moons, E.; Zhang, F.; Anthopoulos, T.D.; Inganäs, O.; Wang, E. Over 14% efficiency all-polymer solar cells enabled by a low bandgap polymer acceptor with low energy loss and efficient charge separation. *Energy Environ. Sci.,* **2020**, *13*(12), 5017-5027.
[http://dx.doi.org/10.1039/D0EE01828G]

[111] Keshtov, M.L.; Kuklin, S.A.; Agrawal, A.; Dahiya, H.; Chen, F-C.; Sharma, G.D. Ternary polymer solar cells based on wide bandgap and narrow bandgap non-fullerene acceptors with an efficiency of 16.40 % and a low energy loss of 0.53 eV. *Mater. Today Energy,* **2021**, *21*100843
[http://dx.doi.org/10.1016/j.mtener.2021.100843]

CHAPTER 11

Physical, Electrical and Dielectric Investigation of Neodymium Doped Lithium Borosilicate Glasses

V.Y. Ganvir[1,*], H.V. Ganvir[1] and **R.S. Gedam[2]**

[1] *Department of Applied Physics, Yeshwantrao Chavan College of Engineering, Nagpur - 441110, India*

[2] *Department of Physics, Visvesvaraya National Institute of Technology, Nagpur - 440010, India*

Abstract: In the present research work, melt-quench technique was employed for synthesis of Nd_2O_3 doped lithium borosilicate glasses having general system $30Li_2O$-$(70-x)$ $[1/7SiO_2:6/7B_2O_3]$-xNd_2O_3. Electrical conductivity of produced samples was tested in frequency band of 2mHz to 20MHz at 423K to 673K, using Impedance Analyser. Impedance data was used for scaling which shows that the process of conduction is based on the composition and not on the temperature. The inclusion of neodymium oxide in the lithium borosilicate glass affects molar volume, density and various physical properties like Ion concentration, ionic radius. The electrical modulus data obtained from impedance analyser was utilized to study relaxation behaviour of the samples. In the temperature band, 423-673 K, the variance of the dielectric loss (Tan δ), dielectric constant (ε') and ac conductivity (σ') with frequency was measured using impedance spectroscopy and discussed at length.

Keywords: Borosilicate glasses, Dielectric, Density, Electrical Conductivity, Impedance spectroscopy.

INTRODUCTION

The search for the quickest ionic conductors has resulted in a widespread study of lithium-based compounds by several researchers in recent years [1 - 3]. As lithium is more electropositive, its compounds are primarily desirable because of their massive energy densities and strong open circuit potential [3]. Therefore, lithium-based 'glasses' and 'glass ceramic's are highly suited for solid-state batteries with high energy capacity. Owing to their uses in different industrial and scientific contexts, the rare earth oxide doped lithium borosilicate glasses are rigorously investigated.

---

[*] **Corresponding author V.Y. Ganvir:** Department of Applied Physics, Yeshwantrao Chavan College of Engineering, Nagpur - 441110, India; Tel: +91-9881367366; E-mail: vyganvir@gmail.com

Dibya Prakash Rai (Ed.)

Nowadays, extensive studies on rare earth oxides doped glasses is carried out, as their presence increases optical bandgap, optical properties, laser amplification *etc* [3 - 6]. The ionic conductivity of the glass-ceramics & glasses has been evaluated primarily by industry and research societies, both theoretically and experimentally [6, 7].

Nd_2O_3 is an incredibly valuable 'rare-earth' for processing separate optical and physical properties of glass-ceramics &glasses [8, 9]. $Nd^{3+}$is commonly used in glass-ceramics &glasses as the luminescence activator [10].

Currently glasses containing Nd_2O_3 have been thoroughly studied with focus on optical properties. However, a little methodical work is devoted on the electrical features of Nd_2O_3doped lithium borosilicate glasses [9, 10]. To utilise these materials for battery usage, it is essential to examine the electrical conduction mechanism of the borosilicate glasses. 'IS' *i.e.* Impedance spectroscopy being a very effective tool for characterisation of certain material's and their interfaces electrical properties [6, 7].

Thus, herein we report a comprehensive impedance spectroscopy research on Nd_2O_3 doped lithium borosilicate glasses with an emphasis on conductivity mechanism and modulus formalism which are subsequently linked to different physical properties.

MATERIAL AND METHODS

The glass system '30 Li_2O: (70 - x Nd_2O_3) {1/7 SiO_2: 6/7 B_2O_3} were prepared with precise melt quench technique where 'x' was varied from 0 to 2 in the stage of 0.5 mol percentage. For glass sample preparations, the high purity starting raw materials such as SiO_2, Li_2CO_3, B_2O_3 and Nd_2O_3 (Quality 99.9 per cent) were used. These additives were thoroughly blended in acetone for around 1 hour in adequate amounts, and then this mixture was placed for 3 hours in a crucible made of platinum in an electric furnace at 1223-1273 K. Then, this melt was transferred to an aluminium mould kept at room temperature to get the desired form of the bulk glass samples.

The quenched bulk glasses were then immediately transferred to an annealing furnace at 588 K for 3 hours to remove the thermal tension that had developed during quenching, and then allowed to cool to ambient temperature. The sample name along with their composition variation is given in Table **1**.

Table 1. Compositions of prepared samples.

Sample Id	Li_2CO_3	B_2O_3	SiO_2	Nd_2O_3
0 BSLI	30.00	60.00	10.00	0
0.5 BSLIND	30.00	59.57	9.93	0.50
1.0 BSLIND	30.00	59.14	9.86	1.00
1.5 BSLIND	30.00	58.71	9.79	1.50
2.0 BSLIND	30.00	58.29	9.71	2.00

The X-ray diffractograms of the synthesized samples were recorded using X'pert pro-PANalytical. SHIMADZU SMK-401 density measuring instruments (Archimedes principle based) was employed to estimate the density of prepared glasses where toluene is used as dipping solvent.

For impedance measurements, the synthesized glass sample were cut and polished into cylinders of about 2.5 to 3 mm thickness. The proper electrical contact with the sample holder's silver electrode and glass samples were ensured by providing coating of silver paint on the parallel sides of glass samples. Impedance measurements were taken as a function of temperature for all samples in the 20mHz to 20MHz frequency band, in the temperature range of 423-673 K using a 'high-resolution dielectric analyser (Novocontrol Make)'. This data was utilized to investigate the electrical characteristics of Nd^{3+} doped lithium borosilicate glasses.

RESULTS AND DISCUSSION

The X-ray diffractograms of prepared glass samples are shown in Fig. (**1**). This X-ray spectrum is characterised by broad humps and a lack of prominent peaks associated with any of the sample components. This demonstrates the amorphous nature of the material [11].

Fig. (**2**) shows the FTIR spectra for Nd_2O_3 containing lithium borosilicate glasses. The presence of distinct vibrational bands in the moulded glass samples is confirmed by the observed transmission spectra, confirming the development of BO_4 structural units. In Nd_2O_3 doped lithium borosilicate glasses, the band at 740 cm^{-1} can result from the overlapping contributions of multiple borate units [4, 12]. The B-O bond stretching in structural units from the di borate group is ascribed to the band at 943 cm^{-1} [4, 12]. The symmetric stretching vibration of BO_4 units is thought to be the source of the 1294 cm^{-1} band [4, 12]. However, there are a few slight differences in band intensities and peak positions amongst these samples. At 694 cm^{-1}, the FTIR band is caused by the combined vibrations of BO_4 groups. The band 1329 cm^{-1} changes to the low-wavenumber side, whereas the band 894 cm^{-1} shifts to the high-wavenumber side. This movement of the bands toward lower or higher wave numbers is caused by a change in the relative strength of the chemical bonds caused by the change in glass composition [4, 12]. Additionally, this shift in band location indicates the structural alterations and increase of BO$_4$ units [4, 12].

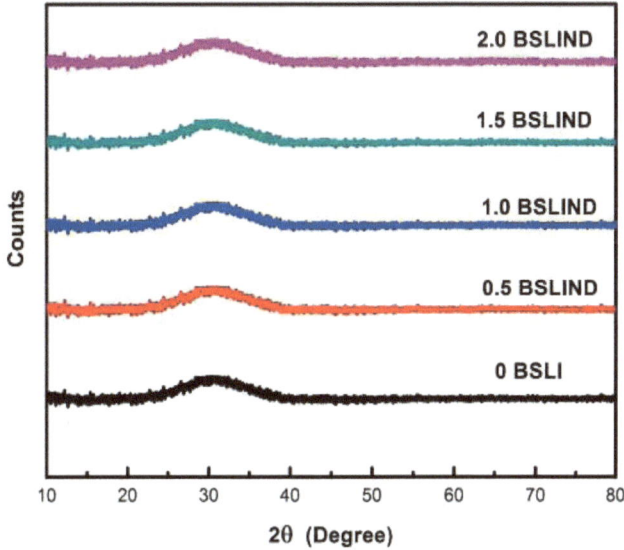

Fig. (1). XRD spectra of sodium borosilicate glasses doped with Nd$_2$O$_3$.

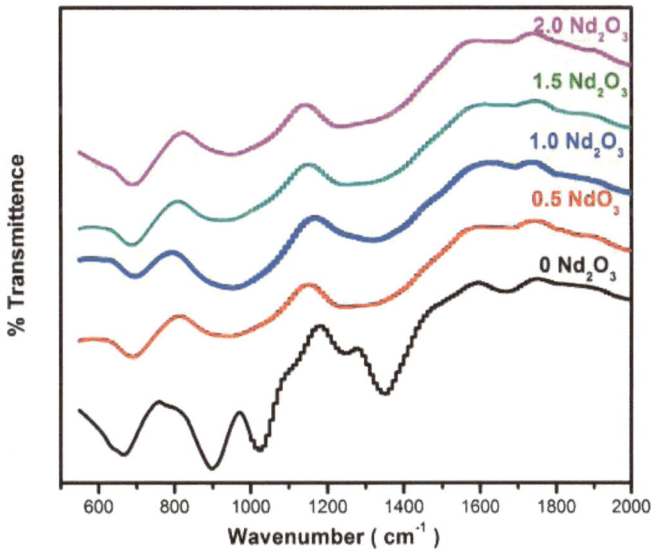

Fig. (2). FTIR spectra of Nd$_2$O$_3$ - lithium borosilicate glasses.

The dc conductivity derived from the impedance investigation at a variable temperature is given in Fig. (**3**), which is pertinent to the Arrhenius equation.

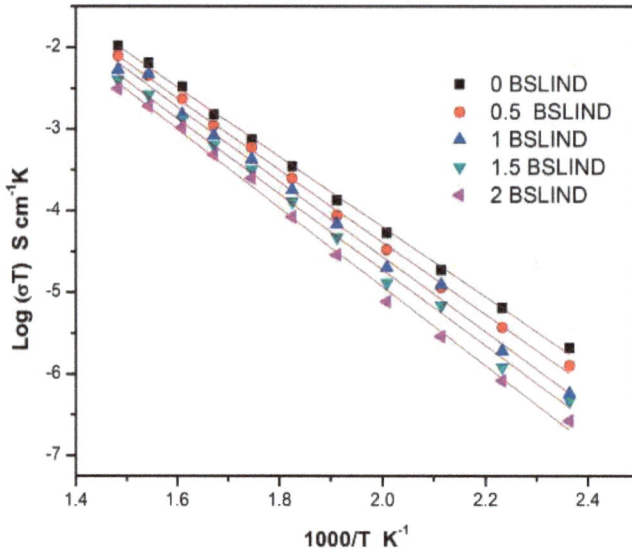

Fig. (3). Conductivity variation with temperature for prepared glasses.

$$\sigma = \sigma_0 \exp\left(-\frac{E_a}{k_b T}\right) \qquad (1)$$

Where σ_0, T, E_a and k_B are pre-exponential component, Temperature, activation energy and Boltzmann constant respectively.

Each sample's activation energy was calculated using equation 1 and corresponding line slopes. The estimated activation energies (E_a) were then plotted against the conductivity at 573 K for various mole percent Nd_2O_3, as shown in Fig. (**4**).

This figure illustrates that with the inclusion of Nd_2O_3, the conductivity of the prepared samples is reduced and the activation energy rises. The decline in conductivity can be associated with increased activation energy due to addition of Nd_2O_3. As lithium concentration is uniform in all samples of system, the reduction in conductivity for mobile Li+ ions are attributed to a decrease in the accessible vacant sites. The decrease at these sites may be ascribed to the conversion of BO_3 units to BO_4^- units, as shown by the FTIR findings of the investigated glass. Similar results are observed in rare earth doped vanado-tellurite &borate glass system [13, 14]. Additionally, the decrease in conductivity is fully advocated by the increase in density and molar volume, showing the glass system's rigidity.

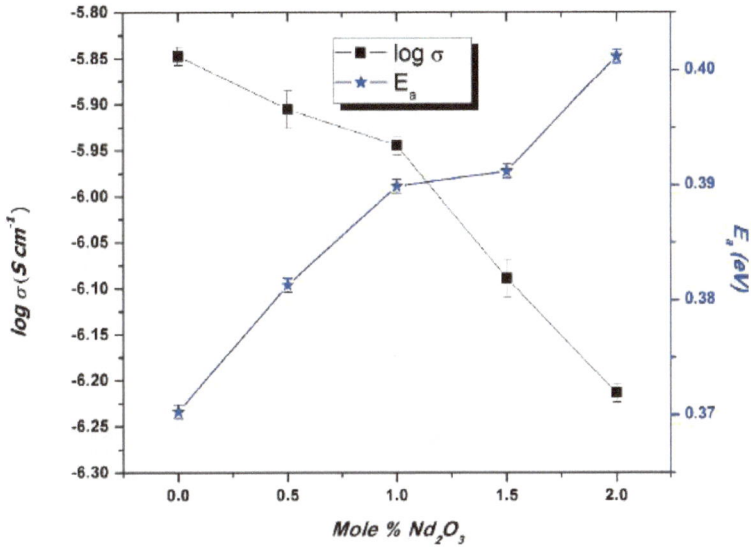

Fig. (4). Conductivity and activation energy with function of Nd_2O_3.

Fig. (5) displays variations in molar volume (V_m) and density (ρ) for Nd_2O_3 comprising glasses.

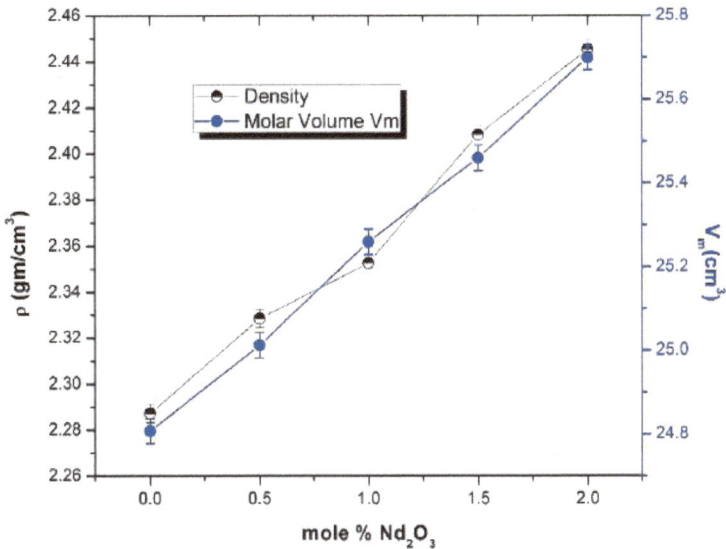

Fig. (5). Density and Molar volume as a function of Nd_2O_3 concentration.

The molar volume (V_m) and density (ρ) are observed to rise with an increase in Nd_2O_3 concentration. This rise in molar volume and density is due to the more molecular weight (336.48 gm/ mol) of Nd_2O_3 than other additives and large Nd^{3+}

ion radius. This large radius of Nd^{3+}ion causes expansion of glassy structures thereby increasing the molar volume. Also, the Nd^{3+} ions are incorporated into the gaps around the units that open the glassy structure. These results are in substantial alignment with the reported results, where molar volume and glass density rises with REOs [12, 13]. Further, several physical parameters such as ions concentration (N), polaron radius (r_p), molar volume (V_m), field strength (F) and internuclear distance (r_i) were determined using the density values. The physical properties play a significant role in explaining atomic arrangements in the glass network and provide useful knowledge regarding glass structure. The obtained values of the physical parameters for the prepared glass system are given respectively in Table **2**.

Table 2. Physical properties of the studied glasses.

Parameters	Unit	0.0	0.50	1.00	1.50	2.00
ρ	(g/cm³)	2.2873	2.3284	2.3526	2.4083	2.4451
V_m	(cm³)	24.8081	25.0122	25.2593	25.4596	25.6996
(N) (× 10²²)	(ions/cm³)	-	1.2072	2.3845	3.5486	4.6872
r_i	(Å)	-	4.3592	3.4743	3.0431	2.7735
r_p	(Å)	-	1.7565	1.3999	1.2262	1.1175
F (× 10¹⁷)	(cm⁻²)	-	1.9448	3.0616	3.9908	4.8044

From Table **2**, it is revealed that the Nd^{3+} ions (N) concentration rises with increase in Nd_2O_3content whereas the internuclear distance (r_i) and polaron radius (r_p) decrease with rise in Nd^{3+} ions (N) concentration. Internuclear distance (r_i) between Nd-O decreasesdue to congestion of glassy network by Nd^{3+}interstices. Thus the strength of Nd-Obond increases. Growing Nd-O bond strength results in greater field intensity (F) around Nd^{3+}ions resulting in increased glass density with the addition of Nd_2O_3 [12 - 14].

The relaxing phenomenon of the prepared samples can be understood by the application of electric modulus formalism. Fig. (**6**) illustrates the change of the real portion of the electric modulus with frequency for 2 mol% Nd_2O_3 glass samples.

This figure illustrates that M' is 0 at low frequencies owing to the absence of restoring forces for mobile Li^+ ions, and M' increases as frequency increases [15, 16]. Further M' achieves the maximal value conforming to $M\infty = (\varepsilon_\infty)^{-1}$ owing to relaxation processes [15, 16]. The remaining samples in the investigated series exhibit a similar qualitative pattern.

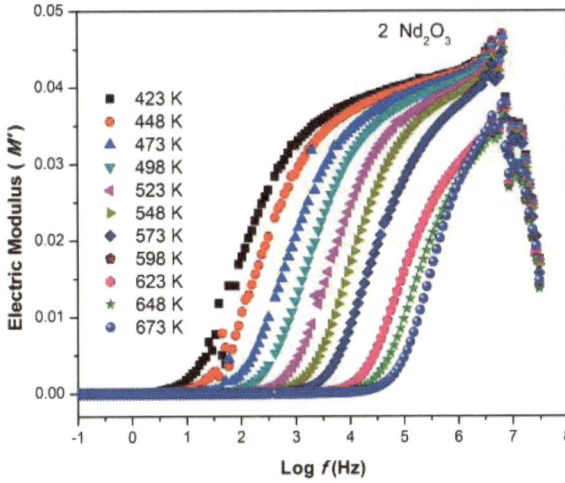

Fig. (6). Real portion of electric modulus (M') with function of frequency and temperature for 2-Nd$_2$O$_3$ sample.

The variance with the frequency of the imaginary component of the equivalent electric modulus M" has been shown in Fig. (**7**).

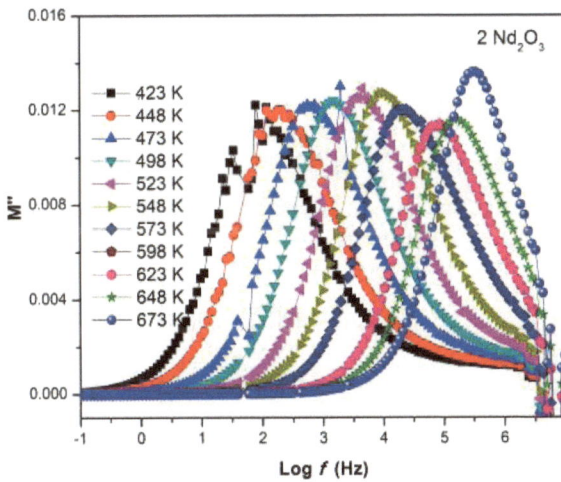

Fig. (7). Imaginary component of electric modulus (M") with function of frequency and temperature for 2 mol % Nd$_2$O$_3$ sample.

As can be seen from this graph, M"$_{Max}$ switches to higher frequencies as the temperature increases. f$_p$ denotes the center of relaxation peak and is also used to compute the relaxation time τ' ($\tau = 1/2\pi fp$) under the constraint $\omega_c\tau = 1$. The estimated relaxation time for this series's 2 mole percent Nd$_2$O$_3$ is shown versus 1000/T in Fig. (**8**).

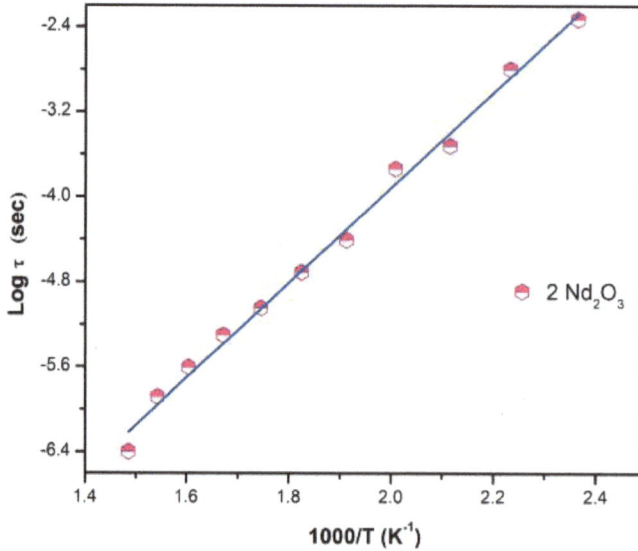

Fig. (8). Change of relaxation time with function of temperature for 2 mol% Nd_2O_3 sample.

The activation energies, $E_{a(DC)}$ and $E_{a(\tau)}$, for studied glass samples were calculated using the Arrhenius equation and are given in Table 3. The parallelism between $E_{a(DC)}$ and $E_{a(\tau)}$ implies that charge carriers must surpass the same energy barrier during conduction and relaxation [13, 14, 16 - 21]

Table 3. Values of $E_{a(DC)}$ and $E_{a(\tau)}$ for studied glass samples.

Nd_2O_3 Mole %	$E_{a(DC)}$ (eV)	$E_{a(\tau)}$(eV)
0.0	0.3703	0.3649
0.5	0.3822	0.3789
1.0	0.3999	0.3814
1.5	0.4007	0.3839
2.0	0. 4080	0.3883

The outcome of a normalized plot using *Log* (σ'/σ_{dc}) as the Y-axis parameter and *Log* $(f/\sigma_{dc}T)$ as the X-axis parameter for a 2 mol% Nd_2O_3 glass sample is given in Fig. (**9**), where data for various temperatures converge on a single master curve [16 - 22]. Other glass samples in the series exhibit a similar nature graph.

Fig. (**10**) depicts a superimposed plot of scaled data for the imaginary portion of the modulus. The Y axis has been scaled to (M''/M''_{Max}) in this image, while the X axis has been scaled to *Log* (f/f_p). As shown in Figs. (**9** and **10**), dynamic processes operating at various frequencies need almost identical thermal activation, implying that the conduction mechanism is temperature independent [17 - 22].

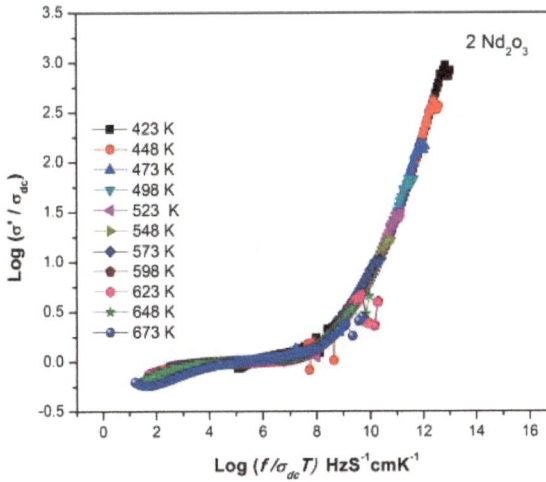

Fig. (9). Conductivity scaling data for 2mol % Nd_2O_3 glass sample for different temperatures.

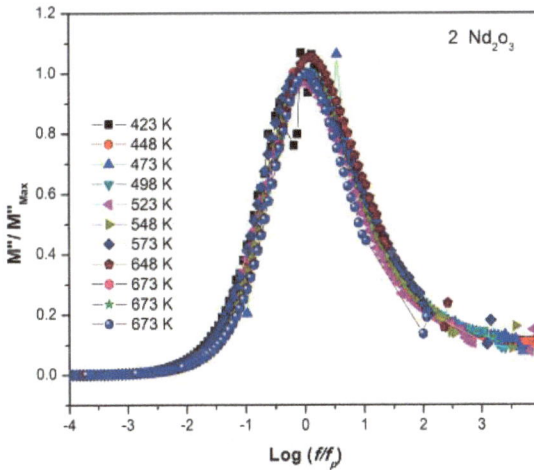

Fig. (10). Imaginary portion of electric modulus M''/M''$_{max}$ with Log f/fp for 2 mol % Nd_2O_3 glass sample at different temperatures..

To investigate the impact of composition on the conduction mechanism, an additional mol percent component was included. For prepared glass samples, the Y-axis is scaled using $Log (\sigma' / \sigma_{dc})$ and $Log (fx/ \sigma_{dc}T)$ as the X-axis. As shown in Fig. (**11**), the data does not create a single master curve. Thus, establishes that the conduction mechanism is composition-dependent, rather than temperature-dependent [15 - 18].

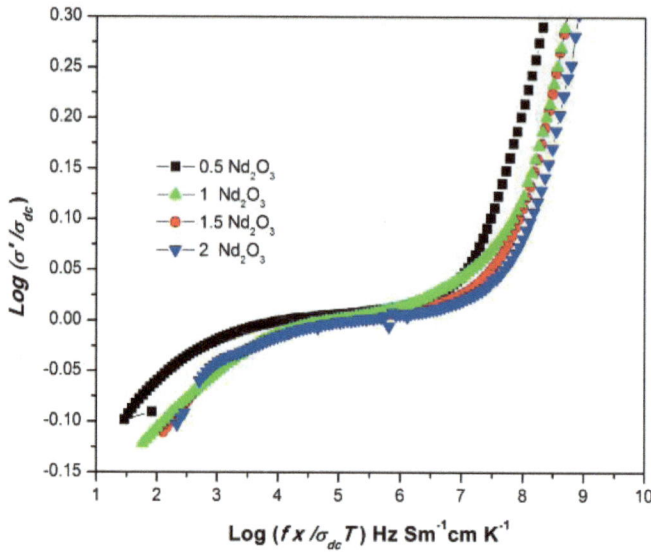

Fig. (11). Scaling data for different mol % Nd_2O_3 glass samples at 573 K..

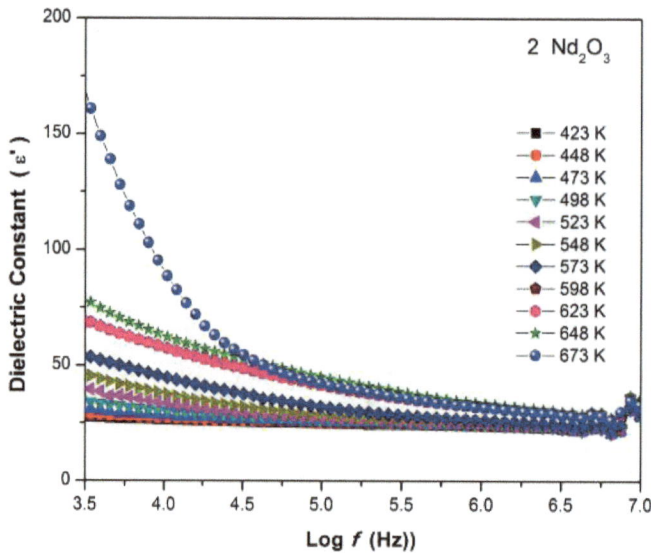

Fig. (12). Change of dielectric constant (ε') with a function of frequency for 2 mol % Nd_2O_3 sample for different temperature.

Figs. (**12** and **13**) illustrate the change of the dielectric constant (ε') and the dielectric loss (Tan δ) with logarithm of frequency for 2 mol % Nd_2O_3 glass sample (2.0 BSLIND). It can be noticed from these figures that ε' and Tan δ decreases with increase in frequency and decrease in temperature. Additionally, the dielectric constant increases with decreasing frequency and increasing temperature, which is typical for oxide glasses [15, 16]. Increases in ε' and Tan δ with increasing temperature in the low frequency zone may be attributed to a reduction in bond energy. With increasing temperature, dipolar polarisation is influenced by i) an increase in orientational vibration owing to weakening intermolecular forces and (ii) a disruption in orientational vibration due to thermal activation [15, 16]. Accumulation of charge at high energy barrier sites as a consequence of charge carriers hopping from low energy barrier sites leads in increased net polarisation, which leads to low frequency region's high dielectric constant [15 - 20]. At high frequencies, however, the difficulty of charge carriers to spin quickly and adequately results in a reduction in charge carrier oscillation, leads to decline of dielectric constant and dielectric loss. Other glass samples exhibit a similar qualitative trend.

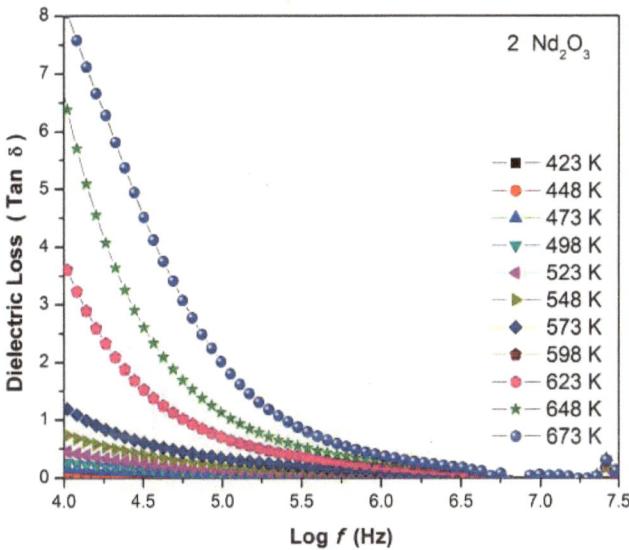

Fig. (13). Change of dielectric loss (Tan δ) versus log of frequency for 2 mol % Nd_2O_3 sample for different temperature.

Fig. **(14)** illustrates the ac conductivity as a function of frequency for 2 mol % Nd_2O_3 glass sample at various temperatures (2.0 BSLIND).

As shown in this graph, ac conductivity rises with increasing temperature at all frequencies, which is attributable to thermal activation. AC conductivity is temperature independent at low frequencies and high temperatures due to the absence of a change in the slope. However, at low temperatures and high frequencies, the slope changes, indicating that AC conductivity is frequency dependent. This change in the frequency independence of AC conductivity to frequency dependence indicates the origin of the conductivity relaxation phenomena [15, 16]. Conductivity is frequency independent in the low frequency range owing to the random distribution of ionic charge, while conductivity exhibits dispersion at higher frequencies [14 - 16]. Related qualitative activity with other glass objects is detected.

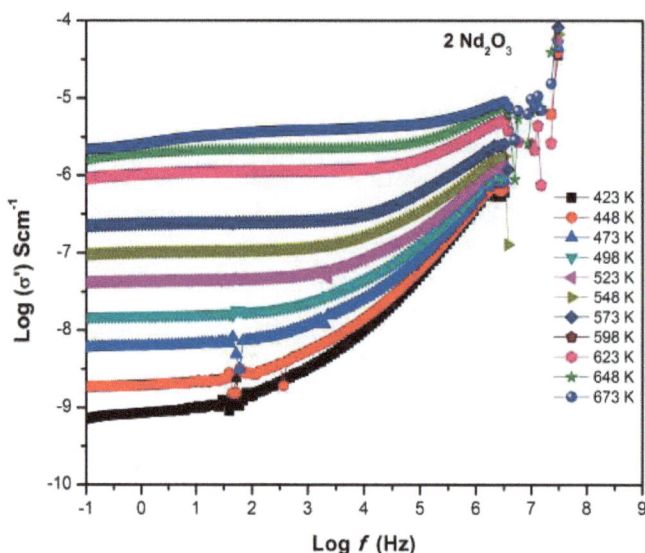

Fig. (14). AC conductivity versus frequency for 2 mol % Nd_2O_3 sample.

CONCLUSION

The melt quench technique was used to produce lithium borosilicate glasses doped with Nd_2O_3. XRD examination revealed that the produced glass samples were amorphous in nature. The development of units is confirmed by the FTIR spectrum. The density (ρ) of glasses rises when Nd_2O_3 is added. The increase in density is due to the fact that Nd_2O_3 has a greater molecular weight than the other glass components. The estimated values for different physical parameters, including the concentration of the Nd^{3+} ion (N), mean spacing between the rare earth ions (r_i), polaron radius (r_p) and field strength (F) of Nd-O, support the change density (ρ) and molar volume (V_M). The inclusion of Nd_2O_3 reduces the electrical conductivity of lithium borosilicate glasses. Increase in density (ρ) supports this reduction in conductivity. The normalised plot of *Log (σ'/σ_{dc}) vs Log (f/ $\sigma_{dc}T$)* and *M''/M''$_{Max}$vs Log (f/f$_{Max}$)* demonstrates that the dynamic processes operating at various frequencies need almost the same thermal activation energy. In other words, conduction mechanism is independent of temperature. The figure of *Log (σ'/σ_{dc}) vs Log (fx/ $\sigma_{dc}T$)* at 573 K demonstrates that the conduction method is compositional dependent, not temperature dependent. Temperature and frequency lead to the progressive reduction in the dielectric constant (ϵ') and dielectric loss (Tan δ) of investigated glasses, due to electron hopping and bond energy fluctuation in the net polarisation.

CONSENT FOR PUBLICATION

Not applicable.

CONFLICT OF INTEREST

The authors declare no conflict of interest, financial or otherwise.

ACKNOWLEDGEMENTS

Declared none.

REFERENCES

[1]   Muralidharan, P. Sol–gel synthesis, structural and ion transport studies of lithium borosilicate glasses. *Solid State Ion., 2004, 166*(1-2), 27-38.
        [http://dx.doi.org/10.1016/j.ssi.2003.10.011]

[2]   Muralidharan, P.; Venkateswarlu, M.; Satyanarayana, N. AC conductivity studies of lithium borosilicate glasses: synthesized by sol–gel process with various concentrations of nitric acid as a catalyst. *Mater. Chem. Phys., 2004, 88*(1), 138-144.
        [http://dx.doi.org/10.1016/j.matchemphys.2004.06.032]

[3]   Deshpande, A.V.; Paighan, N.S. **2009**.
        [http://dx.doi.org/10.1088/1757-899X/2/1/012051]

[4]   Pawar, P.P.; Munishwar, S.R.; Gautam, S.; Gedam, R.S. Physical, thermal, structural and optical properties of Dy 3+ doped lithium alumino-borate glasses for bright W-LED. *J. Lumin., 2017, 183*, 79-88.
        [http://dx.doi.org/10.1016/j.jlumin.2016.11.027]

[5]   Lakshminarayana, G.; Buddhudu, S. Spectral analysis of Sm3+ and Dy3+: B2O3–ZnO–PbO glasses. *Physica B, 2006, 373*(1), 100-106.
        [http://dx.doi.org/10.1016/j.physb.2005.11.143]

[6]   Pereira, R.; Gozzo, C.B.; Guedes, I.; Boatner, L.A.; Terezo, A.J.; Costa, M.M. Impedance spectroscopy study of SiO2–Li2O:Nd2O3 glasses. *J. Alloys Compd., 2014, 597*, 79-84.
        [http://dx.doi.org/10.1016/j.jallcom.2014.01.151]

[7]   Ganvir, V.Y.; Gedam, R.S. Influence of Sm $_2$ O $_3$ addition on electrical properties of lithium borosilicate glasses. *Integr. Ferroelectr., 2017, 185*(1), 102-108.
        [http://dx.doi.org/10.1080/10584587.2017.1370346]

[8]   Anjaiah, J.; Laxmikanth, C. *Optical Properties of Neodymium Ion Doped Lithium Borate Glasses,* **2015**, *5*, 173-183.

[9]   Li, H.; Li, L.; Vienna, J.D.; Qian, M.; Wang, Z.; Darab, J.G.; Peeler, D.K. Neodymium(III) in alumino-borosilicate glasses. *J. Non-Cryst. Solids, 2000, 278*(1-3), 35-57.
        [http://dx.doi.org/10.1016/S0022-3093(00)00327-6]

[10]   Dymshits, O.S.; Zhilin, A.A.; Savostjanov, V.A.; Chuvaeva, T.I. The structure of luminescence centers of neodymium in glasses and transparent glass-ceramics of the Li2Q--Al & -SiO , system. **1996**.

[11]   Pawar, P.P.; Munishwar, S.R.; Gedam, R.S. Physical and optical properties of Dy 3+ /Pr 3+ Co-doped lithium borate glasses for W-LED. *J. Alloys Compd., 2016, 660*, 347-355.

[http://dx.doi.org/10.1016/j.jallcom.2015.11.087]

[12]   Ramteke, D.D.; Annapurna, K.; Deshpande, V.K.; Gedam, R.S. Effect of Nd3+ on spectroscopic properties of lithium borate glasses. *J. Rare Earths,* **2014**, *32*(12), 1148-1153.
       [http://dx.doi.org/10.1016/S1002-0721(14)60196-4]

[13]   Gedam, R.S.; Ramteke, D.D. Influence of CeO2 addition on the electrical and optical properties of lithium borate glasses. *J. Phys. Chem. Solids,* **2013**, *74*(10), 1399-1402.
       [http://dx.doi.org/10.1016/j.jpcs.2013.04.022]

[14]   Prashant Kumar, M.; Sankarappa, T. DC conductivity of rare earth ions doped vanado-tellurite glasses. *J. Non-Cryst. Solids,* **2009**, *355*(4-5), 295-300.
       [http://dx.doi.org/10.1016/j.jnoncrysol.2008.11.004]

[15]   Ramteke, D.D.; Gedam, R.S. Study of Li2O–B2O3–Dy2O3 glasses by impedance spectroscopy. *Solid State Ion.,* **2014**, *258*, 82-87.
       [http://dx.doi.org/10.1016/j.ssi.2014.02.006]

[16]   Bahgat, A.A Study of dielectric relaxation in Na-doped Bi–Pb–Sr–Ca–Cu–O glasses, Journal of Non-Crystalline Solids. **1998**.
       [http://dx.doi.org/10.1016/S0022-3093(97)00482-1]

[17]   Abdel-Khalek, E.K.; Bahgat, A.A. Optical and dielectric properties of transparent glasses and nanocrystals of lithium niobate and lithium diborate in borate glasses, Phys. B. *Condens. Matter,* **2010**, *405*, 1986-1992.

[18]   Roling, B. Scaling properties of the conductivity spectra of glasses and supercooled melts. *Solid State Ion.,* **1998**, *105*(1-4), 185-193.
       [http://dx.doi.org/10.1016/S0167-2738(97)00463-3]

[19]   Roling, B.; Happe, A.; Funke, K.; Ingram, M.D. Carrier Concentrations and Relaxation Spectroscopy: New Information from Scaling Properties of Conductivity Spectra in Ionically Conducting Glasses. *Phys. Rev. Lett.,* **1997**, *78*(11), 2160-2163.
       [http://dx.doi.org/10.1103/PhysRevLett.78.2160]

[20]   P., Muralidharan; M., Venkateswarlu; N., Satyanarayana AC conductivity studies of lithium borosilicate glasses: Synthesized by sol-gel process with various concentrations of nitric acid as a catalyst *Mater. Chem. Phys,* **1993**, *88*, 138-144.
       [http://dx.doi.org/10.1088/0022-3727/26/7/019]

[21]   Ganvir, V Y; Ganvir, H V; Gedam, R S Effect of Dy_2O_3 on Electrical Conductivity, Dielectric Properties and Physical Properties in Lithium Borosilicate Glasses, Integrated Ferroelectrics. , 1-11.
       [http://dx.doi.org/10.1080/10584587.2019.1674947]

[22]   Ganvir, V.Y.; Gedam, R.S. Effect of La_2O_3 addition on structural and electrical properties of sodium borosilicate glasses. *Mater. Res. Express,* **2017**, *4*(3), 035204.
       [http://dx.doi.org/10.1088/2053-1591/aa66e4]

CHAPTER 12

Comprehensive Quantum Mechanical Study of Structural Features, Reactivity, Molecular Properties and Wave Function-Based Characteristics of Capmatinib

Renjith Thomas[1,*] and **T. Pooventhiran**[1]

[1] *Department of Chemistry, St Berchmans College, Changanaserry, Kerala, India*

Abstract: Lung cancer is one of the major classes of cancer affecting men. Capmatinib is developed and approved as a medicine to fight non-small lung cancer. Even though the compound is developed for the management of cancer, it may be of several other applications. To tune the property of molecules to fit diverse functions, the study of the electronic structure and other quantum mechanical properties is very important. Literature survey indicates that the detailed structure and reactivity profile of this compound was not reported. We use molecular modeling using DFT and TD-DFT methods using B3LYP/CAM-B3LYP/aug-cc-pVDZ level to study the structure, reactivity, and other Physico-chemical properties of this compound. The optical properties of the compound are compared with standard materials. Different useful indices from molecular wave function analysis present the reactivity and stability of the compound in detail. A qualitative study of non-covalent interactions was also reported along with many useful local information entropy studies, which showed that the molecule is stable with low uncertainty of electrons in spatial distribution.

Keywords: Capmatinib, DFT, Local information entropy, MESP, NCI, Structure.

INTRODUCTION

Lung cancer is a common type of cancer in men, and among the various types of lung cancer is non-small cell lung cancer (NSCLC), which is typically detected at an advanced stage [1 - 3]. There are many subtypes of this disease [4]. Several reported studies indicate that receptor tyrosine kinase (RTK) MET acts as an oncogene and is regarded as a significant target for cancer drug improvement [5]. MET signals provide basic cell differentiation and cell growth, including embryonic development cycles, wound cure, and tissue regeneration [6].

---
* **Corresponding author Renjith Thomas:** Department of Chemistry, St Berchmans College, Changanaserry, Kerala, India; E-mail: renjith@sbcollege.ac.in

Dibya Prakash Rai (Ed.)

Alterations of MET lead to activation of mutations, overexpression, gene amplification, and translocations [7]. Anti-MET antibodies are failed to obtain high levels in medical trials, while erlotinib, as well as the monoclonal anti-MET antibody onartuzumab combination therapies in patients with NSCLC and MET overexpression, have produced insufficient results in a phase III analysis based on immunohistochemistry (IHC) analysis [8]. Small-molecule MET inhibitors are of different types. Type I block and type **Ib** inhibitors like ATP binding can prevent phosphorylation and activation of the receptor, and tepotinib, capmatinib, savolitinib, and AMG-337 are highly specific for MET when compared to type **Ia** inhibitors like crizotinib [9]. Glesatinib, cabozantinib, and merestinib are ATP competitive inhibitors that interact with a hydrophobic pocket adjacent to the ATP binding sites [10]. They also interact with allosteric sites instead of the ATP binding sites.

The benign tumors like cutaneous neurofibroma, plexiform neurofibroma, and glioma are the biallelic inactivation of the tumor-suppressor gene neurofibromatosis type-1 in glial cells in the skin, along a nerve plexus or in the brain, respectively [11]. The bidirectional (on-targeted and off-targeted) interactions with the NSCLC increase also reported therapeutic resistance [12]. Some mechanisms, including gene mutation, amplification, rearrangement, and protein overexpression, cause abnormal mesenchymal-epithelial transition (MET) pathway in lung cancer [13]. Capmatinib hepatoprotective effect were shown to be mediated by lowering the excessive formation of lipid peroxidation and nitrosative stress products caused by acetaminophen [14]. Capmatinib controls or destroys human [15 - 24] and animal (mouse) [25, 26] MET-targets by oncogene activation of receptor tyrosine kinases (RTKs). The MET inhibitors can treat with NSCLC harboring MET exon 14 skipping in mouse xenograft [27]. The intracranial activity of osimertinib with capmatinib in a patient with EGFR was reported [28].

Capmatinib is an orally administered, bioavailable, highly selective small molecule that has been shown to effectively inhibit the MET pathway both in vitro and *in vivo* [7, 29]. Capmatinib literature shows that no serious studies have been investigated in the analysis stability of the structure as well as electronic and reactivity properties in detail till date. This book chapter is an attempt to satisfy my new research direction. The geometry optimization of capmatinib was done. The TD-DFT study was done along with its non-linear optics properties. The NBO analysis, ALIE analysis, ELF analysis, LOL analysis, MESP analysis, RDG analysis, LIE analysis, and NCI analysis were done explicitly.

METHODOLOGY

Gaussian-09 software [30] was used for optimization of capmatinib by the method DFT- B3LYP [31 - 35] with basis set of aug-cc-pVDZ [36 - 41], and IR spectra simulation, frontier molecular orbital properties, and natural bonding orbital analysis were frequency calculations done by above basis set. We used RCAM-B3LYP functional [42, 43] and aug-cc-pVDZ for TD-DFT. GaussSum [44] to see the results and get data [44] for UV-Visible spectrum with molecular orbital contributions. Capmatinib molecule has more than two reaction sites, such as on fluorine, carbons, and protons in fluorophenyl-, oxygen, carbons, protons, and nitrogen in N-methylacetamid-, nitrogens, carbons, and protons in imidazotriazin-, carbon and hydrogens in acetyl, and nitrogen, carbons, and protons in quinolin-group. Capmatinib molecule reaction sites found by multiwavefunction software by analyzing total electrostatic potentials, average localized ionization energy (ALIE), electron localization functions (ELF), localized orbital locator (LOL), reduced density gradient (RDG), localized entropy interaction (LE), localized electron locator (LEL), and non-covalent interactions (NCI) [45].

RESULT AND DISCUSSIONS

Structure of Capmatinib

Capmatinib molecule is optimized to get the minimum energy state with DFT by B3LYP/aug-cc-pVDZ. Fig. (1) shows the geometry structure for capmatinib.

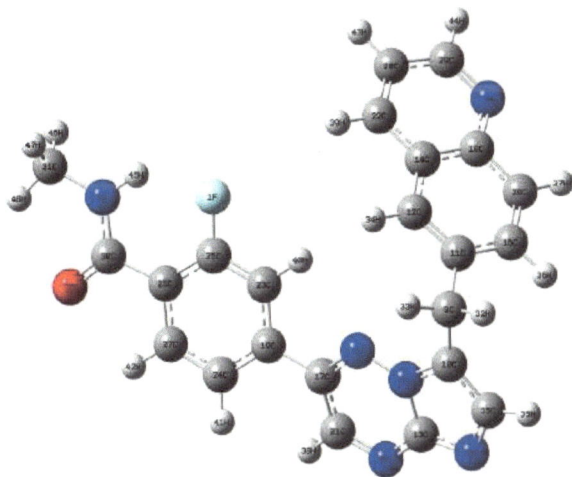

Fig. (1). Optimised geometry of capmatinib.

The bond lengths for 1F-26C, 2O-30C, 8N-30C, 8N-31C, 8N-45H, 25C-30C, 17C-19C, 3N-4N, 3N-10C, 3N-13C, 4N-17C, 5N-13C, 5N-15C, 6N-13C, 6N-21C, 9C-10C, 9C-11C, 14C-18C, 7N-18C, and 7N-29C having 1.37, 1.23, 1.36, 1.45, 1.01, 1.52, 1.48, 1.34, 1.37, 1.42, 1.32, 1.33, 1.36, 1.34, 1.32, 1.50, 1.52, 1.43, 1.37, and 1.32 Å in orderly. The bond angles for 30C-8N-31C, 31C-8--45H, 2O-30C-8N, 2O-30C-25C, 4N-3N-10C, 4N-3N-13C, 3N-4N-17C, 3N-13--6N, 13C-6N-21C, 3N-13C-5N, 13C-5N-15C, 10C-9C-11C, and 18C-7N-29C having 122.19, 118.63, 123.10, 119.70, 127.51, 124.90, 114.55, 119.69, 116.09, 110.65, 104.75, 114.16, and 117.49^0 respectively.

Nature Bond Orbital (NBO) Study of Capmatinib

The molecular stability of capmatinib determined by the nature of bonding orbital study explained by intramolecular electron motions are very important. The hyperconjugation [46 - 49] interactions explained by natural bonding orbital study also explain the electron's occupancies. The stability of the capmatinib molecule explained by delocalization energy provides valuable information and occupancy. NBO calculations were performed using the NBO suite, which is included with the Gaussian-09 software suite [50].

The descending order of atomic orbitals (AOs) by the occupancies is valence orbitals < core < Rydberg [51 - 55]. The delocalization of electrons depends upon their occupancies, the fluorine atom's maximum occupancy is 1.9118 in valence 2pz orbital from other electron-donating atoms, the oxygen atom's maximum occupancy is 1.5058 in valence 2pz orbital from its valence 2s orbital, the nitrogen atom'-s maximum occupancies are more than 1.5000 in their valence in 2py, and 2pz orbitals form their self-valence 2s orbitals, and carbon atoms got maximum occupancies is more than 1.5000 in their valence 2pz orbital form their self-valence 2s orbitals. Capmatinib molecule have g 48 atoms which contribute for 519 NAOs. The NAOs numbers are 8, 22, 32, 46, 64, 74, 90, 102, 118, 130, 146, 160, 172, 188, 200, 216, 228, 244, 256, 272, 284, 300, 314, 326, 342, 354, 370, 384, 396, 410, 424, 435, 440, 445, 450, 455, 460, 465, 470, 475, 480, 485, 490, 495, 500, 505, 510, and 515 for the atoms with number F1, O2, N3-8, C9-31, and H32-48. The nature atomic orbital numbers are 4, 20, 34, 50, 60, 78, 92, 104, 120, 132, 144, 158, 174, 186, 204, 214, 230, 242, 258, 270, 286, 298, 312, 328, 340, 356, 368, 382, 398, 408, and 515 for the atoms with number F1, O2, N3-8, and C9-31.

Table **1** shows the natural populations of charges analysis of capmatinib. Capatinib molecule having the total natural charges, natural populations of core, valence, Rydberg population and total population are 0, 61.9710, 151.3595, 0.6695, and 214 respectively.

Table 2. Summary of Natural Population Charges Analysis of Capmatinib.

Atom	No.	Natural Charge	Natural Population				Atom	No.	Natural Charge	Natural Population			
			Core	Valence	Rydberg	Total				Core	Valence	Rydberg	Total
F	1	-0.3585	1.9999	7.3550	0.0036	9.3585	C	26	0.4523	1.9983	3.5217	0.0278	5.5477
O	2	-0.6443	1.9998	6.6358	0.0086	8.6443	C	27	-0.1613	1.9990	4.1461	0.0163	6.1613
N	3	-0.2111	1.9990	5.1953	0.0168	7.2111	C	28	-0.2689	1.9990	4.2561	0.0138	6.2689
N	4	-0.2512	1.9993	5.2288	0.0230	7.2512	C	29	0.0729	1.9991	3.9041	0.0238	5.9271
N	5	-0.5185	1.9994	5.4957	0.0234	7.5185	C	30	0.7085	1.9992	3.2427	0.0496	5.2915
N	6	-0.4377	1.9993	5.4146	0.0238	7.4377	C	31	-0.4133	1.9994	4.4031	0.0108	6.4133
N	7	-0.4789	1.9994	5.4572	0.0223	7.4789	H	32	0.2551	0	0.7415	0.0034	0.7449
N	8	-0.6634	1.9993	5.6546	0.0095	7.6634	H	33	0.2514	0	0.7445	0.0040	0.7486
C	9	-0.4798	1.9991	4.4641	0.0166	6.4798	H	34	0.2274	0	0.7692	0.0034	0.7726
C	10	0.1398	1.9990	3.8361	0.0252	5.8602	H	35	0.2251	0	0.7732	0.0018	0.7749
C	11	-0.0261	1.9989	4.0102	0.0171	6.0261	H	36	0.2309	0	0.7659	0.0032	0.7690
C	12	-0.2021	1.9988	4.1880	0.0153	6.2021	H	37	0.2424	0	0.7536	0.0040	0.7576
C	13	0.5932	1.9991	3.3715	0.0362	5.4069	H	38	0.2218	0	0.7758	0.0024	0.7782
C	14	-0.0889	1.9988	4.0750	0.0151	6.0889	H	39	0.2301	0	0.7663	0.0035	0.7699
C	15	0.0038	1.9990	3.9748	0.0225	5.9962	H	40	0.2592	0	0.7366	0.0042	0.7408
C	16	-0.2161	1.9989	4.2025	0.0147	6.2161	H	41	0.2353	0	0.7617	0.0030	0.7647
C	17	0.1676	1.9990	3.8059	0.0275	5.8324	H	42	0.2632	0	0.7321	0.0047	0.7368
C	18	0.1935	1.9988	3.7836	0.0241	5.8065	H	43	0.2335	0	0.7635	0.0031	0.7665
C	19	-0.0586	1.9989	4.0430	0.0167	6.0586	H	44	0.2102	0	0.7872	0.0027	0.7898
C	20	-0.1959	1.9989	4.1801	0.0169	6.1959	H	45	0.4117	0	0.5820	0.0063	0.5883
C	21	0.0374	1.9991	3.9404	0.0231	5.9626	H	46	0.2077	0	0.7896	0.0026	0.7923
C	22	-0.1696	1.9989	4.1555	0.0151	6.1696	H	47	0.2079	0	0.7895	0.0026	0.7921
C	23	-0.2621	1.9988	4.2441	0.0191	6.2621	H	48	0.2415	0	0.7539	0.0046	0.7585
C	24	-0.2201	1.9989	4.2075	0.0138	6.2201	**Total**		**0**	**61.9710**	**151.3595**	**0.6695**	**214.0000**
C	25	-0.1972	1.9987	4.1801	0.0183	6.1972							

The natural populations are the natural minimal basis (NMB), and the natural Rydberg basis (RYB) of capmatinib gives the following results. Total core population, valence population, NMB, and RYB are 61.9710, 151.3595, 213.3305, and 0.6695 out of 62, 152, 214, and 214 basis, greater than 99.50, 99.50, 99.50, and lower than 0.50 percentage, respectively.

The contributions towards the core, and valence Lewis are 61.97104, and 145.9425 out of 62, and 152 basis, is more than 99.50 percentage respectively. Therefore total L_contribution is 207.9135 out of 214 basis, which is 97.156 percentage. Contribution for valence n-L and Rydberg n-L orbitals are 5.6054, and 0.4810 out of 214 basis, which is 2.62, and 0.23 percentage respectively. Rydberg n-L orbital is 0.4810 out of 214 basis, and 0.23 percentage, therefore n-L contribution is 6.08646 out of 214 basis, and 2.84 percentage [53, 56 - 58].

Table **2** displays NBOs of capmatinib clarified through Perturbation theory (IInd order) analysis of Fock-matrixs. The transition occurs from the number of donor NBOs are 8, 10, 107, 462, 468, 471, 471, 487, 497, 510, and 510 donor orbitals NBOs are σ (2) N4 - C17, σ (2) N5 - C13, LP (1) N8, σ *(2) N4 - C17, σ *(2) N5 - C13, σ *(2) N7 - C29, σ *(2) N7 - C29, σ *(2) C14 - C18, σ *(2) C19 - C23, σ *(2) C25 - C26, and σ *(2) C25 - C26 to acceptor NBOs numbers are 102, 102, 457, 497, 480, 487, 502, 491, 507, 457, and 507 orbitals are LP (1) N3, LP (1) N3, σ *(2) O2 - C30, σ *(2) C19 - C23, σ *(2) C10 - C15, σ *(2) C14 - C18, σ *(2) C22 - C28, σ *(2) C16 - C20, σ *(2) C24 - C27, σ *(2) O2 - C30, and σ *(2) C24 - C27 by the energies are 75.02, 125.08, 68.69, 56.09, 91.55, 161.98, 135.12, 243.22, 279.71, 242.67, and 200.99 kcal/mol respectively. From the above result highest energy absorbs electron delocalized from σ *(2) C19-C2) bonding orbital to σ *(2) C24-C27 antibonding orbital by 279.71 kcal/mol.

Time-dependent Density Functional Theory (TD-DFT) Study of Capmatinib

The electronic transition was studied by time TDFT with CAM-B3LYP/au--ccpVDZ in a water implicit solvent atmosphere IEFPCM solvation model [59, 60]. Figs. (**2** and **3**) show the UV–Vis spectrum.

Table 2. Fock Matrix of the compound to find delocalization energy.

NBOs	Donor NBO (i)	NBOs	Acceptor NBO (j)	E(2) kcal/mol	E(j)-E(i) a.u.	F(i,j) a.u.
8	BD (2) N4-C17	102	LP (1) N3	75.02	0.05	0.09
10	BD (2) N5-C13	102	LP (1) N3	125.08	0.02	0.07
107	LP (1) N8	457	BD*(2) O2-C30	68.69	0.27	0.12
462	BD*(2) N4-C17	497	BD*(2) C19-C23	56.09	0.04	0.07
464	BD*(2) N5-C13	480	BD*(2) C10-C15	91.55	0.02	0.06
471	BD*(2) N7-C29	487	BD*(2) C14-C18	161.98	0.02	0.08
471	BD*(2) N7-C29	502	BD*(2) C22-C28	135.12	0.02	0.08
487	BD*(2) C14-C18	491	BD*(2) C16-C20	243.22	0.01	0.08
497	BD*(2) C19-C23	507	BD*(2) C24-C27	279.71	0.01	0.08
510	BD*(2) C25-C26	457	BD*(2) O2-C30	242.67	0.01	0.07
510	BD*(2) C25-C26	507	BD*(2) C24-C27	200.99	0.02	0.08

UV-VIS Spectrum

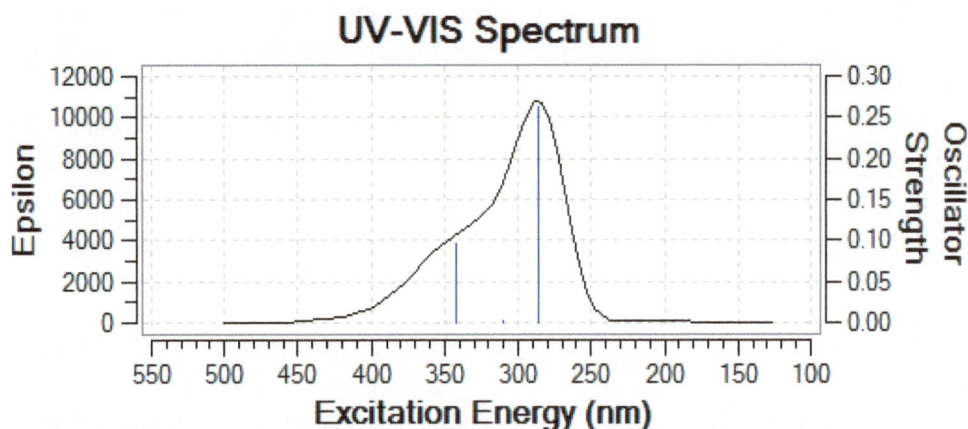

Fig. (2). Ultraviolet-visible spectrum of capmatinib.

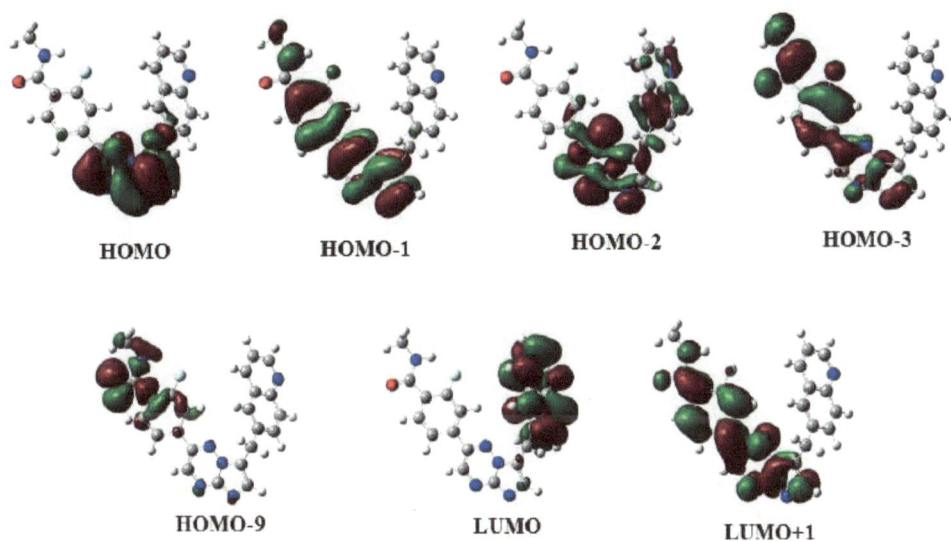

| HOMO | HOMO-1 | HOMO-2 | HOMO-3 |

| HOMO-9 | LUMO | LUMO+1 |

Fig. (3). Major and minor contributions orbitals of capmatinib.

The Capmatinib molecule has two important electronic transitions. The HOMO occurs at imidazotriazin- ring with methyl- groups, and LUMO occurs at quinolin- ring. The numerical values are 100.1250 kcal/mol, 285.5574 nm, 0.2662 for energy, wavelength, and oscillator strength, of first electronic transition respectively. Capmatinib having the major contribution to these transitions are from HOMO third, and second to LUMO are self, and self with 23, and 59 percentages. The minor contributions to these transitions of HOMO are from ninth, and self to LUMO are self, and first with 2, and 3 percent respectively. For the second transition, energy, wavelength, and oscillator strength are 83.6757 kcal/mol, 341.6930 nm, and 0.0971 with major contribution is from higher occupied molecular orbital self to L-UMO with 82 percentage, and the minor contributions of HOMO from third, first, and self to LUMO are self, self, and first with 4, 5, and 3 percentages [54, 55, 61 - 69].

Average Localized Ionization Energy (ALIE), Electron Localized Function (ELF), and the Localized Orbital Locator (LOL) Studies of Capmatinib

The reactivity of the capmatinib molecule is explained by ALIE, ELF [70 - 72], and LOL [73, 74] studies. Fig. (**4**) shows the reactive sites of ALIE, ELF, and LOL of capmatinib.

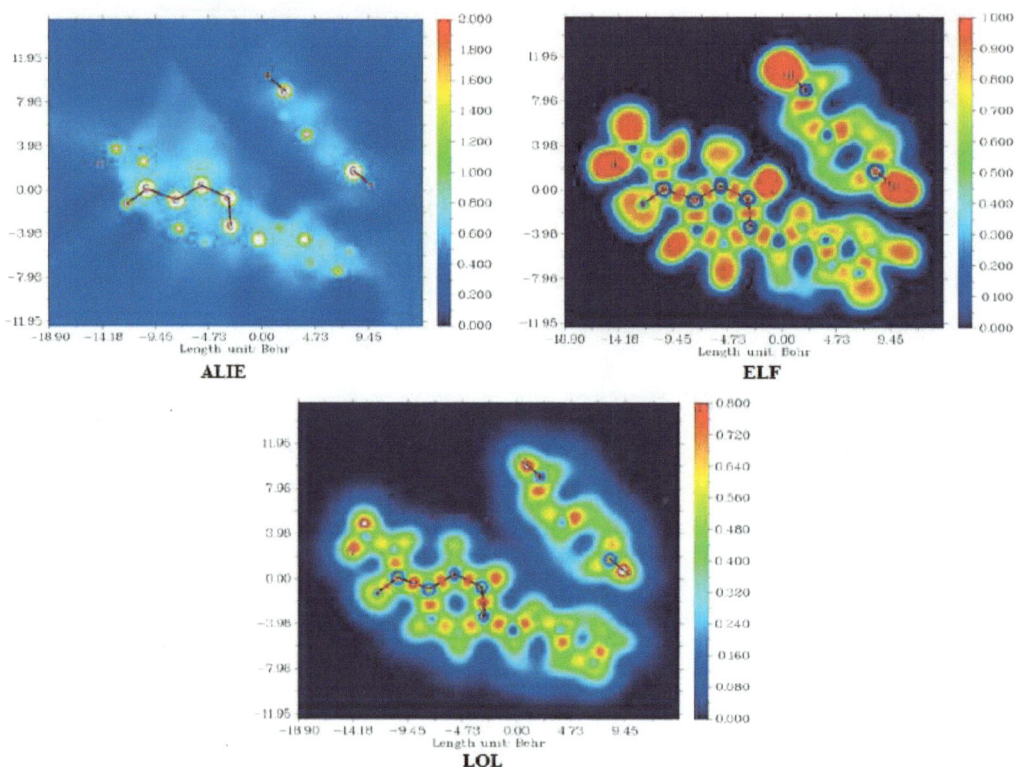

Fig. (**4**). ALIE, ELF, and LOL of capmatinib.

Capmatinib has ALIE, ELF, and LOL regions from blue to red as the scale values between 0.00 and 2.00, 0.00 to 1.00, and 0.00 to 0.08 the molecule within -16.55 to 11.82, -17.52 to 13.21 and -17.52 to 13.21 $Bohr^3$ respectively.

The ALIE study identified the LIE need for electronic excitations [62, 63, 65, 75 - 78]. The color blueish-green appears at N-methyacetamid-, fluorophenyl-, imidazotriazin- and quinolin- group, which indicates delocalization of electrons, blue indicates lone-pair of electrons having oxygen and nitrogen in N-methylacetamid-, nitrogen atoms in imidazotriazin- and quinolin- group, and red indicates the core orbitals of electrons in capmatinib. From ELF study, explains the probability of electrons sites. The color red is the most probability of electrons, which are stable on all the protons at N-methylacetamid-, fluorophenyl-, imidazotriazin-, quinolin- groups, lone-pair of electrons in oxygen, and nitrogen atoms within a molecule, and fluorin- atom in fluorophenyl- group shows strongly localized electrons. The blue mentions high delocalized electrons in the molecule positions at carbon-to-nitrogen-to-carbon in N-methylacetamid-, fluorophenyl-, imidazotriazin-, and quinolin- groups. From the LOL study, the blue, and red color shows weak, and strong pi-delocalized orbitals in capmatinib, also having adjacent mobile orbitals, and delocalization of electrons occurs on oxygen, nitrogen, and carbon atoms in N-methylacetamid-, fluorophenyl-, imidazotriazin-, quinolin- groups.

Local Information Entropy (LIE) of Capmatinib

The uncertainty is directly proportional to entropy, which is the probability of distributions, and explains the stability of capmatinib. The LIE and uncertainty of electrons in spatial distribution are having a directly proportional relationship [79 - 81]. Fig. (5) shows the local information entropy of capmatinib.

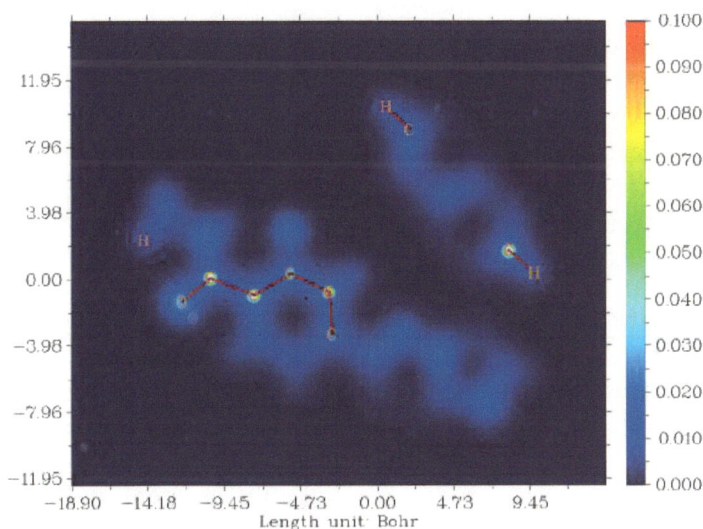

Fig. (5). Local information entropy of capmatinib.

From the color blue to red, the scale value from 0.00 to 0.10, and the range between -17.52 and 13.21 Bohr [3]. Capmatinib showed very low-LIE values at hydrogen, carbon, nitrogen, oxygen, and fluorine atoms.

Molecular Electrostatic Potentials (MESP) from Electronic Charges and Nuclear Charges of Capmatinib

The electrophilic and nucleophilic centers can predict electrostatic potential by the charge distribution of the molecules [64, 67, 82 - 88]. The potential of molecules can be identified due to the presence of electronic as well as nuclear charges.

Fig. (6) shows the MESP of capmatinib by electronic charges. Capmatinib has a molecular range from -17.46 to 13.12 Bohr [3].

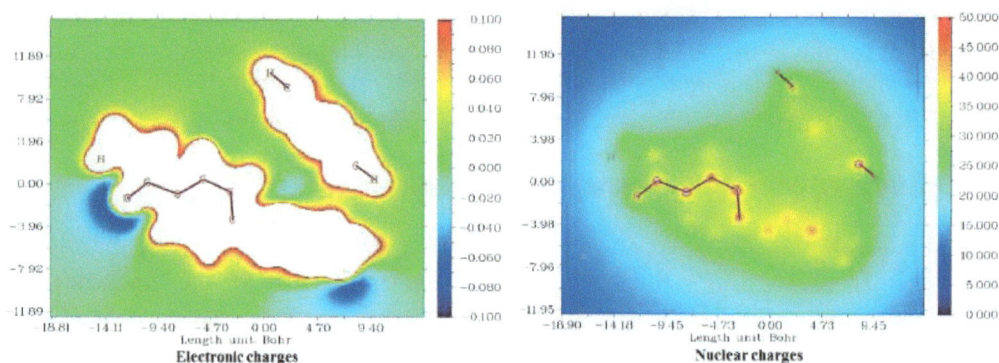

Fig. (6). Molecular electrostatic potentials of capmatinib.

The electron rich-sites of capmatinib shows blue appears at oxygen and nitrogen atoms (electrophilic region) in N-methylacetamid-, imidazotriazin-, and quinolin-groups, therefore electrophiles easily attack those sites: like the electron-poor sites show red appears at all the protons in the amide group (nucleophilic region), N-methylacetamid-, fluorophenyl-, imidazotriazin-, acetyl-, and quinolin- groups, therefore nucleophiles easily attack those sites.

Fig. (6) shows the MESP of capmatinib by nuclear charges. Capmatinib shows from the color blue to red, scale from 0.00 to 50.00, and range from -17.52 to 13.21 Bohr [3], respectively. The charges between 19 and 21, bluish-green on all the protons in N-methylacetamid-, fluorophenyl-, imidazotriazin-, acetyl-, and quinolin- groups, and these sites easily undergo substitution reactions possible, the charges between 47 and 50, red on carbonyl oxygen, and nitrogen in N-methylacetamid-, carbons in fluorophenyl- and quinolin- rings, and these sites easily undergo addition reactions and the charges from 25 to 38, greenish-yellow in color on all the elements except for those atoms, are not involved in any addition or substitution reactions.

Non-covalent Interactions (NCI), and Reduced Density Gradient (RDG) of Capmatinib

NCI always explains the stabilizes the molecule [52, 66, 68, 89, 90]. This is an important application of the bioactive nature of the molecule. The electronic density is directly proportional to the stability [32, 74]. Which is the probability to greater mass is equal to the stability of the molecule (sites). Fig. (7) shows the NCI, and RDG for capmatinib.

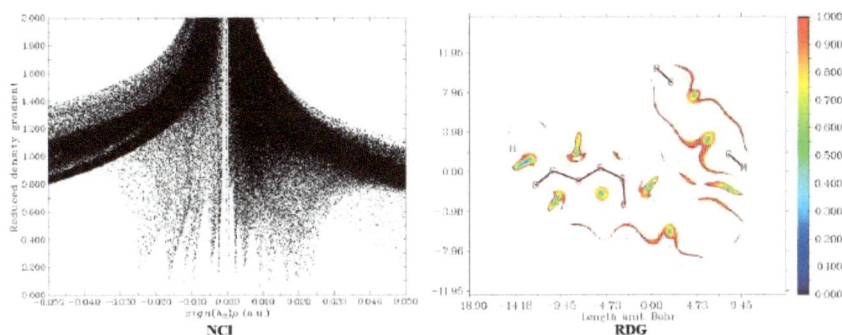

Fig. (7). Noncovalent interactions and reduced density gradient of capmatinib.

NCI of capmatinib, for a graph, plotted energy against an RDG. The strong H-bond appears from $-0.25*10^{-1}$ and $-0.02*10^{-1}$ a.u. from hydrogen atoms in fluorophenyl- group to carbonyl-oxygen in N-methylacetamid- group, weak H-bond appears from $-0.02*10^{-1}$ to $0.02*10^{-1}$ a.u. from protons to carbonyl_oxygen in N-methylactamid- group and weak repulsion interactions range from $0.02*10^{-1}$ to $0.45*10^{-1}$ a.u. having imidazotriazin- and quinolin- groups, and fluorine in fluorophenyl- and N-methylacetamid- groups.

RDG of capmatinib, from blue to red as the scale of the probability value from 0.00 to 1.00, and the molecule within -17.52 to 13.21 Bohr [3]. The red shows the high RDG from 0.90 to 1.00 on the elements nitrogen in N-methylacetamid-, fluorin- and carbons in fluorophenyl-, nitrogen, and carbons in imidazotriazin-, acetyl- carbon, and carbons and nitrogens in quinolin- groups. The small RDG show in colors green and blue are reduced densities are from 0.50 to 0.65 and 0.00 to 0.50 mingle with high reduced density gradients shown in Fig. (7). Therefore, the red in color area particles are probably more stable than the blue area of the capmatinib molecule.

CONCLUDING REMARKS

The book chapter studies show the structure and notable physical properties of Capmatinib. The stability described by intra-molecular charge delocalization was studied with NBO. UV-Visible spectra show a significant peak at 285.55 nm with 0.2616 oscillator strength. The energy, and reaction site properties like MESP from electronic and nuclear charges, ALIE, ELF, LOL, LIE RDG, and NCI are clarified in detail. It is found to be high at fluorine, carbons, and protons in fluorophenyl-, oxygen, carbons, protons, and nitrogen in N-methylacetamid-, nitrogens, carbons, and protons in imidazotriazin-, carbon and hydrogens in acetyl-, and nitrogen, carbons, and protons in quinolin- group. This structural information enables researchers to effectively design novel similar drugs.

CONSENT FOR PUBLICATION

Not applicable.

CONFLICT OF INTEREST

The authors declare no conflict of interest, financial or otherwise.

ACKNOWLEDGMENTS

Authors thanks St Berchmans College for encouragement.

REFERENCES

[1]   Walters, S.; Maringe, C.; Coleman, M.P.; Peake, M.D.; Butler, J.; Young, N.; Bergström, S.; Hanna, L.; Jakobsen, E.; Kölbeck, K.; Sundstrøm, S.; Engholm, G.; Gavin, A.; Gjerstorff, M.L.; Hatcher, J.; Johannesen, T.B.; Linklater, K.M.; McGahan, C.E.; Steward, J.; Tracey, E.; Turner, D.; Richards, M.A.; Rachet, B. Lung cancer survival and stage at diagnosis in Australia, Canada, Denmark, Norway, Sweden and the UK: a population-based study, 2004-2007. *Thorax*, **2013**, *68*(6), 551-564.
        [http://dx.doi.org/10.1136/thoraxjnl-2012-202297] [PMID: 23399908]

[2]   Planchard, D.; Popat, S.; Kerr, K.; Novello, S.; Smit, E.F.; Faivre-Finn, C.; Mok, T.S.; Reck, M.; Van Schil, P.E.; Hellmann, M.D.; Peters, S. Metastatic Non-Small Cell Lung Cancer: ESMO Clinical Practice Guidelines for Diagnosis, Treatment and Follow-Up. *Ann. Oncol.*, **2018**, *29* Suppl. 4, iv192-iv237.
        [http://dx.doi.org/10.1093/annonc/mdy275]

[3]   Siegel, R.L.; Miller, K.D.; Jemal, A. Cancer statistics, 2019. *CA Cancer J. Clin.*, **2019**, *69*(1), 7-34.
        [http://dx.doi.org/10.3322/caac.21551] [PMID: 30620402]

[4]   Ettinger, D.S.; Wood, D.E.; Aisner, D.L.; Akerley, W.; Bauman, J.; Chirieac, L.R.; D'Amico, T.A.; DeCamp, M.M.; Dilling, T.J.; Dobelbower, M.; Doebele, R.C.; Govindan, R.; Gubens, M.A.; Hennon, M.; Horn, L.; Komaki, R.; Lackner, R.P.; Lanuti, M.; Leal, T.A.; Leisch, L.J.; Lilenbaum, R.; Lin, J.; Loo, B.W.J., Jr; Martins, R.; Otterson, G.A.; Reckamp, K.; Riely, G.J.; Schild, S.E.; Shapiro, T.A.; Stevenson, J.; Swanson, S.J.; Tauer, K.; Yang, S.C.; Gregory, K.; Hughes, M. Non-Small Cell Lung Cancer, Version 5.2017, NCCN Clinical Practice Guidelines in Oncology. *J. Natl. Compr. Canc. Netw.*, **2017**, *15*(4), 504-535.
        [http://dx.doi.org/10.6004/jnccn.2017.0050] [PMID: 28404761]

[5]   Comoglio, P.M.; Trusolino, L.; Boccaccio, C. Known and novel roles of the MET oncogene in cancer: a coherent approach to targeted therapy. *Nat. Rev. Cancer,* **2018**, *18*(6), 341-358.
[http://dx.doi.org/10.1038/s41568-018-0002-y] [PMID: 29674709]

[6]   Smyth, E.C.; Sclafani, F.; Cunningham, D. Emerging molecular targets in oncology: clinical potential of MET/hepatocyte growth-factor inhibitors. *OncoTargets Ther.,* **2014**, *7*, 1001-1014.
[http://dx.doi.org/10.2147/OTT.S44941] [PMID: 24959087]

[7]   Baltschukat, S.; Engstler, B.S.; Huang, A.; Hao, H.X.; Tam, A.; Wang, H.Q.; Liang, J.; DiMare, M.T.; Bhang, H.C.; Wang, Y.; Furet, P.; Sellers, W.R.; Hofmann, F.; Schoepfer, J.; Tiedt, R. Capmatinib (INC280) Is Active Against Models of Non-Small Cell Lung Cancer and Other Cancer Types with Defined Mechanisms of MET Activation. *Clin. Cancer Res.,* **2019**, *25*(10), 3164-3175.
[http://dx.doi.org/10.1158/1078-0432.CCR-18-2814] [PMID: 30674502]

[8]   Spigel, D.R.; Edelman, M.J.; O'Byrne, K.; Paz-Ares, L.; Mocci, S.; Phan, S.; Shames, D.S.; Smith, D.; Yu, W.; Paton, V.E.; Mok, T. Results From the Phase III Randomized Trial of Onartuzumab Plus Erlotinib Versus Erlotinib in Previously Treated Stage IIIB or IV Non-Small-Cell Lung Cancer: METLung. *J. Clin. Oncol.,* **2017**, *35*(4), 412-420.
[http://dx.doi.org/10.1200/JCO.2016.69.2160] [PMID: 27937096]

[9]   Vansteenkiste, J.F.; Van De Kerkhove, C.; Wauters, E.; Van Mol, P. Capmatinib for the treatment of non-small cell lung cancer. *Expert Rev. Anticancer Ther.,* **2019**, *19*(8), 659-671.
[http://dx.doi.org/10.1080/14737140.2019.1643239] [PMID: 31368815]

[10]   Reungwetwattana, T.; Liang, Y.; Zhu, V.; Ou, S.I. The race to target MET exon 14 skipping alterations in non-small cell lung cancer: The Why, the How, the Who, the Unknown, and the Inevitable. *Lung Cancer,* **2017**, *103*, 27-37.
[http://dx.doi.org/10.1016/j.lungcan.2016.11.011] [PMID: 28024693]

[11]   Brosseau, J.P.; Liao, C.P.; Le, L.Q. Translating current basic research into future therapies for neurofibromatosis type 1. *Br. J. Cancer,* **2020**, *123*(2), 178-186.
[http://dx.doi.org/10.1038/s41416-020-0903-x] [PMID: 32439933]

[12]   Rotow, J.; Bivona, T.G. Understanding and targeting resistance mechanisms in NSCLC. *Nat. Rev. Cancer,* **2017**, *17*(11), 637-658.
[http://dx.doi.org/10.1038/nrc.2017.84] [PMID: 29068003]

[13]   Drilon, A.; Cappuzzo, F.; Ou, S.I.; Camidge, D.R. Targeting MET in Lung Cancer: Will Expectations Finally Be MET? *J. Thorac. Oncol.,* **2017**, *12*(1), 15-26.
[http://dx.doi.org/10.1016/j.jtho.2016.10.014] [PMID: 27794501]

[14]   Saad, K.M.; Shaker, M.E.; Shaaban, A.A.; Abdelrahman, R.S.; Said, E. The c-Met inhibitor capmatinib alleviates acetaminophen-induced hepatotoxicity. *Int. Immunopharmacol.,* **2020**, *81*(February), 106292.
[http://dx.doi.org/10.1016/j.intimp.2020.106292] [PMID: 32062076]

[15]   Duplaquet, L.; Kherrouche, Z.; Baldacci, S.; Jamme, P.; Cortot, A.B.; Copin, M.C.; Tulasne, D. The multiple paths towards MET receptor addiction in cancer. *Oncogene,* **2018**, *37*(24), 3200-3215.
[http://dx.doi.org/10.1038/s41388-018-0185-4] [PMID: 29551767]

[16]   Fujino, T.; Kobayashi, Y.; Suda, K.; Koga, T.; Nishino, M.; Ohara, S.; Chiba, M.; Shimoji, M.; Tomizawa, K.; Takemoto, T.; Mitsudomi, T. Sensitivity and Resistance of MET Exon 14 Mutations in Lung Cancer to Eight MET Tyrosine Kinase Inhibitors In Vitro. *J. Thorac. Oncol.,* **2019**, *14*(10), 1753-1765.
[http://dx.doi.org/10.1016/j.jtho.2019.06.023] [PMID: 31279006]

[17]   El Husseini, K.; Chaabane, N.; Mansuet-Lupo, A.; Leroy, K.; Revel, M-P.; Wislez, M. Capmatinib-Induced Interstitial Lung Disease: A Case Report. *Curr. Probl. Cancer Case Reports,* **2020**, *2*(September), 100024.
[http://dx.doi.org/10.1016/j.cpccr.2020.100024]

[18]   Rodríguez-Hernández, M.A.; de la Cruz-Ojeda, P.; López-Grueso, M.J.; Navarro-Villarán, E.; Requejo-Aguilar, R.; Castejón-Vega, B.; Negrete, M.; Gallego, P.; Vega-Ochoa, Á.; Victor, V.M.; Cordero, M.D.; Del Campo, J.A.; Bárcena, J.A.; Padilla, C.A.; Muntané, J. Integrated molecular signaling involving mitochondrial dysfunction and alteration of cell metabolism induced by tyrosine kinase inhibitors in cancer. *Redox Biol.,* **2020**, *36*(February), 101510.
[http://dx.doi.org/10.1016/j.redox.2020.101510] [PMID: 32593127]

[19]   Cravero, P.; Vaz, N.; Ricciuti, B.; Clifford, S.E.; DiUbaldi, G.; Drevers, D.; Morton, K.; Rivenburgh, R.E.; Nishino, M.; Awad, M.M. Leptomeningeal Response to Capmatinib After Progression on Crizotinib in a Patient With *MET* Exon 14-Mutant NSCLC. *JTO Clin. Res. Reports,* **2020**, *1*(4), 100072.
[http://dx.doi.org/10.1016/j.jtocrr.2020.100072] [PMID: 34589954]

[20]   Coleman, N.; Hong, L.; Zhang, J.; Heymach, J.; Hong, D.; Le, X. Beyond epidermal growth factor receptor: MET amplification as a general resistance driver to targeted therapy in oncogene-driven non-small-cell lung cancer. *ESMO Open,* **2021**, *6*(6), 100319.
[http://dx.doi.org/10.1016/j.esmoop.2021.100319] [PMID: 34837746]

[21]   Zhong, L.; Li, Y.; Xiong, L.; Wang, W.; Wu, M.; Yuan, T.; Yang, W.; Tian, C.; Miao, Z.; Wang, T.; Yang, S. Small molecules in targeted cancer therapy: advances, challenges, and future perspectives. *Signal Transduct. Target. Ther.,* **2021**, *6*(1), 201.
[http://dx.doi.org/10.1038/s41392-021-00572-w] [PMID: 34054126]

[22]   Ye, Z.; Huang, Y.; Ke, J.; Zhu, X.; Leng, S.; Luo, H. Breakthrough in targeted therapy for non-small cell lung cancer. *Biomed. Pharmacother.,* **2021**, *133*(133), 111079.
[http://dx.doi.org/10.1016/j.biopha.2020.111079] [PMID: 33378976]

[23]   Wu, Y.L.; Smit, E.F.; Bauer, T.M. Capmatinib for patients with non-small cell lung cancer with MET exon 14 skipping mutations: A review of preclinical and clinical studies. *Cancer Treat. Rev.,* **2021**, *95*, 102173.
[http://dx.doi.org/10.1016/j.ctrv.2021.102173] [PMID: 33740553]

[24]   Moreno, V.; Greil, R.; Yachnin, J.; Majem, M.; Wermke, M.; Arkenau, H.T.; Basque, J.R.; Nidamarthy, P.K.; Kapoor, S.; Cui, X.; Giovannini, M. Pharmacokinetics and safety of capmatinib with food in patients with MET-dysregulated advanced solid tumors. *Clin. Ther.,* **2021**, *43*(6), 1092-1111.
[http://dx.doi.org/10.1016/j.clinthera.2021.04.006] [PMID: 34053700]

[25]   Alsahafi, E.; Begg, K.; Amelio, I.; Raulf, N.; Lucarelli, P.; Sauter, T.; Tavassoli, M. Clinical update on head and neck cancer: molecular biology and ongoing challenges. *Cell Death Dis.,* **2019**, *10*(8), 540.
[http://dx.doi.org/10.1038/s41419-019-1769-9] [PMID: 31308358]

[26]   Bonan, N.F.; Kowalski, D.; Kudlac, K.; Flaherty, K.; Gwilliam, J.C.; Falkenberg, L.G.; Maradiaga, E.; DeCicco-Skinner, K.L. Inhibition of HGF/MET signaling decreases overall tumor burden and blocks malignant conversion in Tpl2-related skin cancer. *Oncogenesis,* **2019**, *8*(1), 1-12.
[http://dx.doi.org/10.1038/s41389-018-0109-8] [PMID: 30631034]

[27]   Salgia, R.; Sattler, M.; Scheele, J.; Stroh, C.; Felip, E. The promise of selective MET inhibitors in non-small cell lung cancer with MET exon 14 skipping. *Cancer Treat. Rev.,* **2020**, *87*(April), 102022.
[http://dx.doi.org/10.1016/j.ctrv.2020.102022] [PMID: 32334240]

[28]   Gautschi, O.; Diebold, J. Intracranial Activity of Osimertinib Plus Capmatinib in a Patient With *EGFR* and *MET*-Driven Lung Cancer: Case Report. *JTO Clin. Res. Reports,* **2021**, *2*(4), 100162.
[http://dx.doi.org/10.1016/j.jtocrr.2021.100162] [PMID: 34590012]

[29]   Liu, X.; Wang, Q.; Yang, G.; Marando, C.; Koblish, H.K.; Hall, L.M.; Fridman, J.S.; Behshad, E.; Wynn, R.; Li, Y.; Boer, J.; Diamond, S.; He, C.; Xu, M.; Zhuo, J.; Yao, W.; Newton, R.C.; Scherle, P.A. A novel kinase inhibitor, INCB28060, blocks c-MET-dependent signaling, neoplastic activities, and cross-talk with EGFR and HER-3. *Clin. Cancer Res.,* **2011**, *17*(22), 7127-7138.
[http://dx.doi.org/10.1158/1078-0432.CCR-11-1157] [PMID: 21918175]

[30]   Frisch, M.J.; Trucks, G.W.; Schlegel, H.B.; Scuseria, G.E.; Robb, M.A.; Cheeseman, J.R.; Scalmani, G.; Barone, V.; Mennucci, B.; Petersson, G.A.; Nakatsuji, H.; Caricato, M.; Li, X.; Hratchian, H.P.; Izmaylov, A.F.; Bloino, J.; Zheng, G.; Sonnenberg, J.L.; Hada, M.; Ehara, M.; Toyota, K.; Fukuda, R.; Hasegawa, J.; Ishida, M.; Nakajima, T.; Honda, Y.; Kitao, O.; Nakai, H.; Vreven, T.; Montgomery, J.A., Jr; Peralta, J.E.; Ogliaro, F.; Bearpark, M.; Heyd, J.J.; Brothers, E.; Kudin, K.N.; Staroverov, V.N.; Kobayashi, R.; Normand, J.; Raghavachari, K.; Rendell, A.; Burant, J.C.; Iyengar, S.S.; Tomasi, J.; Cossi, M.; Rega, N.; Millam, J.M.; Klene, M.; Knox, J.E.; Cross, J.B.; Bakken, V.; Adamo, C.; Jaramillo, J.; Gomperts, R.; Stratmann, R.E.; Yazyev, O.; Austin, A.J.; Cammi, R.; Pomelli, C.; Ochterski, J.W.; Martin, R.L.; Morokuma, K.; Zakrzewski, V.G.; Voth, G.A.; Salvador, P.; Dannenberg, J.J.; Dapprich, S.; Daniels, A.D.; Farkas, O.; Foresman, J.B.; Ortiz, J.V.; Cioslowski, J.; Fox, D.J. *Gaussian09 Revision D.01*; Gaussian, Inc.: Wallingford, CT, **2013**.

[31]   Becke, A.D. Density-functional Thermochemistry. III. The Role of Exact Exchange. *J. Chem. Phys.,* **1993**, *98*(7), 5648-5652.
[http://dx.doi.org/10.1063/1.464913]

[32]   Schmider, H.L.; Becke, A.D. Chemical Content of the Kinetic Energy Density. *J. Mol. Struct. THEOCHEM*, **2000**, *527*(1), 51-61.https://doi.org/https://doi.org/10.1016/S0166-1280(00)00477-2
[http://dx.doi.org/10.1016/S0166-1280(00)00477-2]

[33]   Becke, A.D. Density-functional exchange-energy approximation with correct asymptotic behavior. *Phys. Rev. A Gen. Phys.,* **1988**, *38*(6), 3098-3100.
[http://dx.doi.org/10.1103/PhysRevA.38.3098] [PMID: 9900728]

[34]   Becke, A.D.; Johnson, E.R. A density-functional model of the dispersion interaction. *J. Chem. Phys.,* **2005**, *123*(15), 154101.
[http://dx.doi.org/10.1063/1.2065267] [PMID: 16252936]

[35]   Becke, A. D. Perspective: Fifty Years of Density-Functional Theory in Chemical Physics. *J. Chem. Phys.,* **2014**, *140*(18), 18A301.
[http://dx.doi.org/10.1063/1.4869598]

[36]   Dunning, T.H., Jr Gaussian Basis Sets for Use in Correlated Molecular Calculations. I. The Atoms Boron through Neon and Hydrogen. *J. Chem. Phys.,* **1989**, *90*(2), 1007-1023.
[http://dx.doi.org/10.1063/1.456153]

[37]   Frisch, M.J.; Pople, J.A.; Binkley, J.S. Self-consistent Molecular Orbital Methods 25. Supplementary Functions for Gaussian Basis Sets. *J. Chem. Phys.,* **1984**, *80*(7), 3265-3269.
[http://dx.doi.org/10.1063/1.447079]

[38]   Longuet-Higgins, H.C.; Pople, J.A. Electronic Spectral Shifts of Nonpolar Molecules in Nonpolar Solvents. *J. Chem. Phys.,* **1957**, *27*(1), 192-194.
[http://dx.doi.org/10.1063/1.1743666]

[39]   Krishnan, R.; Binkley, J.S.; Seeger, R.; Pople, J.A. Self-consistent Molecular Orbital Methods. XX. A Basis Set for Correlated Wave Functions. *J. Chem. Phys.,* **1980**, *72*(1), 650-654.
[http://dx.doi.org/10.1063/1.438955]

[40]   Rassolov, V.A.; Ratner, M.A.; Pople, J.A.; Redfern, P.C.; Curtiss, L.A. 6-31G* Basis Set for Third-Row Atoms. *J. Comput. Chem.,* **2001**, *22*(9), 976-984.
[http://dx.doi.org/10.1002/jcc.1058]

[41]   Ditchfield, R.; Hehre, W.J.; Pople, J.A. Self-Consistent Molecular-Orbital Methods. IX. An Extended Gaussian-Type Basis for Molecular-Orbital Studies of Organic Molecules. *J. Chem. Phys.,* **1971**, *54*(2), 724-728.
[http://dx.doi.org/10.1063/1.1674902]

[42]   Yanai, T.; Tew, D.P.; Handy, N.C. A New Hybrid Exchange–Correlation Functional Using the Coulomb-Attenuating Method (CAM-B3LYP). *Chem. Phys. Lett.,* **2004**, *393*(1), 51-57.https://doi.org/https://doi.org/10.1016/j.cplett.2004.06.011
[http://dx.doi.org/10.1016/j.cplett.2004.06.011]

[43]   Okuno, K.; Shigeta, Y.; Kishi, R.; Miyasaka, H.; Nakano, M. Tuned CAM-B3LYP Functional in the Time-Dependent Density Functional Theory Scheme for Excitation Energies and Properties of Diarylethene Derivatives. *J. Photochem. Photobiol. Chem.,* **2012**, *235*, 29-34.
[http://dx.doi.org/10.1016/j.jphotochem.2012.03.003]

[44]   O'Boyle, N.M.; Tenderholt, A.L.; Langner, K.M. cclib: a library for package-independent computational chemistry algorithms. *J. Comput. Chem.,* **2008**, *29*(5), 839-845.
[http://dx.doi.org/10.1002/jcc.20823] [PMID: 17849392]

[45]   Lu, T.; Chen, F. Multiwfn: a multifunctional wavefunction analyzer. *J. Comput. Chem.,* **2012**, *33*(5), 580-592.
[http://dx.doi.org/10.1002/jcc.22885] [PMID: 22162017]

[46]   Reed, A.E.; Curtiss, L.A.; Weinhold, F. Intermolecular Interactions from a Natural Bond Orbital, Donor-Acceptor Viewpoint. *Chem. Rev.,* **1988**, *88*(6), 899-926.
[http://dx.doi.org/10.1021/cr00088a005]

[47]   Weinhold, F. Natural bond orbital analysis: a critical overview of relationships to alternative bonding perspectives. *J. Comput. Chem.,* **2012**, *33*(30), 2363-2379.
[http://dx.doi.org/10.1002/jcc.23060] [PMID: 22837029]

[48]   Dunnington, B.D.; Schmidt, J.R. Generalization of Natural Bond Orbital Analysis to Periodic Systems: Applications to Solids and Surfaces via Plane-Wave Density Functional Theory. *J. Chem. Theory Comput.,* **2012**, *8*(6), 1902-1911.
[http://dx.doi.org/10.1021/ct300002t] [PMID: 26593824]

[49]   Glendening, E.D.; Landis, C.R.; Weinhold, F. Natural Bond Orbital Methods. *WIREs Comput. Wiley Interdiscip. Rev. Comput. Mol. Sci.,* **2012**, *2*(1), 1-42.
[http://dx.doi.org/10.1002/wcms.51]

[50]   Glendening, E.D.; Reed, A.E.; Carpenter, J.E. NBO 3.1. Theoretical Chemistry Institute, University of Wisconsin, Madison 2003.

[51]   Al-Otaibi, J.S.; Mary, Y.S.; Mary, Y.S.; Panicker, C.Y.; Thomas, R. Cocrystals of Pyrazinamide with P-Toluenesulfonic and Ferulic Acids: DFT Investigations and Molecular Docking Studies. *J. Mol. Struct.,* **2019**, *1175*, 916-926.
[http://dx.doi.org/10.1016/j.molstruc.2018.08.055]

[52]   Al-Otaibi, J.S.; Mary, Y.S.; Armaković, S.; Thomas, R. Hybrid and Bioactive Cocrystals of Pyrazinamide with Hydroxybenzoic Acids: Detailed Study of Structure, Spectroscopic Characteristics, Other Potential Applications and Noncovalent Interactions Using SAPT. *J. Mol. Struct.,* **2020**, *1202*, 127316.
[http://dx.doi.org/10.1016/j.molstruc.2019.127316]

[53]   Thadathil, D.A.; Varghese, S.; Akshaya, K.B.; Thomas, R.; Varghese, A. An Insight into Photophysical Investigation of (E)-2-Fluoro-N'-(1-(4-Nitrophenyl)Ethylidene)Benzohydrazide through Solvatochromism Approaches and Computational Studies. *J. Fluoresc.,* **2019**, *29*(4), 1013-1027.
[http://dx.doi.org/10.1007/s10895-019-02415-y] [PMID: 31309390]

[54]   Alsalme, A.; Pooventhiran, T.; Al-Zaqri, N.; Rao, D.J.; Thomas, R. Structural, Physico-Chemical Landscapes, Ground State and Excited State Properties in Different Solvent Atmosphere of Avapritinib and Its Ultrasensitive Detection Using SERS/GERS on Self-Assembly Formation with Graphene Quantum Dots. *J. Mol. Liq.,* **2021**, *322*, 114555.
[http://dx.doi.org/10.1016/j.molliq.2020.114555]

[55]   Pooventhiran, T.; Al-zaqri, N.; Alsalme, A.; Bhattacharyya, U.; Thomas, R. Structural Aspects, Conformational Preference and Other Physico-Chemical Properties of Artesunate and the Formation of Self-Assembly with Graphene Quantum Dots : A Fi Rst Principle Analysis and Surface Enhancement of Raman Activity Investigation. *J. Mol. Liq.,* **2021**, *325*, 114810.
[http://dx.doi.org/10.1016/j.molliq.2020.114810]

[56] Mary, Y.S.; Mary, Y.S.; Resmi, K.S.; Thomas, R. DFT and molecular docking investigations of oxicam derivatives. *Heliyon,* **2019**, *5*(7), e02175.
[http://dx.doi.org/10.1016/j.heliyon.2019.e02175] [PMID: 31388594]

[57] Pooventhiran, T.; Thomas, R.; Bhattacharyya, U.; Sowrirajan, S.; Irfan, A.; Rao, D.J. Structural Aspects, Reactivity Analysis, Wavefunction Based Properties, Cluster Formation with Helicene and Subsequent Detection from Surface Enhancement in Raman Spectra of Triclabendazole Studies Using First Principle Simulations. *Vietnam J. Chem.,* **2021**, *59*(6), 887-901.
[http://dx.doi.org/10.1002/vjch.202100067]

[58] Surendar, P.; Pooventhiran, T.; Rajam, S.; Irfan, A.; Thomas, R. Schiff Bases from α-Ionone with Adenine, Cytosine, and l-Leucine Biomolecules: Synthesis, Structural Features, Electronic Structure, and Medicinal Activities. *J. Comput. Biophys. Chem.,* **2021**, *0*(0), 2250001.
[http://dx.doi.org/10.1142/S2737416522500016]

[59] Majumdar, D.; Agrawal, Y.; Thomas, R.; Ullah, Z.; Santra, M.K.; Das, S.; Pal, T.K.; Bankura, K.; Mishra, D. Syntheses, Characterizations, Crystal Structures, DFT/TD-DFT, Luminescence Behaviors and Cytotoxic Effect of Bicompartmental Zn (II)-dicyanamide Schiff Base Coordination Polymers: An Approach to Apoptosis, Autophagy and Necrosis Type Classical Cell Death. *Appl. Organomet. Chem.,* **2020**, *34*(1), e5269.
[http://dx.doi.org/10.1002/aoc.5269]

[60] Majumdar, D.; Das, S.; Thomas, R.; Ullah, Z.; Sreejith, S.S.; Das, D.; Shukla, P.; Bankura, K.; Mishra, D. Syntheses, X-Ray Crystal Structures of Two New Zn (II)-Dicyanamide Complexes Derived from H2vanen-Type Compartmental Ligands: Investigation of Thermal, Photoluminescence, in Vitro Cytotoxic Effect and DFT-TDDFT Studies. *Inorg. Chim. Acta,* **2019**, *492*, 221-234.
[http://dx.doi.org/10.1016/j.ica.2019.04.041]

[61] Srikanth, K.E.; Veeraiah, A.; Pooventhiran, T.; Thomas, R.; Solomon, K.A.; Soma Raju, C.J.; Latha, J.N.L. Detailed molecular structure (XRD), conformational search, spectroscopic characterization (IR, Raman, UV, fluorescence), quantum mechanical properties and bioactivity prediction of a pyrrole analogue. *Heliyon,* **2020**, *6*(6), e04106.
[http://dx.doi.org/10.1016/j.heliyon.2020.e04106] [PMID: 32529077]

[62] Pooventhiran, T.; Bhattacharyya, U.; Rao, D.J.; Chandramohan, V.; Karunakar, P.; Irfan, A.; Mary, Y.S.; Thomas, R. Detailed Spectra, Electronic Properties, Qualitative Non-Covalent Interaction Analysis, Solvatochromism, Docking and Molecular Dynamics Simulations in Different Solvent Atmosphere of Cenobamate. *Struct. Chem.,* **2020**, *31*(6), 2475-2485.
[http://dx.doi.org/10.1007/s11224-020-01607-8]

[63] Al-Zaqri, N.; Pooventhiran, T.; Alsalme, A.; Rao, D.J.; Rao, S.S.; Sankar, A.; Thomas, R. First-Principle Studies of Istradefylline with Emphasis on the Stability, Reactivity, Interactions and Wavefunction-Dependent Properties. *Polycycl. Aromat. Compd.,* **2020**, 1-15.
[http://dx.doi.org/10.1080/10406638.2020.1857273]

[64] Alsalme, A.; Pooventhiran, T.; Al-Zaqri, N.; Rao, D.J.; Rao, S.S.; Thomas, R. Modelling the structural and reactivity landscapes of tucatinib with special reference to its wavefunction-dependent properties and screening for potential antiviral activity. *J. Mol. Model.,* **2020**, *26*(12), 341.
[http://dx.doi.org/10.1007/s00894-020-04603-1] [PMID: 33200284]

[65] Al-Zaqri, N.; Pooventhiran, T.; Alsalme, A.; Warad, I.; John, A.M.; Thomas, R. Structural and physico-chemical evaluation of melatonin and its solution-state excited properties, with emphasis on its binding with novel coronavirus proteins. *J. Mol. Liq.,* **2020**, *318*, 114082.
[http://dx.doi.org/10.1016/j.molliq.2020.114082] [PMID: 32863490]

[66] Alharthi, F.A.; Al-Zaqri, N.; Alsalme, A.; Al-Taleb, A.; Pooventhiran, T.; Thomas, R.; Rao, D.J. Excited-state electronic properties, structural studies, noncovalent interactions, and inhibition of the novel severe acute respiratory syndrome coronavirus 2 proteins in Ripretinib by first-principle simulations. *J. Mol. Liq.,* **2021**, *324*, 115134.
[http://dx.doi.org/10.1016/j.molliq.2020.115134] [PMID: 33390634]

[67]   Al-Zaqri, N.; Pooventhiran, T.; Alharthi, F.A.; Bhattacharyya, U.; Thomas, R. Structural investigations, quantum mechanical studies on proton and metal affinity and biological activity predictions of selpercatinib. *J. Mol. Liq.,* **2021**, *325*, 114765.
[http://dx.doi.org/10.1016/j.molliq.2020.114765] [PMID: 33746318]

[68]   Al-Zaqri, N.; Pooventhiran, T.; Rao, D.J.; Alsalme, A.; Warad, I.; Thomas, R. Structure, Conformational Dynamics, Quantum Mechanical Studies and Potential Biological Activity Analysis of Multiple Sclerosis Medicine Ozanimod. *J. Mol. Struct.,* **2021**, *1227*, 129685.
[http://dx.doi.org/10.1016/j.molstruc.2020.129685]

[69]   Surendar, P.; Pooventhiran, T.; Rajam, S.; Bhattacharyya, U.; Bakht, A.; Thomas, R. Quasi Liquid Schiff Bases from Trans -2-Hexenal and Cytosine and l -Leucine with Potential Antieczematic and Antiarthritic Activities : Synthesis, Structure and Quantum Mechanical Studies. *J. Mol. Liq.,* **2021**, *334*, 116448.
[http://dx.doi.org/10.1016/j.molliq.2021.116448]

[70]   Fuster, F.; Sevin, A.; Silvi, B. Topological Analysis of the Electron Localization Function (ELF) Applied to the Electrophilic Aromatic Substitution. *J. Phys. Chem. A,* **2000**, *104*(4), 852-858.
[http://dx.doi.org/10.1021/jp992783k]

[71]   Fuentealba, P.; Chamorro, E.; Santos, J.C. Chapter 5 Understanding and Using the Electron Localization Function. In: *In Theoretical Aspects of Chemical Reactivity*; Toro-Labbé, A.B.; T.T., C.C.; Santos, J.C., Eds.; Elsevier, **2007**; 19, pp. 57-85.
[http://dx.doi.org/10.1016/S1380-7323(07)80006-9]

[72]   Gibbs, G.V.; Cox, D.F.; Boisen, M.B., Jr; Downs, R.T.; Ross, N.L. The Electron Localization Function: A Tool for Locating Favorable Proton Docking Sites in the Silica Polymorphs. *Phys. Chem. Miner.,* **2003**, *30*(5), 305-316.
[http://dx.doi.org/10.1007/s00269-003-0318-2]

[73]   Jacobsen, H. Localized-Orbital Locator (LOL) Profiles of Chemical Bonding. *Can. J. Chem.,* **2008**, *86*(7), 695-702.
[http://dx.doi.org/10.1139/v08-052]

[74]   Tsirelson, V.; Stash, A. Analyzing experimental electron density with the localized-orbital locator. *Acta Crystallogr. B,* **2002**, *58*(Pt 5), 780-785.
[http://dx.doi.org/10.1107/S0108768102012338] [PMID: 12324690]

[75]   Hossain, M.; Thomas, R.; Mary, Y.S.; Resmi, K.S.; Armaković, S.; Armaković, S.J.; Nanda, A.K.; Vijayakumar, G.; Alsenoy, C.V. Understanding Reactivity of Two Newly Synthetized Imidazole Derivatives by Spectroscopic Characterization and Computational Study. *J. Mol. Struct.,* **2018**, *1158*, 176-196.
[http://dx.doi.org/10.1016/j.molstruc.2018.01.029]

[76]   Murray, J.S.; Seminario, J.M.; Politzer, P.; Sjoberg, P. Average Local Ionization Energies Computed on the Surfaces of Some Strained Molecules. *Int. J. Quantum Chem.,* **1990**, *38*(S24), 645-653.
[http://dx.doi.org/10.1002/qua.560382462]

[77]   Armaković, S.; Armaković, S.J.; Vraneš, M.; Tot, A.; Gadžurić, S. Determination of Reactive Properties of 1-Butyl-3-Methylimidazolium Taurate Ionic Liquid Employing DFT Calculations. *J. Mol. Liq.,* **2016**, *222*, 796-803.
[http://dx.doi.org/10.1016/j.molliq.2016.07.094]

[78]   Rikalo, A.; Nikolić, M.; Alanov, M.; Vuković, A.; Armaković, S.J.; Armaković, S.A. DFT and MD Study of Reactive, H2 Adsorption and Optoelectronic Properties of Graphane Nanoparticles – An Influence of Boron Doping. *Mater. Chem. Phys.,* **2020**, *241*, 122329.https://doi.org/https://doi.org/10.1016/j.matchemphys.2019.122329
[http://dx.doi.org/10.1016/j.matchemphys.2019.122329]

[79]   Alipour, M.; Badooei, Z. Toward Electron Correlation and Electronic Properties from the Perspective of Information Functional Theory. *J. Phys. Chem. A,* **2018**, *122*(31), 6424-6437.

[http://dx.doi.org/10.1021/acs.jpca.8b05703] [PMID: 30052445]

[80]   Rong, C.; Wang, B.; Zhao, D.; Liu, S. Information-Theoretic Approach in Density Functional Theory and Its Recent Applications to Chemical Problems. *WIREs Comput. Wiley Interdiscip. Rev. Comput. Mol. Sci.,* **2020**, *10*(4), e1461.
[http://dx.doi.org/10.1002/wcms.1461]

[81]   Zhou, X-Y.; Rong, C.; Lu, T.; Zhou, P.; Liu, S. Information Functional Theory: Electronic Properties as Functionals of Information for Atoms and Molecules. *J. Phys. Chem. A,* **2016**, *120*(20), 3634-3642.
[http://dx.doi.org/10.1021/acs.jpca.6b01197] [PMID: 27115776]

[82]   Politzer, P.; Laurence, P.R.; Jayasuriya, K. Molecular electrostatic potentials: an effective tool for the elucidation of biochemical phenomena. *Environ. Health Perspect.,* **1985**, *61*, 191-202.
[http://dx.doi.org/10.1289/ehp.8561191] [PMID: 2866089]

[83]   Politzer, P.; Murray, J.S. The Fundamental Nature and Role of the Electrostatic Potential in Atoms and Molecules. *Theor. Chem. Acc.,* **2002**, *108*(3), 134-142.
[http://dx.doi.org/10.1007/s00214-002-0363-9]

[84]   Politzer, P.; Murray, J.S. Molecular Electrostatic Potentials and Chemical Reactivity. *Rev. Comput. Chem.,* **1991**, 273-312.https://doi.org/doi:10.1002/9780470125793.ch7

[85]   Politzer, P.; Lane, P.; Concha, M.C. Atomic and Molecular Energies in Terms of Electrostatic Potentials at Nuclei. *Int. J. Quantum Chem.,* **2002**, *90*(1), 459-463.
[http://dx.doi.org/10.1002/qua.10105]

[86]   Breneman, C.M.; Martinov, M. The Use of Electrostatic Potential Fields in QSAR and QSPR. *In Molecular Electrostatic Potential,* Murray, J.S.; Sen, K.B.T.T.; C.C., **1996**, *3*, 143-179.https://doi.org/https://doi.org/10.1016/S1380-7323(96)80043-4

[87]   Politzer, P.; Murray, J.S. Electrostatic Potentials at the Nuclei of Atoms and Molecules. *Theor. Chem. Acc.,* **2021**, *140*(1), 7.
[http://dx.doi.org/10.1007/s00214-020-02701-0]

[88]   Bulat, F.A.; Toro-Labbé, A.; Brinck, T.; Murray, J.S.; Politzer, P. Quantitative analysis of molecular surfaces: areas, volumes, electrostatic potentials and average local ionization energies. *J. Mol. Model.,* **2010**, *16*(11), 1679-1691.
[http://dx.doi.org/10.1007/s00894-010-0692-x] [PMID: 20361346]

[89]   Karshikoff, A. Non-Covalent Interactions in Proteins. **2006**.https://doi.org/doi:10.1142/p477

[90]   Boto, R.A.; Piquemal, J.P.; Contreras-García, J. Revealing Strong Interactions with the Reduced Density Gradient: A Benchmark for Covalent, Ionic and Charge-Shift Bonds. *Theor. Chem. Acc.,* **2017**, *136*(12), 1-9.
[http://dx.doi.org/10.1007/s00214-017-2169-9]

SUBJECT INDEX

A

Absorption 3, 18, 21, 23, 29, 30, 53, 65, 68, 69, 81, 110, 111, 182, 188, 192, 214, 227
  intensities 69
  medication 53
  optical 65, 111
  spectra 69, 110, 214
  spectrum 68, 188
Acceptor LUMO 208, 220
Acetaminophen 253
Acid 6, 52, 53,
  amino 6
  Deoxyribonucleic 6
  nucleic 52, 53
Activated reactive evaporation 85
Activation 65, 66, 67, 96, 241, 242, 245, 246, 253
  energy 65, 66, 67, 241, 242, 245
  oncogene 253
Activity 6, 11, 180, 181
  immune 6
  photo-induced 181
ALIE analysis 253
Alzheimer's disease 9
Amphiphilic block copolymers 52
Amplifiers 109, 118
  integrated optical 118
Anaesthesia 55
Anisotropic 79
  nanomaterials 79
  nanoparticles 79
Anisotropy 120
  magneto-crystalline 120
Antibody onartuzumab combination therapies 253
Antifungal Activity 82
Anti-reflection coating (ARC) 63
Application(s) 1, 2, 3, 4, 5, 7, 8, 9, 10, 49, 50, 51, 54, 56, 59, 206
  electronic 206

of nanomedicines 3, 10
Archimedes principle 239
Atherosclerosis 7
Atmosphere, oxidizing 114
Atomic 59, 103, 122, 146, 212, 214, 255
  force microscopy (AFM) 59, 103
  orbitals (AOs) 122, 146, 212, 214, 255
Atoms 41, 57, 126, 127, 135, 136, 146, 151, 152, 153, 211, 212, 214, 255, 256
  electron-donating 255
  electronegative 214
Augmented plane wave (APW) 34, 120, 145, 146, 148, 149, 150, 151, 155, 165, 166
  method 146
Autoclave 93

B

Ball milling technique 89, 90, 91
Bandgap energy 25, 192
Band(s) 35, 41, 63, 68, 69, 75, 114, 115, 144, 179, 239, 240
  bending absorption 68
  distinct vibrational 239
  electronic energy 35
  intensities 239
  luminescence 115
Behaviours 22, 26, 28, 36, 57, 69, 143
  topological insulator 36
Bloch 126
  hypothesis 126
Bloch function 37
  energy-independent 37
Bloch's theorem 124
Blood transfusion 3
Boltzmann constant 241
Borosilicate glasses 237, 238
Bragg's diffraction 101
Brillouin-zone coordination techniques 165

X

Z

www.ingramcontent.com/pod-product-compliance
Lightning Source LLC
Chambersburg PA
CBHW050814220326
41598CB00006B/204